Phytochemicals in the Management of Rheumatoid Arthritis

Rheumatoid arthritis (RA) is a chronic inflammatory disease that primarily affects joints and bones. Disease-modifying anti-rheumatic drugs, biologics, and NSAIDs are mostly acclaimed for symptomatic relief, but these may have adverse outcomes. Extensive research has been conducted to find alternative and safe treatment options for RA. *Phytochemicals in the Management of Rheumatoid Arthritis* examines advanced knowledge about pathogenesis, diagnosis, and conventional treatment of RA and the application of phytochemicals in the management and prophylaxis of RA through modulation of molecular targets such as matrix metalloproteinases, nuclear factor kappa B, arachidonic acid, RANKL, and JAK/STAT.

This book contains research on the mechanisms of action, dosage, and molecular pathways affected by phytochemicals to treat RA and emphasizes the targeted-drug delivery and novel dosage forms of phytochemicals to improve therapeutic efficacy and abate adverse effects. Additionally, it provides information on clinical studies and the use of photodynamic therapy for the management of RA. It will equip postgraduate students, health professionals, practitioners, pharmaceutical researchers, and biomedical scientists with valuable evidence about the anti-arthritic foods, nutrients, and phytochemicals required for treating RA.

Phytochemicals in the Management of Rheumatoid Arthritis

Edited by
Ammara Saleem, Muhammad Furqan Akhtar,
and Alam Zeb

CRC Press
Taylor & Francis Group
Boca Raton London New York

CRC Press is an imprint of the
Taylor & Francis Group, an **informa** business

Designed cover image: Shutterstock

First edition published 2027
by CRC Press
2385 NW Executive Center Drive, Suite 320, Boca Raton FL 33431

and by CRC Press
4 Park Square, Milton Park, Abingdon, Oxon, OX14 4RN

CRC Press is an imprint of Taylor & Francis Group, LLC

ISBN: 9781032860862 (hbk)
ISBN: 9781032878911 (pbk)
ISBN: 9781003535201 (ebk)

DOI: 10.1201/9781003535201

Typeset in Times
by codeMantra

We dedicate this book to all who seek truth through research and wisdom through learning.

Contents

Preface

Rheumatoid arthritis (RA) is a systemic, autoimmune, and chronic inflammatory disease that primarily affects joints and bones. It mostly affects patients during the fourth or fifth decades of life. It causes severe pain and inflammation of bones and joints, bone erosion, cartilage destruction, and extra-articular events. Disease-modifying anti-rheumatic drugs, biologics, and NSAIDs are mostly acclaimed for symptomatic relief, but these are not devoid of adverse outcomes. Extensive research has been conducted to find alternative and safe treatment options for RA. This book comprehends and delivers advanced knowledge about the pathogenesis, diagnosis, and conventional treatment of RA and the application of phytochemicals in the management and prophylaxis of RA through modulation of molecular targets such as matrix metalloproteinases, nuclear factor kappa B, arachidonic acid, RANKL, and JAK/STAT.

This book will deliver cutting-edge research regarding the bioavailability, mechanism of action, dosage, and molecular pathways affected by phytochemicals to treat RA. Moreover, this book will emphasize on the targeted-drug delivery and novel dosage forms of phytochemicals to improve therapeutic efficacy and abate adverse effects. This book will additionally provide information regarding clinical studies and the use of photodynamic therapy for the management of RA. This book will equip the postgraduate students, health professionals, practitioners, pharmaceutical researchers, and biomedical scientists with valuable evidence about the anti-arthritic foods, nutrients, and phytochemicals for treating RA.

Currently, there is no book available on the usage of phytochemicals and alternative therapies for the management of RA or arthritis that necessitates the significance and need of this book for readers, researchers, and professionals. Additionally, this comprehensive book will provide complete and up-to-date information about important phytochemicals and photoactive chemicals effective against arthritis, preclinical studies, clinical studies, and safety aspects of phytochemicals; the problems associated with these therapies; the regulatory aspects of phytomedicines (nutraceuticals); the advanced/targeted-drug delivery of phytochemicals in RA; and the understanding of which is compulsory for beginners to the expert level in the field.

This book comprises of four sections (Sections A–D) and has a total of 16 chapters that are distributed in different sections.

Section A covers the introductory portion and includes four chapters. This section provides the reader with in-depth knowledge of the phytochemicals and RA. This section forms the basis for understanding the concepts, which are presented in the later sections. Chapter 1 is very important. It is helpful for the beginners and equips them with the background to the main phytochemical classes that are used in the management of RA and their role in various other diseases, major sources of phytochemicals, and their importance as a treatment moiety. Chapter 2 is extremely important. It is helpful for the beginners and prepares them with the basic concept of inflammation, arthritis, and RA and also the pathogenesis and treatment options for RA. Chapter 3 provides knowledge about available pharmaceutical drug classes for the treatment of RA and their mechanism of action, and special emphasis is given to their adverse drug reactions. Chapter 4 provides valuable information to the researchers and academia about the use of disease-modifying anti-rheumatic drugs for treating RA alone and in combination with other drugs and phytochemicals. Is a combination of two phytochemicals or phytochemicals with any anti-rheumatic drug beneficial or not? Moreover, previous literature regarding this matter and critical analysis is provided.

Section B includes seven chapters that cover the signaling mechanisms involved in the pathogenesis of RA and targets used by the phytochemicals to treat or modify the disease. This section comprises seven chapters. This section is especially helpful for the readers in research, academia, and professionals from pharmaceutical industries due to the comprehensive coverage of phytochemical/drug targets for treating RA. Moreover, preclinical studies and clinical studies are included in this section. Chapter 5 provides an introduction to nuclear factor kappa B (NF-κB) and matrix metalloproteinase (MMP) pathways and the role of NF-κB and MMPs in inflammatory diseases. It describes the role of NF-κB and MMPs in the pathogenesis ofRA and phytochemicals targeting these pathways and preclinical studies. Chapter 6 provides information about the arachidonic acid pathway and phytochemicals targeting the arachidonic acid pathway to treat RA.

Chapter 7 provides an introduction to JAK/STAT pathway and its role in the pathogenesis of arthritis. It provides information about phytochemicals targeting JAK/STAT pathway to treat RA. Chapter 8 provides an introduction to PI3K/Akt/mTOR signaling pathway and its role in the pathogenesis of arthritis. It provides information about phytochemicals targeting PI3K/Akt/mTOR signaling pathway to treat RA. Chapter 9 provides comprehensive information to the reader about the introduction of transforming growth factor and inducible nitric oxide synthase (iNOS) and their role in inflammatory diseases and in the pathogenesis of RA. It describes phytochemicals targeting transforming growth factor and inducible nitric oxide synthase and preclinical studies.

Chapter 10 provides an introduction to vascular endothelial growth factor and RANKL pathway and their role in inflammatory diseases and in the pathogenesis of RA. It provided information about phytochemicals targeting VEGF and RANKL pathways and preclinical studies. Chapter 11 provides valuable information about clinical studies of phytomedicines for the management of RA.

Section C provides valuable information to the readers regarding the use of phytochemicals as nutraceuticals and the toxicity of phytochemicals. This section contains two chapters. Chapter 12 provides an introduction to nutraceuticals, the importance of nutraceuticals, and the role of nutraceuticals in the treatment of various diseases including RA. Chapter 13 provides valuable literature to readers about the toxic/adverse effects of the plant-derived compounds (phytocompounds) on health. Future studies, suggestions, and guidelines for toxicity study of phytochemicals are discussed to avoid/control these adverse effects.

Section D covers novel dosage forms of phytochemicals and the use of photodynamic therapy for the treatment of RA. This helps researchers understand and find new therapeutic approaches and strategies to develop advanced dosage forms for the treatment of RA. This section comprises three chapters.

Chapter 14 provides an introduction to various targeted dosage forms that are used in numerous research studies to cope with the problems of drug adverse reaction, solubility, bioavailability, target site delivery, etc. and enhance therapeutic effectiveness. Chapter 15 provides valuable and comprehensive knowledge about phytochemicals, which are used as a potent photosensitizer and employed in photodynamic therapy (PDT) for the treatment of arthritis. This chapter provides information to researchers and pharmaceutical industries about the design and use of photoactive extracts/phytochemicals for the treatment of arthritis. Chapter 16 provides readers with valuable and comprehensive knowledge about constraints and future prospects of phytochemical products. This chapter provides information to researchers and pharmaceutical industries about the pharmaceutical, regulatory, and marketing requirements of anti-arthritic phytochemical drugs and nutraceuticals and the importance of encouraging mediculture.

Ammara Saleem
Muhammad Furqan Akhtar
Alam Zeb

About the Editors

Ammara Saleem is an Assistant Professor in the Department of Pharmacology, Faculty of Pharmaceutical Sciences, Government College University Faisalabad, Faisalabad, Pakistan. She has published more than 117 research papers in renowned journals with an impact factor of more than 440. She has 11 years of teaching and research experience. Her specific research areas are inflammatory diseases, metabolic disorders, and neurological and immunological disorders. She has published various research articles regarding medicinal plants for the treatment of inflammatory and immune disorders. She is an editorial board member in journals such as *BioMed Research International* and the *Canadian Respiratory Journal*. She has reviewed for different well-reputed international scientific journals such as *Molecules*, *Food*, and *Frontiers in Pharmacology*.

Muhammad Furqan Akhtar is an Associate Professor in the Riphah Institute of Pharmaceutical Sciences, Riphah International University, Lahore Campus, Pakistan. He has 14 years of teaching and research experience. He is an author of more than 150 peer-reviewed ISI-indexed articles with a cumulative impact factor of 500. He has supervised more than 49 postgraduate pharmacology students and serves as an academic/associate editor of several prestigious international journals such as *International Journal of Endocrinology*, *PLoS One*, *Evidence-Based Complementary and Alternative Medicine*, *Frontiers in Pharmacology*, and *Heliyon*. His research work focuses on preclinical exploration of neurological, endocrine, metabolic, and immune diseases such as infertility, polycystic ovarian syndrome, obesity, drug-induced infertility, diabetes, and arthritis by drug-repurposing, nutraceuticals, and natural and synthetic drugs in animal models of disease and in vitro methods. More specifically, he carries out research studies to investigate the role of potential drugs in ameliorating oxidative stress, genotoxicity, and inflammation through a range of molecular and biochemical techniques so as to elucidate their mechanism of action. At larger scales, he is interested in the treatment of neurological and endocrine diseases, immune-stimulation, phytomedicines, and ethnopharmacology. He also investigates the potential health hazardous effects, especially genotoxicity, endocrine disruption, and neurodegenerative diseases associated with the environmental pollutants such as microplastics and herbicides, so as to develop strategies to overcome these problems. Through these studies, he has investigated, compared, and corroborated a range of nutraceuticals, herbal remedies, and repurposed drugs for the management of various chronic ailments.

Alam Zeb is a Professor of Biochemistry, at the Department of Biochemistry, University of Malakand. He previously worked in The Bioactive Lab, Center for Desert Agriculture, King Abdullah University of Science and Technology, Thuwal, Saudi Arabia. He has beeen serving for 23 years at the same university teaching chemistry and biochemistry courses to undergraduate, graduate, and postgraduate students of biotechnology and biochemistry. He has received his PhD with distinction from the Institute of Biochemistry, Technical University of Graz, Austria. He has published more than 100 research articles regarding the subject in various international peer-reviewed journals. He had authored two books, and the recent one is a textbook entitled *Phenolic Antioxidants in Foods: Chemistry, Biochemistry and Analysis*, published in 2021. The second textbook is entitled *Food Frying: Chemistry, Biochemistry and Safety*, published in 2019. This book has been ranked fourth among the best top ten books of all times in the field of chemistry and safety by Bookauthority.org. He had supervised several PhD and MPhil research students during his service at the University of Malakand. He is a member of the editorial board of several international journals, including *Frontiers in Nutrition*, *Journal of Food Processing and Preservation*, and *Current Functional Foods*. He has been a productive scientist of Pakistan since 2009 and has received research productivity awards from the Pakistan Council for Science and Technology, Pakistan, for several consecutive years, representing his contribution in the field of biochemistry.

Contributors

Atika Afzal
Faculty of Pharmacy
The University of Lahore
Lahore, Pakistan

Qadeer Ahmad
Department of Pharmacology
Faculty of Pharmaceutical Sciences
Government College University
Faisalabad, Pakistan

Anil Ahsan
Molecular Pharmacology & Experimental Therapeutics at Mayo Clinic Rochester
Rochester, Minnesota

Sajid Ali
Department of Chemistry, Ångström Laboratory
Uppsala University
Uppsala, Sweden

Mahdia Arshad
Faculty of Pharmacy
Hamdard University
Islamabad, Pakistan

Rabia Arshad
Faculty of Pharmacy
The University of Lahore
Lahore, Pakistan

Ghulam Md Ashraf
Neuroscience and Neuropsychiatry Group
Research Institute for Medical and Health Sciences
University of Sharjah
Sharjah, United Arab Emirates

Tanzeela Awan
Bakhtawar Amin College of Pharmaceutical Sciences
Bakhtawar Amin Medical and Dental College
Multan, Pakistan

Mirza Muhammad Faran Ashraf Baig
Department of Chemistry
School of Science
The Hong Kong University of Science and Technology
Hong Kong, China

Ejaz Basheer
Department of Pharmacognosy
Faculty of Pharmaceutical Sciences
Government College University Faisalabad
Faisalabad, Pakistan

Sampath Chinnam
Department of Chemistry
M.S. Ramaiah Institute of Technology
Bengaluru, India

Muhammad Ejaz-ul-Haq
Riphah Institute of Pharmaceutical Sciences
Riphah International University
Faisalabad, Pakistan

Noor-ul-Eman
Department of Pharmacology
Faculty of Pharmaceutical Sciences
Government College University
Faisalabad, Pakistan

Ayesha Fatima
Bakhtawar Amin College of Pharmaceutical Sciences
Bakhtawar Amin Medical and Dental College
Multan, Pakistan

Mehak Fatima
Department of Pharmacology
Faculty of Pharmaceutical Sciences
Government College University
Faisalabad, Pakistan

Humaira Gul
Department of Pharmacology
Faculty of Pharmaceutical Sciences
Government College University
Faisalabad, Pakistan

Rukhsar Imran
Faculty of Pharmacy
The University of Lahore
Lahore, Pakistan

Muhammad Irfan
Department of Pharmaceutics
Faculty of Pharmaceutical Sciences
Government College University
Faisalabad, Pakistan

Faraza Javaid
College of Pharmacy
Quaid-e-Azam Educational Complex
Sahiwal, Pakistan

Chanda Javed
Department of Medical Imaging Technology
Shifa Tameer e Milat University
Islamabad, Pakistan

Joham Javed
Faculty of Pharmacy
The University of Lahore
Lahore, Pakistan

Sana Kashan
Institute of Molecular Biology and Biotechnology (IMBB)
The University of Lahore
Lahore, Pakistan

Syed Haroon Khalid
Faculty of Pharmacy
Universiti Technolgi Mara (UiTM)
Selangor, Malaysia
and
Department of Pharmaceutics
Faculty of Pharmaceutical Sciences
Government College University
Faisalabad, Pakistan

Muhammad Imran Khan
Riphah Institute of Pharmaceutical Sciences
Riphah International University
Lahore, Pakistan

Shaneel Kousar
Faculty of Pharmacy
The University of Lahore
Lahore, Pakistan

Ayesha Mahmood
Faculty of Pharmacy
The University of Lahore
Lahore, Pakistan

Muntaha Maqbool
Department of Pharmacology
Faculty of Pharmaceutical Sciences
Government College University
Faisalabad, Pakistan

Hamna Mehmood
Department of Pharmacology
Faculty of Pharmaceutical Sciences
Government College University
Faisalabad, Pakistan

Aisha Mobashar
Faculty of Pharmacy
The University of Lahore
Lahore, Pakistan
and
Equator University of Science and Technology
Maska, Uganda

Aqsa Mushtaq
Faculty of Pharmacy
Hamdard University
Karachi, Pakistan

Muhammad Naeem
Department of Operation Theater Technology
Faculty of Medical Sciences
Government College University Faisalabad
Faisalabad, Pakistan

Samia Gul Niazi
Faculty of Pharmacy
Hamdard University
Islamabad, Pakistan

Shahid Rasool
College of Pharmacy
University of Sargodha
Sargodha, Pakistan

Shakila Sabir
Riphah Institute of Pharmaceutical Sciences
Riphah International University
Faisalabad, Pakistan

Uzma Saher
Department of Pharmacy
The Women University
Multan, Pakistan

Mohammad Saleem
Department of Pharmacology,
University College of Pharmacy,
University of The Punjab,
Lahore, Pakistan

Ramla Shabbir
Faculty of Pharmacy
The University of Lahore
Lahore, Pakistan

Shahida Shaeen
Faculty of Pharmacy
Hamdard University
Islamabad, Pakistan

Aisha Shehzad
Lyallpur Institute of Advanced Studies College of Pharmacy
Faisalabad, Pakistan

Muhammad Asim Shehzad
Department of Surgery
Allied Hospital Faisalabad
Faisalabad, Pakistan

Rida Siddique
Faculty of Pharmaceutical Sciences
Government College University
Faisalabad, Pakistan

Maria Tahir
Department of Pharmacology
Faculty of Pharmaceutical Sciences
Government College University
Faisalabad, Pakistan

Muhammad Talha
Department of Pharmacognosy
Faculty of Pharmaceutical Sciences
Government College University Faisalabad
Faisalabad, Pakistan

Yusuf Tatur
Division of Medical Biochemistry
Department of Basic Medical Sciences
Faculty of Medicine, Recep Tayyip Erdogan University
Rize, Türkiye

Muhammad Najeeb Ullah
Department of Pharmacognosy
Faculty of Pharmaceutical Sciences
Government College University Faisalabad
Faisalabad, Pakistan

Justyna Wozniak
Institute of Clinical Pharmacology
University Hospital
RWTH University
Aachen, Germany

Hafiza Sidra Yaseen
Faculty of Pharmacy
The University of Lahore
Lahore, Pakistan

Sara Zahid
Institute of Molecular Biology and Biotechnology (IMBB)
The University of Lahore
Lahore, Pakistan

Section A

Introduction

1 Introduction to Chemistry of Phytochemicals

Alam Zeb

ABBREVIATIONS

7-HC	7-hydroxycoumarin
IL-1β	Interleukin-6
JNK	c-Jun N-terminal kinase
NF-κB	Nuclear factor κ light-chain enhancer of activated B
RA	Rheumatoid arthritis
TNF-α	Tumor necrosis factor-alpha

1.1 INTRODUCTION

Phytochemicals are bioactive, non-nutritive plant-based substances that can prevent illness. They are nutrients that are not essential but are produced by plants. Consuming a diet rich in phytochemicals promotes health benefits and protects the body against illness. Due to their numerous uses, these bioactive components derived from plants have recently garnered a lot of attention. Medicinal plants provide the largest source of ingredients for traditional medical systems, modern medications, nutraceuticals, dietary supplements, traditional medicines, pharmaceutical intermediates, and chemical entities for synthesized pharmaceuticals. Because of their positive impacts on human health and significant advantages to consumers, phytochemicals are very intriguing and have considerable antioxidant potential.

Food contains phytochemicals with nutraceutical qualities that are crucial for improving human health. These compounds provide defense against various illnesses, such as malignancies; heart attacks; diabetes; high blood pressure; infections from bacteria, viruses, and parasites; inflammatory diseases; mental illnesses; spasmodic conditions; ulcers; arthritis; and other bone-related disorders (Thakur, Singh, & Khedkar, 2020).

An autoimmune condition called rheumatoid arthritis (RA) is characterized by the invasive and erosive nature of synovium tissue to the bone. It is caused by the immune system's malfunction, which damages healthy cells in the joints and induces inflammation, which in turn causes swelling, pain, and bone loss. To aid researchers in the development of phytomedicines against arthritis, Gandhi et al. (2022) recently summarized potential medicinal plants and naturally occurring bioactive compounds against arthritis based on their strong anti-inflammatory properties as shown in several in vitro, in vivo, preclinical, and clinical studies. The sources of phytochemicals were *Punica granatum, Cinnamomum cassia, Curcuma longa, Nigella sativa, Boswellia serrata, Zingiber officinale, Withania somnifera, Tinospora cordifolia, Commiphora wightii, Piper longum, Tribulus terrestris, Camellia sinensis, Glycyrrhiza glabra, Cannabis sativum, Coriander sativum, Saussurea lappa, Aloe barbadensis,* and *Emblica officinalis*. Several phytochemicals, either alone or in a mixture in the form of extract, have been shown to possess anti-RA properties. Thus, only those phytochemicals that are currently known to possess are discussed here.

DOI: 10.1201/9781003535201-2

1.2 TYPES OF PHYTOCHEMICALS

Phytochemicals include various types of compounds, classified as dietary and nondietary. Nondietary compounds, which are toxic, are not discussed in this chapter. Dietary phytochemicals come from both food and nonfood plants. The term *dietary phytochemicals* refers to a wide range of chemical compounds, some of which are bioactive molecules found in nature that appear to have important physiological effects on humans, such as steroids, flavonoids, polyphenols, and saponins. There is no clear classification system available due to the complex structure and properties of these compounds. However, known phytochemicals related to RA are classified as polyphenolic compounds, terpenoids, alkaloids, organo-nitrogen compounds, and organo-sulfur compounds.

1.2.1 Polyphenolic Compounds

Polyphenolic compounds are made of one or more hydroxyl groups directly attached to one or more aromatic rings. The aromatic ring is responsible for making phenol a weak acid by donating an H-atom. In nature, phenolic compounds are biosynthesized by plants (Zeb, 2020). More than 8,000 different phenolic compounds have been identified until now. Phenolic compounds are classified based on several properties. However, there is no clear consensus on the classification of phenolic compounds. Phenolic compounds can be classified into four different means, starting from the most general to the most specific (Zeb, 2021):

1. Flavonoids and nonflavonoids
2. Number of aromatic rings
3. Carbon skeleton, which portrays in a very basic way how the carbon atoms of the molecule are organized (e.g., C6, C6–C1, and C6–C3–C6)
4. Basic chemical structure, which is a description or image that specifies common atoms (e.g., C, H, and O), functional groups (e.g., aromatic rings, hydroxyl groups, and keto groups), saturations (e.g., single bonds and double bonds), types of bonds (e.g., covalent and ionic), and how are they all linked to one another.

Number 3 classification was reported by Harborne and Simmonds (1964) and is still widely used as shown in Table 1.1. However, some authors combined classifications 3 and 4 to form one single classification.

An increasing body of research suggests that dietary polyphenols may be useful as a treatment for RA due to their anti-inflammatory, immunosuppressive, apoptotic, and antioxidant properties (Table 1.2). They may also do this by modulating pathways such as nuclear factor κ light-chain enhancer of activated B cell (NF-κB), c-Jun N-terminal kinase (JNK), interleukin (IL)-1β, IL-6, mitogen-activated protein kinase, and tumor necrosis factor-alpha (TNF-α) (Behl et al., 2022; Gonçalves, Rodrigues, Fonseca, & Silva, 2024, Jabbar et al., 2024). A brief account of some of the classes of phenolic compounds used for RA treatment is given here.

1.2.2 Phenolic Acids

The term phenolic acid is usually ascribed to all phenolic compounds having a carboxylic acid group. Thus, phenolic acids are classified as derivatives of hydroxybenzoic acid and cinnamic acid. In the first case, the presence of a carboxylic acid group on a phenyl ring at a specific position classified it into phenolic acids. Examples of the first class are *p*-hydroxybenzoic acid, salicylic acid, protocatechuic acid, and gallic acid. Phenolic aldehyde is distinguished from hydroxybenzoic acid based on the replacement of the carboxylic acid group by an aldehyde group. The most important

TABLE 1.1
Classification of Phenolic Compounds

Class	Structure	Example(s)
Hydroxybenzoic acids	C6–C1	Hydroxybenzoic acid, gallic acid
Phenolic aldehydes	C6–C1	Vanillin
Acetophenones and phenylacetic acids	C6–C2	Phenylacetic acid
Phenylethanoids	C6–C2	Tyrosol, hydroxytyrosol
Cinnamic acids, cinnamyl aldehydes, cinnamyl alcohols, esters	C6–C3	Cinnamic acid, phenylacetone
Coumarins, isocoumarins, and chromones	C6–C3	1-Benzopyran-2-one
Benzophenones, xanthones	C6–C1–C6	Mangostin
Stilbenes	C6–C2–C6	3,4′,5-Stilbenetriol
Anthraquinones	C6–C2–C6	Emodin
Chalcones, dihydrochalcones	C6–C3–C6	Butein
Aurones	C6–C3–C6	Bracteatin
Flavones	C6–C3–C6	Apigenin
Flavanones	C6–C3–C6	Naringin
Flavonols	C6–C3–C6	2,3-Dihydroflavonol
Flavanols	C6–C3–C6	Catechin, epicatechin
Isoflavons	C6–C3–C6	Diadzein
Neoflavones	C6–C3–C6	Dalbergin
Anthocyanidins	C15	Cyanidin, malvidin
Biflavonyls	C30	Hinokiflavone
Betacyanins	C18	Betanin
Lignans, neolignans, dimers, or oligomers	Dimers	Sesamin; syringaresinol
Lignins	Polymers	Polymer of sinapyl alcohol
Tannins	Oligomers or polymers	Tannic acid, procyanidin B2, punicalagin

TABLE 1.2
Effects of Polyphenols on In Vitro and In Vivo Rheumatoid Arthritis Model System

Class	Compound	Medium	Effects	Reference
Phenolic acids	Gallic acid	Fibroblast-like synoviocytes	Regulation of apoptosis and pro-inflammatory markers	Yoon et al. (2013)
	p-Coumaric acid	Rat	Anti-inflammatory properties	Pragasam et al. (2013)
Phenylethanoids	Hydroxytyrosol	Rat	Markedly reduced the degree of bone resorption, soft tissue swelling, and osteophyte formation	Silva et al. (2015)
Coumarins	7-Hydroxycoumarin	Fibroblast-like synoviocytes & rat	Inhibition of cell proliferation and induction of apoptosis through Wnt/β-catenin pathway	Cai et al. (2022)

(Continued)

TABLE 1.2 (*Continued*)
Effects of Polyphenols on In Vitro and In Vivo Rheumatoid Arthritis Model System

Class	Compound	Medium	Effects	Reference
Stilbenes	Resveratrol	Rats	Reduction in serum levels of IL-1β, C-reactive protein, and prostaglandin E2, as well as nuclear factor κB synovial tissue expression	Fernández-Rodríguez et al. (2021)
Chalcones	Trans-chalcone	Rats	Reduction in IL-1β, IL-6, and IL-17	Jabbar et al. (2024)
Flavonoids	Apigenin	Mice	Downregulation of VEGF and VEGFR	Li et al. (2019)
	Galangin, kaempferol, quercetin, myricetin	RA patients	Suppression of ROS production	Santos et al. (2014)
	Catechins	Fibroblast-like synoviocytes	Hyaluronic acid–epigallocatechin-3-gallate conjugates act as an anti-arthritic agent	Lee et al. (2021)
	Taxofolin	Human bone marrow-derived macrophages and mouse model	Inhibition of osteoclastogenesis through the suppression of the NF-κB signaling pathway	Zhang et al. (2018)
	Genestein	MH7A cells	Inhibition of IL-6-induced VEGF expression	Cheng et al. (2020)
Anthocyanins	Cyanidin	Human osteoarthritis chondrocyte	Modulating the Sirt6/NF-κB signaling	Jiang et al. (2019)

example is vanillin. Cinnamic acids are phenolic compounds with the backbone of C6–C3 and are also known as phenylpropanoids. The first important member of the class is cinnamic acid. Cinnamic acid, p-coumaric acid, caffeic acid, 5-hydroxy ferulic acid, ferulic acid, and sinapic acid are members of cinnamic acids. Some important phenolic acids are shown in Figure 1.1.

Cinnamic acids form an ester with 3, 4, or 5-hydroxyl of L-quinic acid or various other organic acids. Chlorogenic acid (3-caffeoylquinic acid) is an ester of caffeic acid and quinic acid. The isomers of chlorogenic acid are cryptochlorogenic acid (4-caffeoylquinic acid) and neochlorogenic acid (5-caffeoylquinic acid). Chlorogenic acid or its isomers are found in various plant foods such as apples, artichokes, carrots, coffee beans, eggplants, grapes, pears, plums, and potatoes (Santana-Gálvez, Cisneros-Zevallos, & Jacobo-Velázquez, 2017; Santana-Gálvez & Jacobo-Velázquez, 2018).

Phenolic acids have been shown to possess anti-RA properties. Yoon et al. (2013) showed that through the control of apoptosis-related protein expressions, gallic acid therapy was discovered to activate apoptosis in fibroblast-like synoviocytes of RA patients and decrease the expression of pro-inflammatory genes. According to these findings, gallic acid may be utilized as a potential treatment for RA due to its pro-apoptotic and anti-inflammatory properties. The *p*-coumaric acid's anti-inflammatory properties against adjuvant-induced rat arthritis are amply demonstrated by Pragasam, Murunikkara, Sabina, and Rasool (2013).

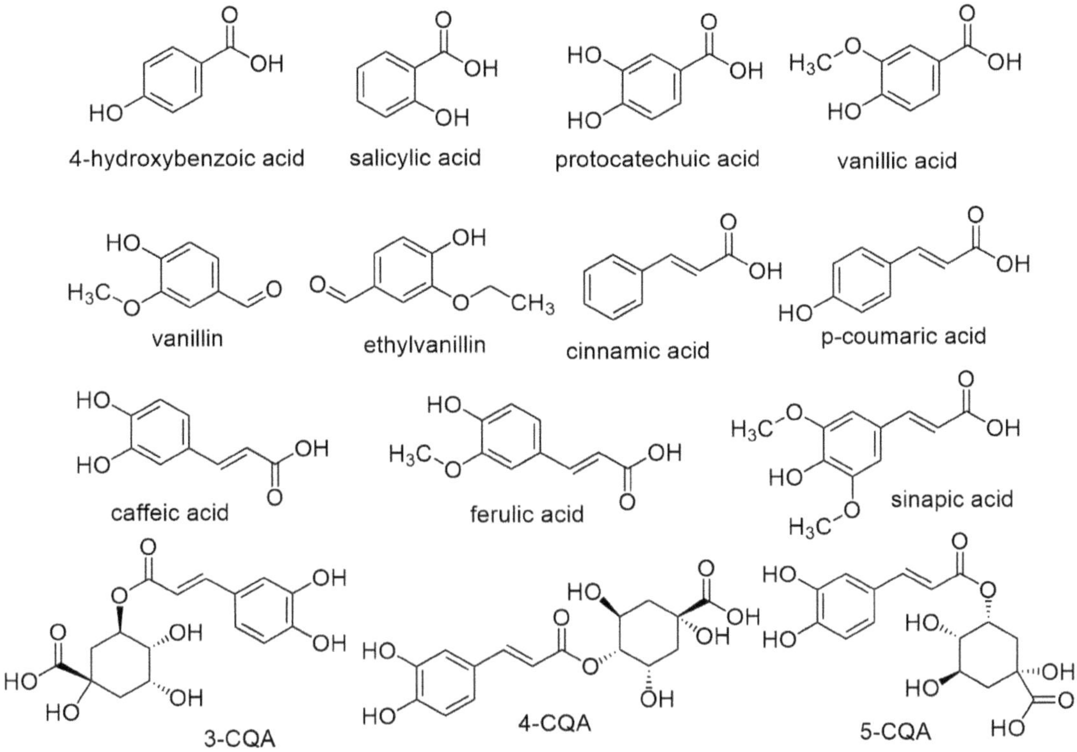

FIGURE 1.1 Chemical structures of important phenolic acids.

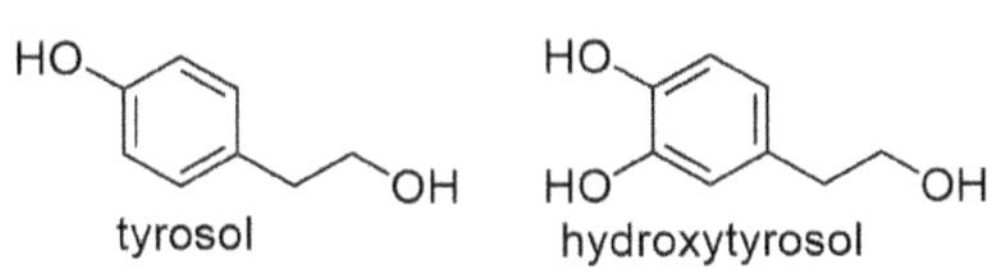

FIGURE 1.2 Chemical structures of important phenylethanoids.

1.2.3 Phenylethanoids

Phenolic compounds with characteristic phenyl and ethanol groups are called phenylethanoids. These compounds are represented by C6–C2. The common examples are tyrosol (4-(2-hydroxyethyl) phenol) and hydroxytyrosol (4-(2-hydroxyethyl)-1,2-benzenediol) as shown in Figure 1.2. Both tyrosol and hydroxytyrosol possess in vitro antioxidant activities (Covas et al., 2003). They are present in wine and olive oils. Silva et al. (2015) showed that hydroxytyrosol supplementation of refined olive oil may be beneficial in RA, having a notable effect on both acute and chronic inflammatory processes.

1.2.4 Coumarins

Coumarins have the structural backbone of C6–C3 consisting of an oxygen heterocycle as part of the molecule. They are also known as 1,2-benzopyrones, consisting of fused benzene and pyrone rings. Coumarins have been classified as simple coumarins, Furanocoumarins, Pyranocoumarins, Bis-, and Triscoumarins, and Coumarinolignans. Coumarin, 7-hydroxycoumarin (7-HC), bergenin, and esculetin are common coumarins present in foods (Figure 1.3). Coumarins are widely distributed in

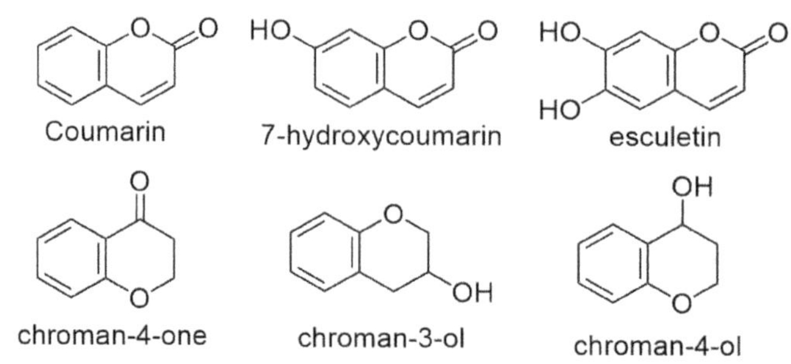

FIGURE 1.3 Chemical structures of important coumarins.

plants and plant foods such as citrus fruits. Coumarins and their derivatives are strong antioxidants in in vitro model systems (Kostova et al., 2011).

The collagen-induced arthritis in rats was lessened in severity by 7-HC (Cai et al., 2022). This was shown by decreased paw edema, arthritic index, joint damage, serum level of collagen type II antibody, and serum and synovium production of IL-1β, IL-6, and TNF-α. Specifically, 7-HC in vivo exhibited pro-apoptotic and anti-proliferative effects on collagen-induced arthritis rat synovial cells, as evidenced by elevated synovial apoptosis index, decreased Bcl-2 protein level, increased Bax level, and cleaved caspase 3 protein.

1.2.5 Stilbenes

Stilbenes are a class of phenolic compounds characterized by the skeleton of C6–C2–C6. They have a 1,2-diphenylethylene nucleus and can be divided into two classes, i.e., monomeric and oligomeric stilbenes. The ethylene bond may be either cis or trans. Some common examples are 4,4′-dihydroxy-*E*-stilbene, 3,4-dihydroxy-3′-methoxystilbene, resveratrol, hexahydroxy-*E*-stilbene, and 3,4,4′-trihydroxy-*E*-stilbene (Figure 1.4). So far, more than 300 derivatives of stilbenes have been reported in both food and nonfood plants. The most important stilbene is resveratrol and its derivatives, which are widely known for their anti-arthritis properties (Karimi et al., 2022). Almonds, for example, contain *trans*-resveratrol, d4-resveratrol, dienestrol, hexestrol, oxyresveratrol, piceatannol, pterostilbene, and resveratrol-3-β-glucoside. Several stilbenes have been found in different wines and their products, legumes, berries, and fruits (Santana-Gálvez & Jacobo-Velázquez, 2018; Vitrac et al., 2005). Resveratrol supplements reduce the crosstalk with inflammation and stimulate the noncanonical autophagy pathway, which in turn controls the synovial hyperplasia in rats (Fernández-Rodríguez et al., 2021).

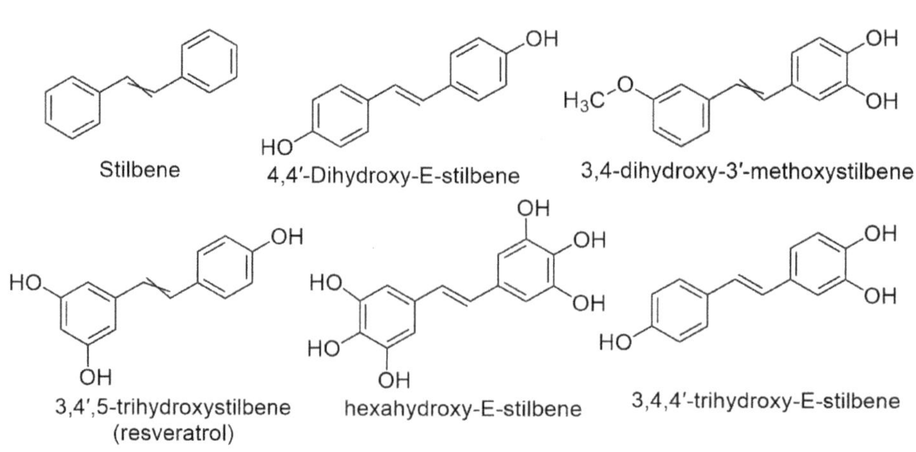

FIGURE 1.4 Chemical structures of important stilbenes.

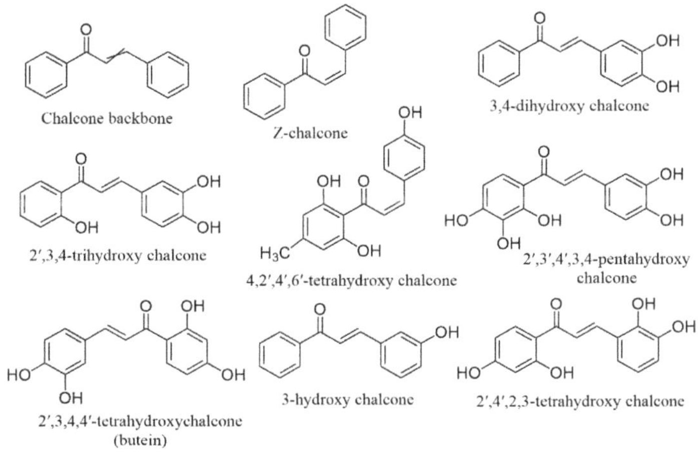

FIGURE 1.5 Chemical structures of chalcones.

1.2.6 Chalcones

Chalcones and their derivatives are phenolic compounds made of the basic skeleton of C6–C3–C6. The C3 bridge connects the two aromatic rings, which may contain a double bond, a functional group, or any substituents as shown in Figure 1.5. Chalcones such as lutein ((E)-1-(2,4-dihydroxy phenyl)-3-(3,4-dihydroxy phenyl) prop-2-en-1-one) are responsible for the yellowish color of flowers. Chalcones are converted to food sweeteners called dihydrochalcone. Chalcones and their derivatives are present in edible oils (Dziedzic & Hudson, 1983), apples, and citrus fruits (Mah, 2020; Tomás-Barberán & Clifford, 2000). Chalcones also form an ester with monosaccharides to form glycosides. Trans-chalcone has demonstrated efficacy in reducing flexion pain and joint stiffness in rats by significantly reducing inflammatory cytokines.

1.2.7 Flavonoids

Flavonoids are phenolic compounds having 15 carbons with a C6–C3–C6 skeleton. The chemical structure consists of three rings, i.e., A, B, and C. Rings A and B are benzene rings, whereas ring C is a pyran. Thus, the backbone is 2-phenyl-3,4-dihydro-2H-chromene as shown in Figure 1.6. Flavonoids are a highly diverse group of phenolic compounds, which are essential components of plant cells. They can be classified based on several factors such as structure, functional groups, properties, and entropy of information theory (Zeb, 2021). The generally accepted classes of flavonoids are flavone, flavanone, neoflavone, isoflavone, flavonol, and anthocyanin.

Flavones: Flavones are a diverse subgroup of flavonoids present in nearly all plant foods. Plant foods are rich in flavones such as 6-hydroxy flavone, apigenin, luteolin, acacetin, diosmetin, chrysoeriol, hispidulin, oroxylin A, chrysin, baicalein, wogonin, and tangeritin. The structures of some of the important flavones are shown in Figure 1.7. Apigenin suppresses osteoclastogenesis, angiogenesis, and synovial hyperplasia, which prevents collagen-induced arthritis in the mice model (Li et al., 2019).

Flavanones: Flavanones are distinguished from flavones in the absence of a double bond in ring-C. The rest of the molecule may have different substituents attached as shown in Figure 1.8. The absence of a double bond (at C2–C3), the presence of a chiral carbon atom at the C2 position, and the absence of substitution at the C3 position of the C-ring mark the structural differences characterizing flavanones as a separate class of flavonoids (Zeb, 2021).

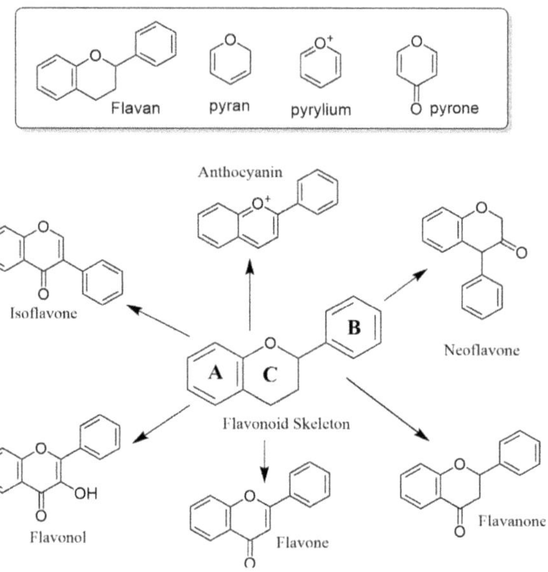

FIGURE 1.6 Chemical structures of backbones and classification of flavonoids.

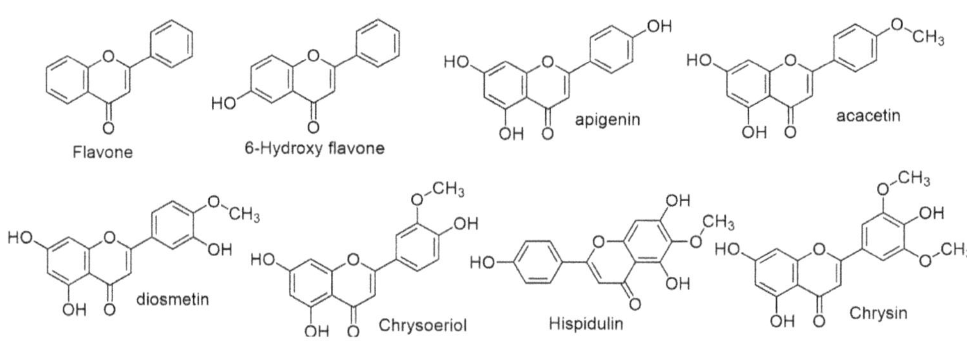

FIGURE 1.7 Chemical structures of flavone backbone and important flavones.

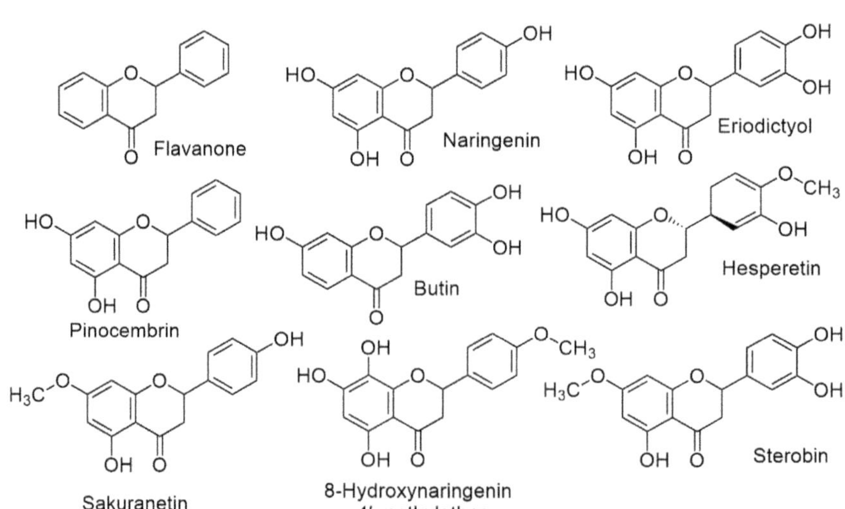

FIGURE 1.8 Chemical structures of flavanone backbone and important flavanones.

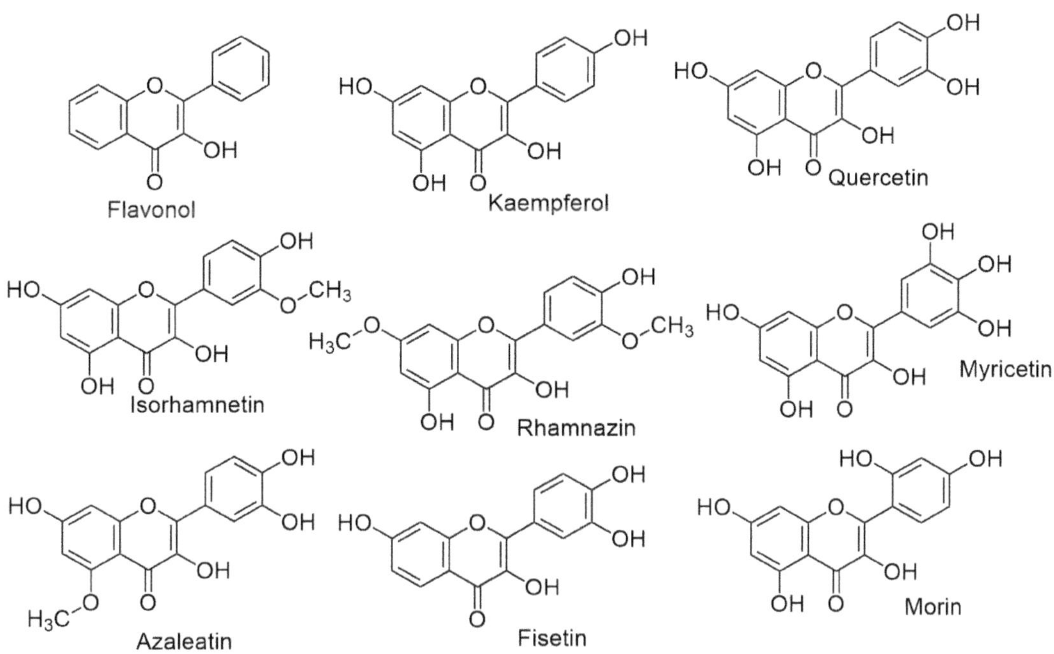

FIGURE 1.9 Chemical structures of flavonol backbone and important flavonols.

There are more than 300 flavanones present in nature and more than a hundred as glycosides. The most recurrent examples of flavanones are those present in citrus fruits. These include naringenin, eriodictyol, pinocembrin, butin, hesperetin, sakuranetin, and sterobin.

Flavonols: Flavonols are one of the most important classes of flavonoids. They have a backbone of 3-hydroxy flavone. The diversity in flavonols is due to their difference in the hydroxyl groups attached to the A and B rings as shown in Figure 1.9. Flavonols are also present in glycosidic forms. They are present in fruits and vegetables (Aherne & O'Brien, 2002). Berries such as cranberry and blueberry are especially rich in flavonols (Kamiloglu, Tomas, & Capanoglu, 2019; Wang, Fong, Singh, Vorsa, & Johnson-Cicalese, 2019). The therapeutic efficiency of flavonols such as galangin, kaempferol, quercetin, and myricetin was determined in RA patients (Santos et al., 2014). The degree of hydroxylation of the flavonol B-ring determined the flavonols' capacity to scavenge hypochlorous acid and chloramines, but not H_2O_2. For patients with active RA, modulating immune complex-mediated neutrophil activation using these drugs offers a safe and promising therapeutic approach to managing inflammation.

Flavanols: Flavanols are distinguished from flavonols by the absence of a ketonic group at the 4-position and a double bond in the C-ring. Catechin, epicatechin, and epigallocatechin are important examples (Figure 1.10) present in plants such as cocoa, red wine, green tea, red grapes, berries, and apples (Hackman et al., 2007). They contain two chiral carbons, resulting in four diastereoisomers. For example, (+)-catechin (2R,3S), (–)-catechin (2S,3R), (–)-catechin (2S,3S), and (–)-catechin (2R,3R) representing four isomers of catechin. Pulse radiolysis studies of the flavanols revealed that they are superior antioxidants to their counterparts flavonols and flavones (Bors & Michel, 1999). The primary catechin found in green tea, epigallocatechin-3-gallate, has garnered much attention as a potential treatment for RA due to its capacity to inhibit synoviocyte proliferation and IL-6 release (Lee, Bae, Ng, Yamashita, & Kurisawa, 2021).

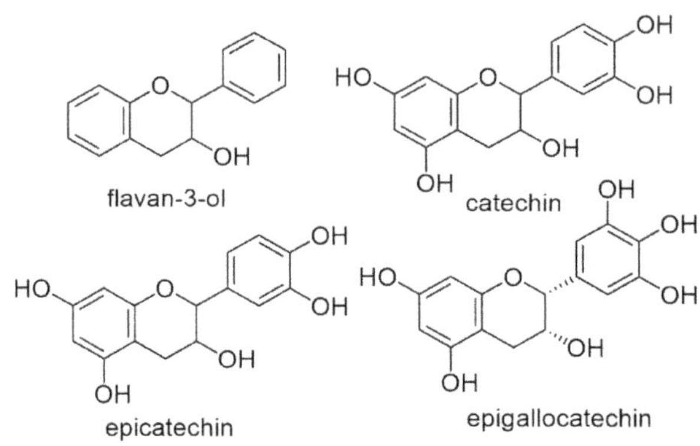

FIGURE 1.10 Chemical structures of flavanols.

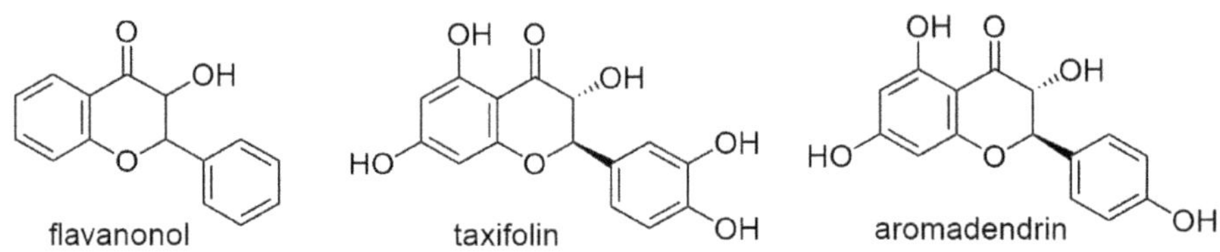

FIGURE 1.11 Chemical structures of flavanonols.

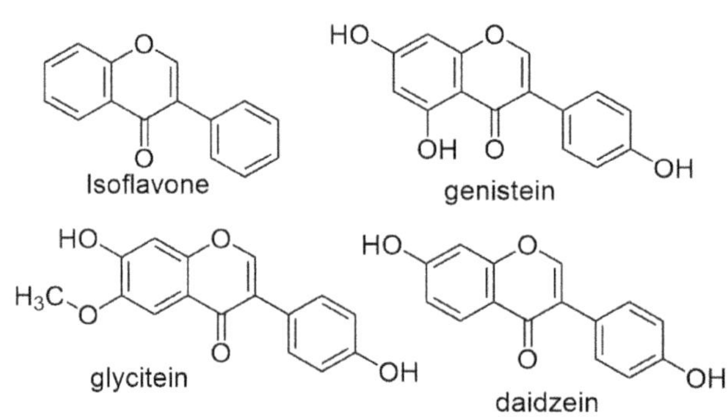

FIGURE 1.12 Chemical structures of isoflavone backbone and important isoflavones.

Flavanonols: They are also known as dihydroflavonols. Like flavanols, each flavanonol has two chiral carbons and will thus have R & S isomers. Toxifolin, also known as dihydro quercetin, has two stereocenters unlike quercetin, which has none. These are (2R,3R) configuration, making it one out of four stereoisomers that form two pairs of enantiomers. However, keto-enol tautomerism is absent in taxifolin and present in quercetin. The absence of a double bond, however, reduces the importance of the hydroxyl group at carbon-3 of the C-ring. Similar is the case of aromadendrin (**2.131**) also known as dihydrokaempferol as shown in Figure 1.11. These flavanonols also form glycosides in plants. Taxifolin suppresses the NF-κB signaling pathway to prevent osteoclastogenesis (Zhang, Wang, Yang, Gao, & Tang, 2018). In a mouse model of calvarial osteolysis, taxifolin inhibited bone loss.

Isoflavones: Isoflavones are those phenolic compounds in which phenyl ring-B is attached to carbon-3 of the ring-C. They have a similar backbone to the C6–C3–C6 skeleton (Figure 1.12). Many of the isoflavones act as phytoestrogens in animals (Fotsis et al., 1993). Genistein is chemically 4,5,7-trihydroxyisoflavone and is present in various foods such

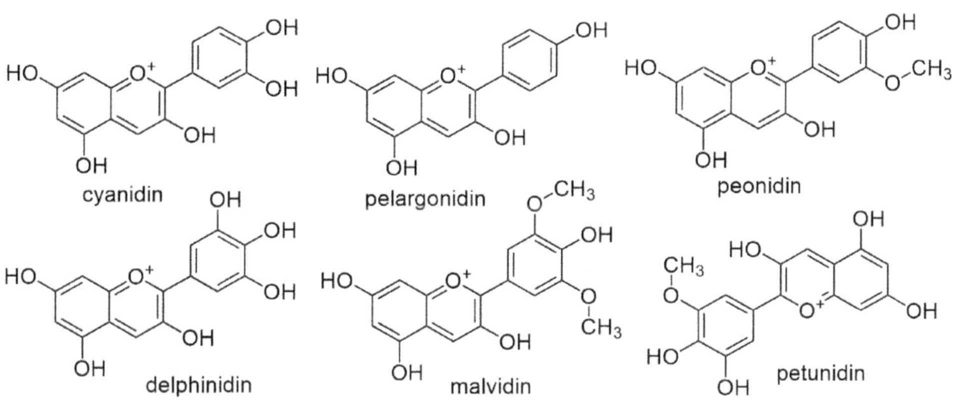

FIGURE 1.13 Chemical structures of anthocyanidins.

as beans and vegetables (Liggins et al., 2000). Daidzein also known as 4′,7-dihydroxyisoflavone is also present in several vegetables (Coward, Smith, Kirk, & Barnes, 1998). Genistein blocks the IL-6/JAK2/STAT3/VEGF signaling pathway, which is responsible for the angiogenesis that results from RA (Cheng et al., 2020).

1.2.8 Anthocyanidins and Anthocyanins

Anthocyanidins are one of the most important classes of phenolic compounds present in plants. The basic skeleton of the anthocyanidins is also C6–C3–C6. The C-ring is a pyrilium cation with a hydroxyl group at position 3. The positive charge on the heterocyclic ring can move around the molecule, making it more stable and contributing to the specific colors. The specific color of some foods may be due to the anthocyanidins such as pelargonidin (orange-red), cyanidin (red), peonidin (rose-red), delphinidin (blue-violet), petunidin (blue-purple), and malvidin (purple) (Figure 1.13). Anthocyanidins rarely occur in free form and usually form glycosides. There are two other classes of anthocyanidins, which are usually derivatives of the parent structures leucoanthocynidins and deoxyanthocyanidins. Anthocyanidin glycosides are also known as anthocyanins. They are mostly 3-glycosides, with the glycone attached by an *O*-glycosidic bond. The glycone is usually glucose, galactose glucuronide, and xylose (Zeb, 2021). The important examples are cyanidin-3-glucoside, malvidin-3-glucoside, peonidin-3-glucoside, delphinidin-3-glucoside, petunidin-3-glucoside, and cyanidin-3-glucoside and their derivatives. Anthocyanins are one of the most important and extensively studied phenolic compounds in plants. They are present in nearly all edible fruits (Einbond, Reynertson, Luo, Basile, & Kennelly, 2004) and possess a significant beneficial role in human health (Khoo, Azlan, Tang, & Lim, 2017). Both anthocyanidins and anthocyanins possess anti-RA properties. In IL-1β-stimulated human osteoarthritis chondrocytes, cyanidin may block the NF-κB pathway. Its action may depend in part on Sirt6 activation, indicating that cyanidin may protect by modulating the Sirt6/NF-κB signaling axis (Jiang et al., 2019). Additionally, the in vivo investigation demonstrated that cyanidin reduced the progression of osteoarthritis in surgically destabilized medial meniscus animal osteoarthritis models.

1.2.9 Terpenoids

Terpenoids are a class of naturally occurring chemicals derived from isoprene, which are made of five carbons. Terpenoids are classified as monoterpenoids, sesquiterpenoids, diterpenoids, sesquiterpenes, triterpenoids, and polyterpenoids based on their molecular skeletons. Terpenoids, one of the most prevalent and varied classes of natural compounds, have garnered a lot of interest lately because of their significant and varied biological properties, such as their strong anti-inflammatory, antioxidant, anticancer, and antiviral properties.

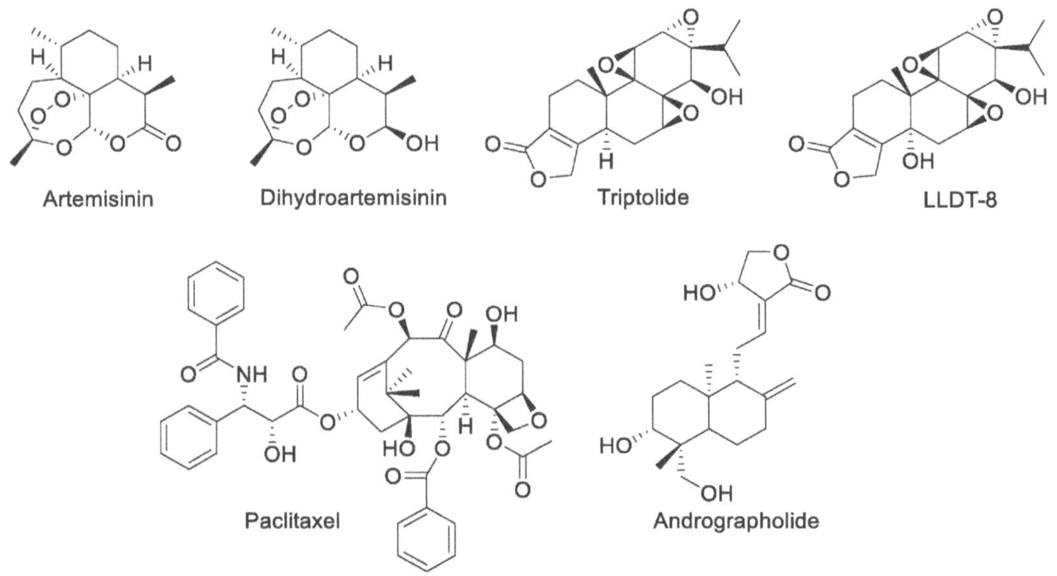

FIGURE 1.14 Chemical structures of important terpenes.

Two isoprene units ($C_{10}H_{16}$) make up monoterpenoids, which exist in acyclic, monocyclic, and bicyclic forms at different oxidation states. Most volatile oils include monoterpenoids, and many of these oils have an aroma. It is commonly known that monoterpenoids play significant roles in the fragrance and medicinal sectors. Several monoterpenoids are pharmacologically active, and they may be further categorized according to their cyclic carbon structures. Sesquiterpenoids ($C_{15}H_{24}$) are chemicals that exist in simple to complex mono- and polycyclic rings and include three isoprene units (Figure 1.14). Triterpenoids, also known as C_{30} compounds ($C_{30}H_{48}$), are widely distributed and originate from around 40 distinct carbon skeletons. Tetraterpenoids include compounds like carotenes ($C_{40}H_{64}$) and are another diverse group that will be discussed separately.

The primary components of essential oils are mono- and sesquiterpenes; the other terpenes are found in rubber, balsams, resins, and waxes. Oleoresin, an approximately equal combination of rosin (C20-diterpene) and turpentine (85% C10-monoterpenes and 15% C15 sesquiterpenes), functions as a poisonous substance to invader insects and associated pathogenic fungus in many conifer species. Other natural compounds also have isoprenoid units in their structure. Terpenoid fragments are therefore also present in indole alkaloids, numerous quinones (vitamin K), alcohols (vitamin E, vitamin A derived from β-carotene), phenols, and isoprenoid alcohols (terpenols or polyprenols). Over 1,000 monoterpenes, 7,000 sesquiterpenes, and 3,000 diterpenes are known to exist, and the quantity is continually rising.

Numerous research studies conducted in the last several decades have demonstrated the anti-inflammatory properties of natural terpenoids and their derivatives. Significantly, terpenoids have demonstrated encouraging characteristics in both preclinical and clinical settings including inflammatory illnesses. The primary active component of the traditional plant *Andrographis paniculate*, for example, was andrographolide (Figure 1.14), which was discovered as a diterpenic molecule (Hancke, Srivastav, Cáceres, & Burgos, 2019).

1.2.10 Carotenoids

Tetraterpene pigments with yellow, orange, red, and purple hues are called carotenoids. Most carotenoids have a 40-carbon backbone and 8 isoprene units. A polyene chain with nine conjugated double bonds and an end group at both ends make up their typical general structures (Maoka, 2020).

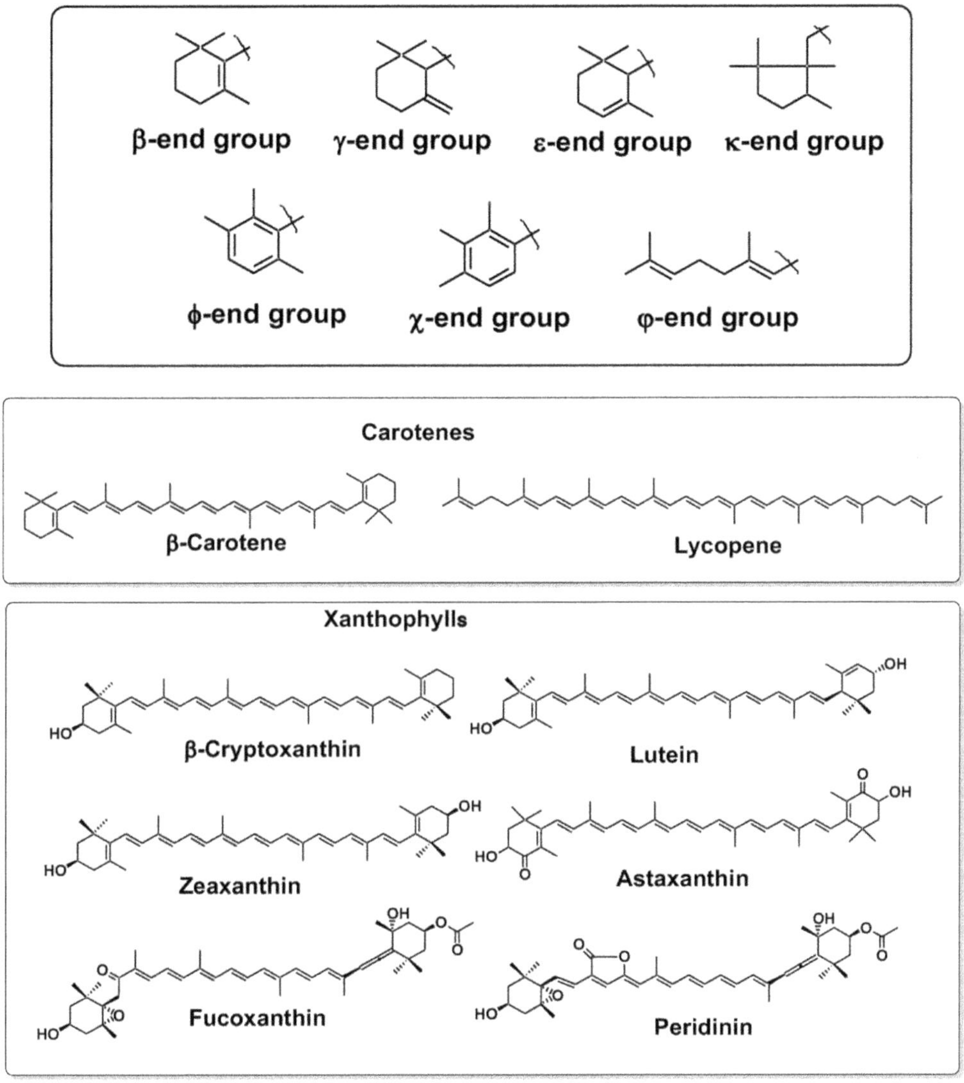

FIGURE 1.15 Chemical structures of carotenoid rings and classification.

Figure 1.15 displays the end groups of carotenoids and the structures of the polyene chain. There are two categories of carotenoids: xanthophylls and carotenes. Hydrocarbons are carotenes, which include lycopene, α-carotene, β-carotene, and ω,ψ-carotene (γ-carotene). In nature, carotenes come in over 50 different varieties. Conversely, xanthophylls are carotenoids that include oxygen atoms such as hydroxy, carbonyl, aldehyde, carboxylic, epoxide, and furanoxide groups (Galanakis, 2019). Examples of these compounds are lutein, zeaxanthin, astaxanthin, fucoxanthin, and peridinin. Certain forms of xanthophylls are found in protein complexes, fatty acid esters, glycosides, and sulfates. Various xanthophyll structures are evident. There are reportedly over 800 different types of xanthophylls in nature.

The majority of carotenoids (C40 carotenoids) have a 40-carbon backbone. Higher carotenoids are those that include a skeleton consisting of 45 or 50 carbons. Certain forms of archaea include about 40 different types of higher carotenoids. Conversely, apocarotenoids are carotenoids made up of carbon skeletons with fewer than 40 carbons. Certain plant and animal species have around 120 different types of apocarotenoids, which are by-products of the breakdown of C40 carotenoids.

The most abundant colors in nature, carotenoids are found in photosynthetic bacteria, some types of fungus and archaea, algae, plants, and mammals.

Common human diets include over 50 different forms of carotenoids, of which about 20 are detectable in blood (either serum or plasma) after being consumed. More than 90% of the total carotenoids are composed of them, with β-carotene, α-carotene, lutein, zeaxanthin, lycopene, and β-cryptoxanthin being the main constituents. The small intestine is where dietary carotenoids are absorbed. Lipase or esterase hydrolyzes xanthophyll esters, which are then absorbed. Provitamin A carotenoids undergo partial conversion into retinal in the small intestine mucosa due to the action of β-carotene-15,15′-dioxygenase. Chylomicrons, which are composed of absorbed carotenoids, are subsequently carried by blood to the liver and other organs.

Carotenoids can inhibit the production of inflammatory cytokines, such as IL-8 or prostaglandin E2, by blocking the translocation of nuclear factor κB to the nucleus and interacting with the nuclear factor κB pathway. They can also block oxidative stress by interacting with the nuclear factor erythroid 2-related factor 2 pathway, enhancing its translocation into the nucleus, and activating phase II enzymes and antioxidants, such as glutathione-S-transferases (Zannou, Pashazadeh, Galanakis, Alamri, & Koca, 2022). These characteristics have been linked recently to the effect of carotenoids on intracellular signaling cascades, thereby influencing gene expression and protein translation. A recent review showed that among all the carotenoids, β-carotene and β-cryptoxanthin showed the strongest health-promoting effects on bones (Kulczyński, Sidor, Brzozowska, & Gramza-Michałowska, 2024).

1.2.11 Glucosinolates

Glucosinolates are anions made up of thiohydroximates with a variable side chain generated from amino acids, an O-linked sulfate residue, and an S-linked β-glucopyranosyl residue. The strong flavors and tastes that cabbage, broccoli, and other Brassica vegetables elicit have made glucosinolates, formerly known as mustard oil glycosides, a part of human life for thousands of years. The chemical structure of the about 120 glucosinolates that have been reported is composed of a variable R group and a β-D-glucopyranose residue that is connected to a (Z)-N-hydroximinosulfate ester by a sulfur atom (Halkier & Gershenzon, 2006). Approximately half of all glucosinolates are classified as aliphatic glucosinolates. These may be categorized according to their straight or branched chains, hydroxyl groups present or absent, and the oxidation state of the sulfur atoms in their side chains (Figure 1.16).

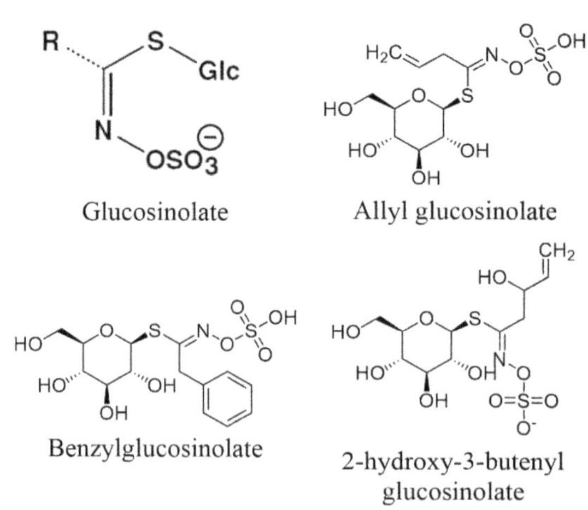

FIGURE 1.16 Chemical structures of glucosinolate backbones and important glucosinolates.

In recent decades, the significance of these secondary metabolites of plants that contain nitrogen and sulfur has increased further due to the discovery of their potential as agents that prevent cancer, compounds that protect crops from damage, and biofumigants in agriculture (Bischoff, 2021). There is proof that their hydrolysis-derived isothiocyanates may have cytoprotective effects by modifying the oxidative stress pathways such as Nrf2/ARE pathway and the inflammatory response NF-kB pathway (Lucarini, Micheli, Di Cesare Mannelli, & Ghelardini, 2022). An extract from the seeds of *Brassica oleracea* L. is thought to be a viable treatment for arthritis (Hamed et al., 2021).

1.2.12 Phytosterols

Phytosterols (PSs) are sterols produced by plants that resemble cholesterol in vertebrate animals both physically and functionally. Chemists have historically used the word *sterol* to refer to substances that have an alcohol moiety added to the fused cyclopentanophenanthrene ring structure of cholesterol. This ring structure is called a steroid nucleus. PSs are made up of a steroid skeleton with a saturated link between the sterol moieties C-5 and C-6. They feature a hydroxyl group connected to the C-3 atom and an aliphatic side chain attached to the C-17 atom (Figure 1.17). PSs can be found in free as well as in bound form. In the latter case, fatty acid, glycoside, or a phenolic moiety is attached to the hydroxy group. PSs are differentiated based on unsaturation, hydroxyl groups, and/or differences in the side chain. The most common plant-based sterols in the human diet are stigmasterol, β-sitosterol, and 4 of 47 campesterol. Campesterol has a C24-methyl group attached to it, while β-sitosterol and stigmasterol have an ethyl group at C-24. Sterol C-22 desaturases are the mechanism by which stigmasterol is produced from sitosterol. Minor components include D-7-avenasterol and brassicasterol.

PSs and their derivatives have been shown to have various pharmacological effects, including the potential to promote human well-being, according to mounting scientific data. One of the most effective ways to lower total and low-density lipoprotein cholesterol levels and so lower the risk of several ailments is through these health advantages (Salehi et al., 2021). Studies showed that in animals with collagen-induced arthritis, β-sitosterol inhibits synovial angiogenesis by reducing endothelial cell migration and proliferation. This reduces bone deterioration and joint edema (Qian et al., 2022). Research demonstrates that stigmasterol, at least partially by blocking the NF-κB pathway, suppresses several pro-inflammatory and matrix degradation mediators often engaged in

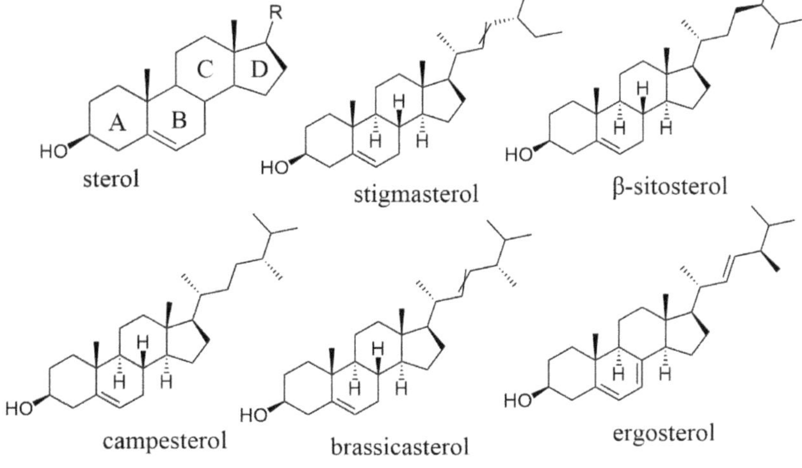

FIGURE 1.17 Chemical structures of steroid nucleus and important phytosterols.

osteoarthritis-induced cartilage deterioration. These encouraging findings support more ex vivo and in vivo studies using stigmasterol (Gabay et al., 2010). Stigmasterol increased antioxidant status, reduced inflammation, and decreased signs of inflamed joints (Ahmad Khan, Sarwar, Rahat, Ahmed, & Umar, 2020).

REFERENCES

Aherne, S. A., & O'Brien, N. M. (2002). Dietary flavonols: chemistry, food content, and metabolism. *Nutrition*, *18*(1), 75–81. https://doi.org/10.1016/S0899-9007(01)00695-5

Ahmad Khan, M., Sarwar, A. H. M. G., Rahat, R., Ahmed, R. S., & Umar, S. (2020). Stigmasterol protects rats from collagen induced arthritis by inhibiting proinflammatory cytokines. *International Immunopharmacology*, *85*, 106642. https://doi.org/10.1016/j.intimp.2020.106642

Behl, T., Mehta, K., Sehgal, A., Singh, S., Sharma, N., Ahmadi, A., Arora, S., & Bungau, S. (2022). Exploring the role of polyphenols in rheumatoid arthritis. *Critical Reviews in Food Science and Nutrition*, *62*(19), 5372–5393. https://doi.org/10.1080/10408398.2021.1924613

Bischoff, K. L. (2021). Glucosinolates. In R. C. Gupta, R. Lall, & A. Srivastava (Eds.), *Nutraceuticals* (Second Edition, pp. 903–909). Academic Press. https://doi.org/10.1016/B978-0-12-821038-3.00053-7

Bors, W., & Michel, C. (1999). Antioxidant capacity of flavanols and gallate esters: pulse radiolysis studies. *Free Radical Biology and Medicine*, *27*(11), 1413–1426. https://doi.org/10.1016/S0891-5849(99)00187-2

Cai, L., Zong, P., Zhou, M.-Y., Liu, F.-Y., Meng, B., Liu, M.-M., Li, Z., & Li, R. (2022). 7-Hydroxycoumarin mitigates the severity of collagen-induced arthritis in rats by inhibiting proliferation and inducing apoptosis of fibroblast-like synoviocytes via suppression of Wnt/β-catenin signaling pathway. *Phytomedicine*, *94*, 153841. https://doi.org/10.1016/j.phymed.2021.153841

Cheng, W.-X., Huang, H., Chen, J.-H., Zhang, T.-T., Zhu, G.-Y., Zheng, Z.-T., Lin, J.-T., Hu, Y.-P., Zhang, Y., Bai, X.-L., Wang, Y., Xu, Z.-W., Song, B., Mao, Y.-Y., Yang, F., & Zhang, P. (2020). Genistein inhibits angiogenesis developed during rheumatoid arthritis through the IL-6/JAK2/STAT3/VEGF signalling pathway. *Journal of Orthopaedic Translation*, *22*, 92–100. https://doi.org/10.1016/j.jot.2019.07.007

Covas, M. I., Miró-Casas, E., Fitó, M., Farré-Albadalejo, M., Gimeno, E., Marrugat, J., & De La Torre, R. (2003). Bioavailability of tyrosol, an antioxidant phenolic compound present in wine and olive oil, in humans. *Drugs under Experimental and Clinical Research*, *29*(5–6), 203–206. https://europepmc.org/abstract/MED/15134375

Coward, L., Smith, M., Kirk, M., & Barnes, S. (1998). Chemical modification of isoflavones in soyfoods during cooking and processing. *The American Journal of Clinical Nutrition*, *68*(6), 1486S–1491S. https://doi.org/10.1093/ajcn/68.6.1486S

Dziedzic, S. Z., & Hudson, B. J. F. (1983). Polyhydroxy chalcones and flavanones as antioxidants for edible oils. *Food Chemistry*, *12*(3), 205–212. https://doi.org/10.1016/0308-8146(83)90007-9

Einbond, L. S., Reynertson, K. A., Luo, X.-D., Basile, M. J., & Kennelly, E. J. (2004). Anthocyanin antioxidants from edible fruits. *Food Chemistry*, *84*(1), 23–28. https://doi.org/10.1016/S0308-8146(03)00162-6

Fernández-Rodríguez, J. A., Almonte-Becerril, M., Ramil-Gómez, O., Hermida-Carballo, L., Viñas-Diz, S., Vela-Anero, Á., Concha, Á., Camacho-Encina, M., Blanco, F. J., & López-Armada, M. J. (2021). Autophagy activation by resveratrol reduces severity of experimental rheumatoid arthritis. *Molecular Nutrition & Food Research*, *65*(2), 2000377. https://doi.org/10.1002/mnfr.202000377

Fotsis, T., Pepper, M., Adlercreutz, H., Fleischmann, G., Hase, T., Montesano, R., & Schweigerer, L. (1993). Genistein, a dietary-derived inhibitor of in vitro angiogenesis. *Proceedings of the National Academy of Sciences*, *90*(7), 2690–2694. https://doi.org/10.1073/pnas.90.7.2690

Gabay, O., Sanchez, C., Salvat, C., Chevy, F., Breton, M., Nourissat, G., Wolf, C., Jacques, C., & Berenbaum, F. (2010). Stigmasterol: a phytosterol with potential anti-osteoarthritic properties. *Osteoarthritis and Cartilage*, *18*(1), 106–116. https://doi.org/10.1016/j.joca.2009.08.019

Galanakis, C. M. (2019). *Carotenoids: Properties, Processing and Applications*. Academic Press.

Gandhi, Y., Kumar, R., Grewal, J., Rawat, H., Mishra, S. K., Kumar, V., Shakya, S. K., Jain, V., Babu, G., Sharma, P., Singh, A., Singh, R., & Acharya, R. (2022). Advances in anti-inflammatory medicinal plants and phytochemicals in the management of arthritis: a comprehensive review. *Food Chemistry Advances*, *1*, 100085. https://doi.org/10.1016/j.focha.2022.100085

Gonçalves, A. C., Rodrigues, S., Fonseca, R., & Silva, L. R. (2024). Potential role of dietary phenolic compounds in the prevention and treatment of rheumatoid arthritis: current reports. *Pharmaceuticals*, *17*(5), 590.

Hackman, R. M., Polagruto, J. A., Zhu, Q. Y., Sun, B., Fujii, H., & Keen, C. L. (2007). Flavanols: digestion, absorption and bioactivity. *Phytochemistry Reviews*, *7*(1), 195. https://doi.org/10.1007/s11101-007-9070-4
Halkier, B. A., & Gershenzon, J. (2006). Biology and biochemistry of glucosinolates. *Annual Review of Plant Biology*, *57*, 303–333. https://doi.org/10.1146/annurev.arplant.57.032905.105228
Hamed, M. A., Aboul Naser, A. F., Aboutabl, M. E., Osman, A. F., Hassan, E. E. S., Aziz, W. M., Khalil, W. K. B., Farghaly, A. A., & El-Hagrassi, A. M. (2021). Bioactive compounds and therapeutic role of *Brassica oleracea* L. seeds in rheumatoid arthritis rats via regulating inflammatory signalling pathways and antagonizing interleukin-1 receptor action. *Biomarkers*, *26*(8), 788–807. https://doi.org/10.1080/1354750X.2021.1999504
Hancke, J. L., Srivastav, S., Cáceres, D. D., & Burgos, R. A. (2019). A double-blind, randomized, placebo-controlled study to assess the efficacy of standardized extract (ParActin®) on pain reduction in subjects with knee osteoarthritis. *Phytotherapy Research*, *33*(5), 1469–1479. https://doi.org/10.1002/ptr.6339
Harborne, J. B., & Simmonds, N. W. (1964). *Biochemistry of Phenolic Compounds*. Academic Press.
Jabbar, Z., Irfan, H. M., Alamgeer, Ullah, A., Mahmoud, M. H., Batiha, G. E., Ahsan, H., Khan, A., & Kashtoh, H. (2024). Trans-Chalcone (1–3-diphenyl-2-propen-1-one) as a therapeutic candidate in joint inflammation via reduction of TNF-α, IL-1β, IL-6, and IL-17 in rodents: An in vivo study by RT-PCR and ELISA analysis. *ACS Omega*, *9*(20), 22123–22135. https://doi.org/10.1021/acsomega.4c00368
Jiang, C., Sun, Z.-M., Hu, J.-N., Jin, Y., Guo, Q., Xu, J.-J., Chen, Z.-X., Jiang, R.-H., & Wu, Y.-S. (2019). Cyanidin ameliorates the progression of osteoarthritis via the Sirt6/NF-κB axis in vitro and in vivo. *Food & Function*, *10*(9), 5873–5885. https://doi.org/10.1039/C9FO00742C
Kamiloglu, S., Tomas, M., & Capanoglu, E. (2019). *Dietary Flavonols and O-glycosides*. Springer Nature Singapore Pte Ltd.
Karimi, A., Azar, P. S., Kadkhodayi, M., Tandorost, A., Vajdi, M., Shoorei, H., & Farhangi, M. A. (2022). A comprehensive insight into effects of resveratrol on molecular mechanism in rheumatoid arthritis: a literature systematic review. *International Journal of Rheumatic Diseases*, *25*(8), 827–843. https://doi.org/10.1111/1756-185X.14356
Khoo, H. E., Azlan, A., Tang, S. T., & Lim, S. M. (2017). Anthocyanidins and anthocyanins: colored pigments as food, pharmaceutical ingredients, and the potential health benefits. *Food & Nutrition Research*, *61*(1), 1361779–1361779. https://doi.org/10.1080/16546628.2017.1361779
Kostova, I., Bhatia, S., Grigorov, P., Balkansky, S., Parmar, V., Prasad, A., & Saso, L. (2011). Coumarins as antioxidants. *Current Medicinal Chemistry*, *18*(25), 3929–3951. https://doi.org/10.2174/092986711803414395
Kulczyński, B., Sidor, A., Brzozowska, A., & Gramza-Michałowska, A. (2024). The role of carotenoids in bone health—a narrative review. *Nutrition*, *119*, 112306. https://doi.org/10.1016/j.nut.2023.112306
Lee, F., Bae, K. H., Ng, S., Yamashita, A., & Kurisawa, M. (2021). Hyaluronic acid–green tea catechin conjugates as a potential therapeutic agent for rheumatoid arthritis. *RSC Advances*, *11*(24), 14285–14294. https://doi.org/10.1039/D1RA01491A
Li, Y., Yang, B., Bai, J.-Y., Xia, S., Mao, M., Li, X., Li, N., & Chen, L. (2019). The roles of synovial hyperplasia, angiogenesis and osteoclastogenesis in the protective effect of apigenin on collagen-induced arthritis. *International Immunopharmacology*, *73*, 362–369. https://doi.org/10.1016/j.intimp.2019.05.024
Liggins, J., Bluck, L. J. C., Runswick, S., Atkinson, C., Coward, W. A., & Bingham, S. A. (2000). Daidzein and genistein contents of vegetables. *British Journal of Nutrition*, *84*(5), 717–725. https://doi.org/10.1017/S0007114500002075
Lucarini, E., Micheli, L., Di Cesare Mannelli, L., & Ghelardini, C. (2022). Naturally occurring glucosinolates and isothiocyanates as a weapon against chronic pain: potentials and limits. *Phytochemistry Reviews*, *21*(2), 647–665. https://doi.org/10.1007/s11101-022-09809-0
Mah, S.H. (2020). Chalcones in diets. In J. Xiao, S. Sarker, & Y. Asakawa (Eds.), *Handbook of Dietary Phytochemicals*. Springer.
Maoka, T. (2020). Carotenoids as natural functional pigments. *Journal of Natural Medicines*, *74*(1), 1–16. https://doi.org/10.1007/s11418-019-01364-x
Pragasam, S. J., Murunikkara, V., Sabina, E. P., & Rasool, M. (2013). Ameliorative effect of p-coumaric acid, a common dietary phenol, on adjuvant-induced arthritis in rats. *Rheumatology International*, *33*(2), 325–334. https://doi.org/10.1007/s00296-012-2394-4
Qian, K., Zheng, X.-X., Wang, C., Huang, W.-G., Liu, X.-B., Xu, S.-D., Liu, D.-K., Liu, M.-Y., & Lin, C.-S. (2022). β-Sitosterol inhibits rheumatoid synovial angiogenesis through suppressing VEGF signaling pathway [Original Research]. *Frontiers in Pharmacology*, *12*, 816477. https://doi.org/10.3389/fphar.2021.816477

Salehi, B., Quispe, C., Sharifi-Rad, J., Cruz-Martins, N., Nigam, M., Mishra, A. P., Konovalov, D. A., Orobinskaya, V., Abu-Reidah, I. M., Zam, W., Sharopov, F., Venneri, T., Capasso, R., Kukula-Koch, W., Wawruszak, A., & Koch, W. (2021). Phytosterols: from preclinical evidence to potential clinical applications [Review]. *Frontiers in Pharmacology*, *11*, 599959. https://doi.org/10.3389/fphar.2020.599959

Santana-Gálvez, J., Cisneros-Zevallos, L., & Jacobo-Velázquez, D. A. (2017). Chlorogenic acid: recent advances on its dual role as a food additive and a nutraceutical against metabolic syndrome. *Molecules*, *22*(3), 358. https://www.mdpi.com/1420-3049/22/3/358

Santana-Gálvez, J., & Jacobo-Velázquez, D. A. (2018). Classification of phenolic compounds. In L. M. Nollet & J. A. Gutierrez-Uribe (Eds.), *Phenolic Compounds in Food: Characterization and Analysis* (pp. 3–20). Taylor & Francis Group, LLC.

Santos, E. O. L., Kabeya, L. M., Figueiredo-Rinhel, A. S. G., Marchi, L. F., Andrade, M. F., Piatesi, F., Paoliello-Paschoalato, A. B., Azzolini, A. E. C. S., & Lucisano-Valim, Y. M. (2014). Flavonols modulate the effector functions of healthy individuals' immune complex-stimulated neutrophils: a therapeutic perspective for rheumatoid arthritis. *International Immunopharmacology*, *21*(1), 102–111. https://doi.org/10.1016/j.intimp.2014.04.014

Silva, S., Sepodes, B., Rocha, J., Direito, R., Fernandes, A., Brites, D., Freitas, M., Fernandes, E., Bronze, M. R., & Figueira, M. E. (2015). Protective effects of hydroxytyrosol-supplemented refined olive oil in animal models of acute inflammation and rheumatoid arthritis. *The Journal of Nutritional Biochemistry*, *26*(4), 360–368. https://doi.org/10.1016/j.jnutbio.2014.11.011

Thakur, M., Singh, K., & Khedkar, R. (2020). Phytochemicals: extraction process, safety assessment, toxicological evaluations, and regulatory issues. In B. Prakash (Ed.), *Functional and Preservative Properties of Phytochemicals* (pp. 341–361). Academic Press. https://doi.org/10.1016/B978-0-12-818593-3.00011-7

Tomás-Barberán, F. A., & Clifford, M. N. (2000). Flavanones, chalcones and dihydrochalcones – nature, occurrence and dietary burden. *Journal of the Science of Food and Agriculture*, *80*(7), 1073–1080. https://doi.org/10.1002/(sici)1097-0010(20000515)80:7<1073::Aid-jsfa568>3.0.Co;2-b

Vitrac, X., Bornet, A., Vanderlinde, R., Valls, J., Richard, T., Delaunay, J.-C., Mérillon, J.-M., & Teissédre, P.-L. (2005). Determination of Stilbenes (δ-viniferin, trans-astringin, trans-piceid, cis- and trans-resveratrol, ε-viniferin) in Brazilian Wines. *Journal of Agricultural and Food Chemistry*, *53*(14), 5664–5669. https://doi.org/10.1021/jf050122g

Wang, Y., Fong, S. K., Singh, A. P., Vorsa, N., & Johnson-Cicalese, J. (2019). Variation of anthocyanins, proanthocyanidins, flavonols, and organic acids in cultivated and wild diploid blueberry species. *Hortscience*, *54*(3), 576–585. https://doi.org/10.21273/hortsci13491-18

Yoon, C.-H., Chung, S.-J., Lee, S.-W., Park, Y.-B., Lee, S.-K., & Park, M.-C. (2013). Gallic acid, a natural polyphenolic acid, induces apoptosis and inhibits proinflammatory gene expressions in rheumatoid arthritis fibroblast-like synoviocytes. *Joint Bone Spine*, *80*(3), 274–279. https://doi.org/10.1016/j.jbspin.2012.08.010

Zannou, O., Pashazadeh, H., Galanakis, C. M., Alamri, A. S., & Koca, I. (2022). Carboxylic acid-based deep eutectic solvents combined with innovative extraction techniques for greener extraction of phenolic compounds from sumac (*Rhus coriaria* L.) *Journal of Applied Research on Medicinal and Aromatic Plants*, *30*, Article 100380. https://doi.org/10.1016/j.jarmap.2022.100380

Zeb, A. (2020). Concept, mechanism, and applications of phenolic antioxidants in foods. *Journal of Food Biochemistry*, *44*(9), e13394. https://doi.org/10.1111/jfbc.13394

Zeb, A. (2021). *Phenolic Antioxidants in Foods: Chemistry, Biochemistry and Analysis*. Springer Nature.

Zhang, H.-Q., Wang, Y.-J., Yang, G.-T., Gao, Q.-L., & Tang, M.-X. (2018). Taxifolin inhibits receptor activator of NF-κB ligand-induced osteoclastogenesis of human bone marrow-derived macrophages in vitro and prevents lipopolysaccharide-induced bone loss in vivo. *Pharmacology*, *103*(1–2), 101–109. https://doi.org/10.1159/000495254

2 Introduction to Rheumatoid Arthritis

Mehak Fatima, Ammara Saleem, Justyna Wozniak, and Muhammad Furqan Akhtar

WHAT WILL YOU LEARN IN THIS CHAPTER?

This chapter will provide a comprehensive review of the concepts of inflammation, arthritis, and rheumatoid arthritis, including pathogenesis and treatment options for rheumatoid arthritis. It also provides knowledge about drug targets, mechanisms of action, and problems related to their use.

LIST OF ABBREVIATIONS

Aa	*Aggregatibacter actinomycetemcomitans*
AAPAs	Antiacetylated protein antibodies
ACP	Anticitrullinated protein
DHODH	Dihydroorotate dehydrogenase
JAK	Janus Kinase
LTP	Lymphoid tyrosine phosphate
NSAIDs	Nonsteroidal anti-inflammatory drugs
OA	Osteoarthritis
PAD	Peptidyl-arginine deaminase
ReA	Reactive arthritis
STAT	Signal transducer and activator of transcription
TNF-α	Tumor necrosis factor-alpha
TRAF1	TNF receptor-associated factor 1

2.1 INFLAMMATION

Inflammation is the body's primary reaction to infection, irritants, or other injuries (Levi, 2017). Currently, inflammation is characterized by five macroscopic pathogenic manifestations, four of which were postulated by Celsus over 2,000 years ago. 'Tumor' is the tissue swelling, 'calor' is the high tissue temperature, 'rubor' is the blood-like redness of the vascularized tissue at the site of inflammation, 'dolor' is the intense feeling of a noxious stimulus, and 'function laesa' is the compromised function of the afflicted organ. All symptoms are considered to result from one main pathophysiological event, which is increased vascular permeability as a direct result of tissue damage (Tracy, 2006).

Despite significant advances in our understanding of inflammation, the complexity of the process and the variability in individual responses mean that some ambiguities remain in the cause-and-effect relationships in pathogenesis and in the strategies used to treat inflammatory disorders.

2.2 MOLECULAR MECHANISM GOVERNING INFLAMMATION

Cells of both the innate and adaptive immune systems express pattern-recognition receptors, specialized transmembrane receptors used by host cells to identify inflammatory stimuli. These germline-encoded receptors are responsible for detecting the presence of pathogenic microbes and the subsequent extent of any cellular injury (Tsuge, Inazumi, Shimamoto, & Sugimoto, 2019).

DOI: 10.1201/9781003535201-3

Inflammation is a complex immune response involving various molecular mechanisms that protect the body against injury, infection, and damage. The process starts when harmful stimuli, such as damaged cells or pathogens, are recognized. This triggers a signaling cascade that activates the production of proinflammatory cytokines, including interleukin (IL)-1β, IL-6, and tumor necrosis factor-alpha (TNF-α), through the activation of transcription factors like nuclear factor (NF)-κB and activator protein-1. These cytokines play a vital role in recruiting immune cells, such as neutrophils and macrophages, to the site of inflammation. Neutrophils function as first responders, releasing granules containing enzymes like neutrophil elastase and myeloperoxidase, which help break down pathogens and damaged tissue. Macrophages phagocytose foreign particles and so release cytokines, which sustain inflammatory response, recruit additional immune cells, and aid in tissue repair and resolution of inflammation (Ke et al., 2019).

The inflammatory response is also characterized by increased vascular permeability, allowing plasma proteins and fluid to leak into the tissue. Stimulation of endothelial cells mediates these processes by expressing adhesion molecules like E-selectin and vascular cell adhesion molecule-1, which facilitate the recruitment of immune cells (Tasneem, Liu, Li, Choudhary, & Wang, 2019). Alongside this, the complement system is activated because of increased blood flow and permeability, further amplifying the inflammatory response. As inflammation progresses, anti-inflammatory cytokines such as IL-10 and transforming growth factor-β are produced, helping in resolving the response and promoting tissue repair. The resolution stage involves the clearance of apoptotic cells and debris, repair of damaged tissue, and restoration of tissue homeostasis. Dysregulation of the molecular mechanism governing inflammation can lead to various diseases, such as atherosclerosis and autoimmune disorders (Gaber, Strehl, & Buttgereit, 2017).

2.3 ARTHRITIS

Arthritis is a common condition affecting one or more joints, causing pain, inflammation, and redness of joints. This condition typically involves inflammation, which occurs when the body's white blood cells and immunological proteins protect against contamination and external influences, such as bacteria and viruses (Gravallese & Firestein, 2023). In some disorders such as autoimmune disorders, the immune system causes inflammation even when no foreign particles are present (Zhang, Jonsson, et al., 2023). Arthritis is a broad term that refers to approximately 100 distinct forms of joint problems and related ailments. It is primarily associated with joint inflammation, which causes symptoms such as pain, stiffness, edema, and restricted mobility. While arthritis can affect individuals of any age, it becomes more common with advancing age. The common types of arthritis are osteoarthritis (OA), rheumatoid arthritis (RA), reactive arthritis (ReA), juvenile idiopathic arthritis (JIA), ankylosing spondylitis, and psoriatic arthritis. To understand the impact and management of arthritis better, we will now delve deeper into these common types.

2.3.1 Osteoarthritis

OA is a prevalent progressive condition of articular cartilage characterized by hypertrophic bone alterations. It causes cartilage to gradually deteriorate and break down, resulting in friction between bones during joint movement, which leads to a phenomenon known as gelling. The gelling phenomenon refers to the worsening of pain with activity, particularly after a period of rest (van den Bosch, 2021). This causes discomfort, stiffness, edema, and limited motion in the afflicted joints. The most commonly affected joints are the hands, knees, hips, and spine, although practically any joint can be impacted (Boer et al., 2021).

Long-term illness management can be extremely helpful, even though OA cannot be cured. Using painkillers to manage discomfort is a crucial first step. There are numerous treatments available, including nonsteroidal anti-inflammatory drugs (NSAIDs), gabapentin, amantadine, and tramadol (opioid analgesic). Physical methods, such as physical therapy and exercise, can also be used to relieve pain (Richard, Driban, & McAlindon, 2023).

2.3.2 Reactive Arthritis (ReA)

It is described as the development of sterile inflammatory arthritis as a result of a distant infection, typically occurring in the gastrointestinal or urogenital tract (Bentaleb et al., 2020). The term 'ReA' was coined by Ahvonen and Sievers in 1969. ReA refers to all infection-triggered arthritis and is classified as spondylarthritis. These disorders are linked to the HLA-B27 genetic marker, lack of rheumatoid factor (RF), familial aggregation, and frequent extra-articular symptoms. Gram-negative, obligatory, or facultative intracellular aerobic bacteria are known to cause this arthritis (Dombret, Skapenko, & Schulze-Koops, 2022). Invasive infections, such as those caused by *Chlamydia trachomatis*, primarily affect the gastrointestinal and urogenital mucosa. During acute arthritis, peripheral blood mononuclear cells produce an imbalance of proinflammatory cytokines, including low levels of TNF-α (Ferreira, Marinho, Vita, Fonseca, & Vasconcelos, 2024). Musculoskeletal symptoms typically appear within 1–2 weeks (or up to 4 weeks) of Chlamydia infection, which may be asymptomatic. ReA mostly affects young adults (30–40 years old) and is uncommon in children (Saricaoglu, Hasanoglu, & Guner, 2021). Male and female patients are equally susceptible to ReA caused by gastrointestinal infections; however, *Chlamydia trachomatis*-induced ReA is more commonly identified in men.

There is no quick fix for ReA, particularly refractory ReA. Clinicians frequently treat ReA patients based on clinical experience. Traditional ReA treatment includes four types of medications: NSAIDs, glucocorticoids, disease-modifying antirheumatic drugs (DMARDs), and antibiotics. A meta-analysis described that antibiotic monotherapy provided no substantial advantage for ReA. However, a double-blind prospective trial described that combination antibiotic therapy was helpful for Chlamydia-induced chronic ReA (Acharya, Smith, Banerjee, Camire, & Vij, 2024).

2.3.3 Juvenile Idiopathic Arthritis (JIA)

It is characterized as an autoimmune disease, targeting specifically the synovium (joint lining) and causing inflammation in the joints. It affects children and adolescents. For a diagnosis of JIA, arthritis must be present for at least 6 weeks in a child younger than 16 years old, with no other identifiable cause for the arthritis. It is the most prevalent chronic rheumatic condition in children, often resulting in significant impairment (Acharya et al., 2024). In high-income countries, the annual incidence is 2–20 cases per 100,000 people, whereas in low-income countries, the incidence ranges from 16 to 150 cases per 100,000 people. Systemic JIA is like adult-onset Still's disease, featuring strong systemic symptoms such as fever, rash, and serositis. The patient's innate immune system is clearly activated, yet there is no consistent link to autoantibodies or human leukocyte antigen (HLA) (Martini et al., 2022).

JIA is further divided into several subtypes based on the characteristics of the arthritis and the presence of certain clinical features such as oligoarthritis, polyarthritis, and enthesites-associated arthritis.

2.4 OLIGOARTHRITIS

It is the most frequent kind of JIA, defined by inflammation in fewer than five joints. It typically affects large joints, such as the knees and ankles, and can be asymmetrical. Asymmetric arthritis, onset before 6 years of age, a higher prevalence in females, the presence of positive antinuclear antibodies, and an increased risk of iridocyclitis are common characteristics in these children (Crayne & Beukelman, 2018). Oligoarthritis primarily affects joints of the lower extremities, with the knee being commonly involved (30%–50% of cases), followed by the ankle. Wrist and ankle arthritis, as well as a high erythrocyte sedimentation rate at onset, are recognized as risk factors for a prolonged course (Boeren, Verstappen, Looijen, de Jong, & van der Helm-van Mil, 2024).

2.5 POLYARTHRITIS

Polyarthritis is a subtype of JIA that affects five or more joints within the first six months of the disease. It can be further divided into two categories based on the presence of RF: (i) RF-negative and (ii) RF-positive.

i. **Polyarthritis (RF-negative)**
   This subtype involves inflammation in five or more joints during the first six months of the disease, without the presence of RF in the blood. It can affect both large and small joints, including the knees, ankles, wrists, fingers, and elbows. This form of polyarthritis is more common in girls and often presents with symptoms such as pain, swelling, and stiffness in the affected joints. Morning stiffness and reduced range of motion are common complaints. Children with RF-negative polyarthritis may also experience systemic symptoms like mild fever and fatigue. The course of the disease can vary, with some children experiencing periods of remission and others having a more chronic, persistent condition. Early and aggressive treatment can help manage symptoms and prevent joint damage.

ii. **Polyarthritis (RF-positive)**
   This subtype also affects five or more joints within the first six months of the disease, but in this case, RF is present in the blood. RF-positive polyarthritis is like adult RA and is characterized by a more severe and potentially erosive disease course. It primarily affects adolescent girls and can involve both large and small joints, leading to pain, swelling, and deformity if not adequately treated. Children with RF-positive polyarthritis are at a higher risk for developing long-term joint damage and disability. Systemic symptoms, such as fatigue and low-grade fever, can also be present. This subtype often requires more intensive treatment, including DMARDs and biologic agents, to control inflammation and prevent joint destruction.

2.6 ENTHESITES-ASSOCIATED ARTHRITIS

In this form, inflammation primarily occurs at the places where tendons and ligaments attach to bones (entheses). It can also affect joints and is linked to conditions such as enthesopathy and spondylitis. Enthesitis-related arthritis mostly affects males over the age of six and is characterized by the presence of both enthesitis and arthritis. The occurrence of enthesitis, particularly at calcaneal insertions of the Achilles tendon, plantar fascia, and tarsal region, is a useful diagnostic feature (Weiss, 2013).

2.7 RHEUMATOID ARTHRITIS (RA)

It is a symmetric polyarticular disease that affects the small diarthrodial joints of hands and feet. It is a multifactorial and complex condition where the body's immune system attacks the joint' lining, leading to stiffness, pain, and swelling. RA mainly involves symmetrical joints, particularly the small joints of hands and feet, and can result in disability and permanent joint damage (Gravallese & Firestein, 2023), significantly reducing the quality of life. RA is more common in females than in males and typically manifests during the prime working years. The symptoms associated with RA are prolonged morning stiffness, fatigue, fever, joint swelling, stiffness, and pain (especially in hands, wrists, elbows, feet, and ankles). Other symptoms include rheumatoid nodules under the skin, loss of strength, sleep disturbances, difficulty in performing daily activities, redness and warmth around affected joints, depression, anxiety, and loss of appetite (Babaahmadi et al., 2023).

2.8 PREVALENCE AND EPIDEMIOLOGY OF RA

In developed countries, the prevalence of RA is between 0.5% and 1% of adults. Research has consistently shown that patients with RA have higher rates of mortality compared to the general population. Depending on demographics, geographic location, and diagnosis methods, the standardized ratio of mortality ranges from 1.28 to 2.98. RA patients are more likely to develop comorbid diseases, which can present with unusual symptoms (Vegas, Drouin, Dray-Spira, & Weill, 2023). Obesity, smoking, and exposure to air pollution are risk factors of RA. Genetic factors contribute to about 50% of the risk of developing RA. Although the exact cause of RA remains unknown, some modifiable lifestyle factors and nonmodifiable risk factors have been identified (Venetsanopoulou et al., 2023).

2.9 PATHOGENESIS OF RA

2.9.1 Genetic Predisposition

The HLA-DRB1 gene, located in the major histocompatibility complex on chromosome six, plays a critical role in the pathogenesis of RA. This gene produces a protein that presents antigens to T cells, triggering an immune response. Most HLA-DRB1 alleles associated with RA encode proteins that share a similar amino acid sequence in the peptide-binding groove, known as shared epitope (SE). This specific sequence on the HLA-DRB1 gene increases the risk of RA and is present in over 80% of RA patients (Khan, Mohan, Muzammil, Alam, & Khayyam, 2024). The sequence alters the way peptides are presented to T cells, leading to an abnormal immune response. Due to the sequence similarity, it is hypothesized that these HLA molecules present particular 'arthritogenic' peptides that may trigger autoimmune reactions in joints. HLA-DRB1 presents peptides from citrullinated proteins (e.g., citrullinated vimentin) to T cells, leading to immune activation due to cross-reactivity with microbial peptides as shown in Figure 2.1. This results in over-activation of pro-inflammatory

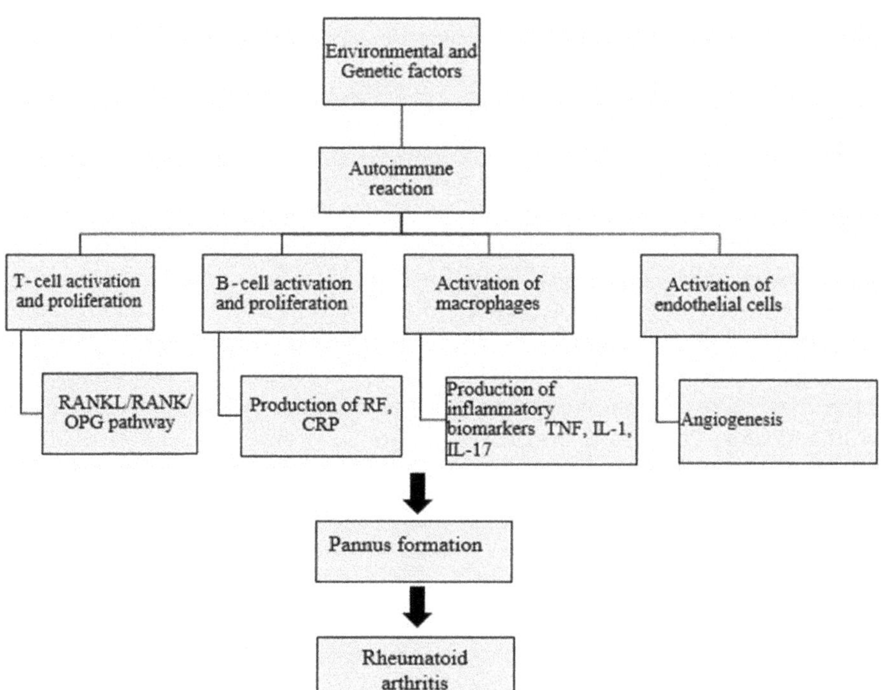

FIGURE 2.1 Pathophysiological cascade of rheumatoid arthritis (RA) development.

T cells and under-activation of regulatory T cells, causing an imbalanced immune response that leads to inflammation and tissue damage. The most common alleles linked with RA are HLA-DRB1 04:01, 04:04, 04:05, and 04:08. These alleles contribute to tissue damage, auto-antibody production (anticitrullinated protein antibodies (ACPAs)), and joint damage (Pisetsky, 2023).

2.9.2 PTPN22: A Key Player in RA Pathophysiology

PTPN22 is a gene that encodes a protein tyrosine phosphatase, known as lymphoid tyrosine phosphatase (LTP). LTP is mostly expressed in T cells (Bai et al., 2023) and plays a key role in T-cell signaling. Variants of PTPN22 have been linked to a higher risk of developing RA. LYP regulates T-cell signaling by dephosphorylation and activation of proteins involved in the T-cell receptor signaling pathway. PTPN22 variants alter T-cell signaling response, leading to the development of autoreactive T cells. This altered signaling response results in increased T-cell activation, enhanced pro-inflammatory cytokines production, impaired function of regulatory T cells, and increased auto-reactive B-cell activation, all contributing to chronic inflammation and RA.

2.9.3 STAT4: A Key Regulator of Cytokine Signaling in RA

STAT4 (Signal Transducer and Activator of Transcription 4) is a protein that plays a key role in cytokine signaling, particularly IL-12 and IL-23 signaling. STAT4 is a key controller of helper T cells Th-1 and Th17 cell differentiation, which are important in the immune response. It also leads to the production of pro-inflammatory cytokines such as IL-6, IL-1, and TNF-α. Alterations in STAT4 signaling contribute to autoimmune diseases including RA. These alterations can result in enhanced Th1 and Th17 cell differentiation, activation of autoreactive T cells, and impaired function of regulatory T cells. This cascade of immune responses causes inflammation and joint damage, leading to the development and progression of RA (Chen, Ning, Sun, & Zhang, 2023).

2.9.4 TRAF1/C5: Encoding Proteins Involved in the TNF Pathway

The TRAF1/C5 gene on chromosome 9 is a risk locus for RA, encoding two important proteins involved in the TNF pathway: TNF receptor-associated factor 1 (TRAF1) and Complement Component 5 (C5), both of which play crucial roles in immune regulation and inflammation. C5 is a vital component of the complement system, part of the innate immune response, and is involved in the formation of the membrane attack complex (MAC), which can lyse pathogens directly. Beyond its role in MAC formation, C5 plays a role in immune regulation by attracting inflammatory cells to sites of infection or injury. This chemotactic function of C5 is important in recruiting immune cells to inflamed synovial tissues in RA. In RA, C5 contributes to the recruitment of inflammatory cells to the synovium, exacerbating inflammation and promoting joint damage. TRAF1 is integral to the TNF receptor signaling pathway, which regulates inflammatory responses. It interacts with TNF receptors to mediate the activation of downstream signaling pathways, leading to the production of pro-inflammatory cytokines, particularly TNF-α. Dysregulation of TRAF1/C5 and the TNF pathway contributes to the progress of chronic swelling and joint damage in RA.

2.10 ENVIRONMENTAL FACTORS

Smoking is a well-established risk factor for RA development, leading to higher expression of the PAD2 (peptidyl-arginine deaminase) enzyme and elevation of citrullination in lungs. An interaction between smoking, HLA-DR SE alleles, and the PTPN22 gene also causes hypercitrullination. This process results in elevated levels of citrullinated antigens, which can interact with T cells and lead to a dysregulation in the immune tolerance. It triggers an autoimmune response and increases the risk of developing RA. Prolonged tobacco use induces several morphological, physiological, biochemical, and enzymatic changes that impair antibacterial defenses, disrupt cellular regulatory activity, and enhance inflammatory responses. Cigarette smoke condensate increases the expression of pro-inflammatory cytokines and matrix-degrading enzymes in RA synovial fibroblasts, contributing to progressive joint destruction (Zhang, Zhang, et al., 2023).

Figure 2.1 illustrates the pathophysiological cascade of RA development, involving immune cell activation, cytokine production, inflammation, joint destruction, and bone erosion, ultimately leading to joint pain, swelling, and loss of function.

2.10.1 Microbiome

The microbiome is a crucial factor in the pathophysiology of many diseases, including RA. Dysbiosis, a disbalance of the microbiome, could influence the development and progression of RA. The microbiome consists of archaea, bacteria, viruses, and eukaryotes that reside within and on the body. While the microbiome itself has not been shown to directly cause arthritis, changes in microbial community may exacerbate or lessen the condition. Dysbiosis of the microbiome may result in bacterial translocation from mucosa to the blood circulation, resulting in the loss of barrier function and local inflammation. Certain components of the bacterial cell wall may molecularly resemble human autoantigens, inducing an immune response that targets the joint (Tsetseri, Silman, Keene, & Dakin, 2023).

In an inflammatory environment, some microbes produce citrullinated antigens by expressing the PAD (protein arginine deaminase) enzyme, which may cross-react with self-peptides, thereby inducing an autoimmune response (Hermans, Lear, Case, & Buckley, 2023). The gingival crevicular fluid from patients with periodontal disease contains citrullinated peptides. Periodontitis is associated with the pathogen. *Aggregatibacter actinomycetemcomitans* (Aa) can express PAD enzymes, resulting in dysregulated citrullination, a process also known as endocitrullination. Endocitrullination triggers an autoimmune response, perpetuating the production of pro-inflammatory cytokines that contributes to joint damage and RA (Thompson et al., 2023).

2.10.2 Role of Autoantibodies in RA

It is estimated that autoantibodies are present in about 50%–80% of RA patients. The 1987 American College of Rheumatology (ACR) classification criteria for RA included RF as a diagnostic marker, despite its limited specificity. Several decades later, more RA-specific ACPAs were discovered (Pisetsky, 2023). ACR-EULAR (European League Against Rheumatism) in 2010 included both RF and ACPA in its classification criteria for RA (Van Hoovels et al., 2018). Antibodies against carbamylated and acetylated proteins, among other post-translationally protein modifications, have been discovered more recently. Seronegative RA patients, who do not have detectable levels of RF and ACPA, tend to have lower inflammatory markers while seropositive RA patients exhibit accelerated radiographic progression and joint destruction. Studying autoantibodies and their properties has greatly improved the knowledge of the pathophysiological mechanisms underlying RA (Mariani et al., 2023).

ACPAs are autoantibodies targeting proteins modified by the citrullination process, where arginine residues are converted to citrulline. The presence of ACPA in the blood is a sensitive and specific marker for RA (Martos et al., 2023). The IgA is mainly found in mucosal areas such as the gut, respiratory tract, and urogenital tract and plays a crucial role in immune defense at these sites. In RA, elevated levels of IgA-ACPA have been observed, suggesting that mucosal immune responses might be involved in the disease's pathogenesis. Synovial fluid from RA-inflamed joints contains citrullinated proteins, suggesting that ACPA may bind to these substances in the joint and worsen local inflammation. By attaching to Fc receptors and activating the complement system, ACPA contributes to inflammation. Carbohydrates, also known as glycoproteins, are affixed to the antibody's Fab and Fc regions. The galactosylation and sialylation of the Fc region of ACPA is lower than that of IgG antibodies to recall antigens. The reduced sialylation promotes osteoclastogenesis through modified FcγR signaling. Lower level of ACPA-IgG sialylation is associated with lower trabecula numbers and decreased bone volume. This specific Fc glycan profile of ACPA influences their capacity to contribute to the pathogenesis of RA. In essence, the glycan modifications on ACPA can affect their inflammatory potential and their role in bone destruction (Sieghart et al., 2023).

2.10.3 Pathogenesis of Anticarbamylated Protein Antibodies

Anticarbamylated protein antibodies (anti-CarP) play a significant role in the pathogenesis of RA. Carbamylation is a chemical process where cyanate, a by-product of myeloperoxidase activity, catalyzes the conversion of lysine residues into homocitrulline. This modification makes proteins immunogenic, triggering an immune response. Carbohydrates, known as glycoproteins, are affixed to the antibody's Fab and Fc regions. The galactosylation and sialylation of the Fc region of anti-CarP antibodies are lower than those of IgG antibodies to recall antigens. Immune cells such as macrophages and dendritic cells are activated by anti-CarP antibodies, contributing to the release of pro-inflammatory cytokines such as TNF-α and IL-1β. This exacerbates inflammation, particularly when carbamylated proteins form immune complexes with anti-CarP antibodies. Anti-CarP antibodies perpetuate the autoimmune response when activated by autoreactive T cells. Epitope spreading occurs when the immune system begins to recognize and respond to multiple carbamylated proteins, amplifying the immune response. Anti-CarP antibodies are closely associated with high disease activity and joint damage in RA (Aripova et al., 2024).

Certain factors, including inflammation, smoking, and renal disease, can elevate cyanide levels and consequently increase carbamylation. Smoking and other genetic or environmental elements may also break the immune tolerance against carbamylated proteins like fibrinogen, further contributing to the pathogenesis of RA.

Figure 2.2 illustrates the role of post-translational modifications (PTMs) in the pathogenesis of RA. PTMs, such as citrullination, carbamylation, and glycosylation, can alter protein structure and function, leading to the generation of neo-epitopes. These modified proteins can trigger autoimmune responses, contributing to the development of RA. The figure highlights the specific PTMs involved in RA pathogenesis and their impact on protein function and immune recognition.

2.10.4 Pathogenesis of Antiacetylated Protein Antibodies

Antiacetylated autoantibodies, also known as antiacetylated protein antibodies (AAPAs), are the autoantibodies modified by acetylation, a PTM where an acetyl group is added to a protein, altering its function and structure. RA susceptibility genes, including PTPN22 and HLA-DRB1, influence the immune system and contribute to the development of AAPAs. Specifically, HLA-DRB1 04:01 is strongly associated with both AAPAs and RA (Studenic et al., 2021).

2.11 TREATMENT OF RA

2.11.1 Disease-Modifying Antirheumatic Drugs

DMARDs are categorized into several types and are used to treat RA and other autoimmune disorders. They work by inhibiting the immune system through various mechanisms, decreasing inflammation and pain and slowing disease progression. Here is an overview of the distinct categories and specific DMARDs: conventional DMARDs, also known as traditional DMARDs, are older medications that decrease the progression and reduce the symptoms of the disease. The main conventional DMARDs include methotrexate (MTX), leflunomide, and sulfasalazine Hydroxychloroquine sulfate is considered a weaker DMARD and is often used for mild RA. In clinical practice, older conventional DMARDs including ciclosporin, azathioprine, sodium aurothiomalate (gold), and

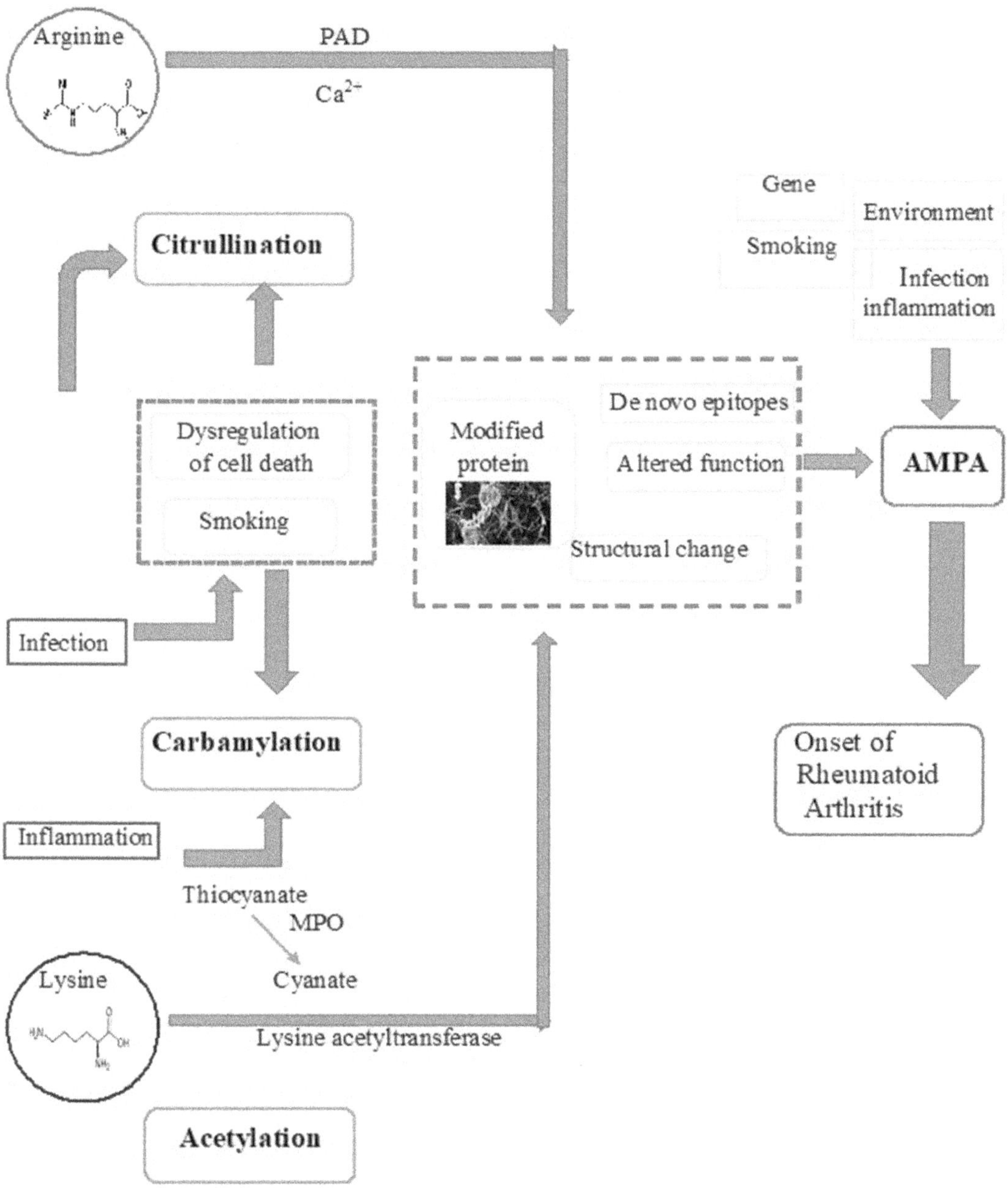

FIGURE 2.2 Post-translational modifications contributing to RA pathogenesis.

penicillamine are no longer frequently used (Smolen et al., 2023). Treatment of RA is shown in Figure 2.3.

MTX is commonly utilized as the primary RA therapy, either alone or in combination with biological and other nonbiological DMARDs. MTX binds to dihydrofolate reductase, inhibiting the conversion of dihydrofolate into tetrahydrofolate, which is a crucial step in purine and pyrimidine synthesis. This inhibition decreases DNA synthesis and cell proliferation, especially in rapidly proliferating immune cells. MTX decreases inflammation by preventing the synthesis of pro-inflammatory cytokines like TNF-α and IL-1β, affecting immune cell function by reducing T-cell activation and

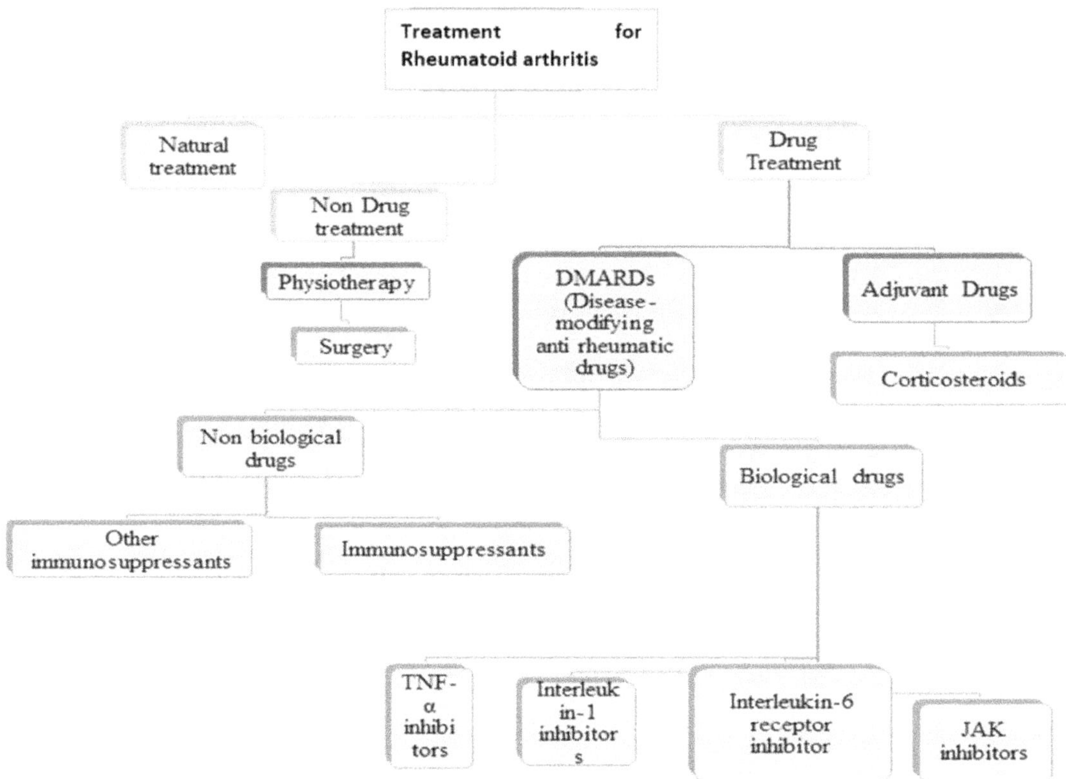

FIGURE 2.3 Current treatment strategies for rheumatoid arthritis.

proliferation and increasing regulatory T-cell activity. It also protects against oxidative stress and cell damage (Khan, Ahmed, et al., 2023).

Leflunomide works by inhibiting the mitochondrial enzyme dihydroorotate dehydrogenase (DHODH), preventing the synthesis of uridine monophosphate and subsequently the production of RNA and DNA. This inhibition affects rapidly dividing cells, such as lymphocytes. Teriflunomide, the active metabolite of leflunomide, inhibits human DHODH and several tyrosine kinases. Teriflunomide has additional antiviral properties (Kjørholt et al., 2024).

Hydroxychloroquine increases lysosomal pH in antigen-presenting cells by interfering with endocytosis and proteolysis. It impacts immune cell functioning, including chemotaxis, phagocytosis, and superoxide production. Hydroxychloroquine accumulates in lysosomes, increasing pH by inhibiting lysosomal acidic proteases. It decreases intracellular processing, glycosylation, and secretion of protein, thereby modulating the immune response (Sen, Riofrio, & Singh, 2024).

2.11.2 Problems Related to DMARDs

DMARDs can cause serious side effects, such as nausea, vomiting, diarrhea, hair loss, and fatigue. These medications may not be effective for all patients and can lose effectiveness over time. High doses can lead to serious side effects, which may limit their use. DMARDs can also cause serious drug interactions. For example, interaction with warfarin can increase the risk of bleeding. Additionally, they can reactivate latent infections, such as tuberculosis and hepatitis B. There is an increased risk of malignancies and fertility issues. Regular blood test and liver monitoring are necessary to manage these risks (Deng, Chen, Shao, Wu, & Wang, 2023).

2.11.3 Biological DMARDs

Biologic DMARDs are a class of medications used to treat autoimmune conditions, including RA. These drugs are composed of proteins that specifically target substances, cells, and processes that lead to tissue damage and inflammation. They block specific proteins involved in inflammation, such as TNF-α, IL-1, and IL-6. Biologic DMARDs are often used in combination with traditional DMARDs, such as MTX, particularly in patients who have not responded to traditional DMARDs or have severe disease. They are also referred to as TNF blockers and anti-TNFs, including adalimumab, certolizumab, golimumab, etanercept, infliximab, and golimumab (Kerschbaumer et al., 2023).

Figure 2.3 illustrates the current treatment strategies for RA, which encompass a range of therapeutic options, including conventional synthetic disease-modifying antirheumatic drugs, biologic DMARDs targeting specific molecules, and targeted synthetic DMARDs, as well as glucocorticoids for short-term inflammation control, all aimed at reducing inflammation, slowing disease progression, and improving patient outcomes.

2.12 TNF-BLOCKING AGENTS

Infliximab was the first TNF-blocking drug to be discovered, followed by adalimumab and etanercept. Infliximab and adalimumab are monoclonal antibodies that specifically target TNF-α, forming an immunoglobulin-like protein to neutralize this pro-inflammatory cytokine. Etanercept, on the other hand, acts as a decoy receptor that binds TNF-α, preventing it from interacting with its natural receptors on cell surfaces. MTX can be administered in conjunction with anti-TNF medicines or separately from it. Despite being more effective than traditional DMARDs in treating some conditions, there are still patients who do not respond to these treatments (Siegmund & Wajant, 2023).

2.13 PROBLEMS RELATED TO THE USE OF TNF-BLOCKING DRUGS

The use of TNF-blocking drugs, while effective, can be associated with serious side effects. Infections are a significant risk associated with these biological agents. Specifically, TB is a critical infection that needs to be checked for when using anti-TNF medications. Exacerbations of hepatitis have been reported in recent data among carriers and patients with chronic hepatitis B. The extended suppression of the immune response may facilitate the reactivation of slow-moving viruses, like the JC polyoma virus, which can lead to progressive multifocal leukoencephalopathy (Patel et al., 2023).

2.14 INTERLEUKIN-1 RECEPTOR INHIBITORS

IL-1 receptor antagonists work by preventing IL-1 from attaching to its receptor, thus eliminating IL-1's pro-inflammatory effects. These inhibitors bind with high affinity to the type I IL-1 receptor (IL-1RI), inhibiting the interaction between IL-1 and its receptor. IL-1 is a key mediator of inflammation and immunological responses. It suppresses the production of pro-inflammatory cytokines and enzymes by occupying the IL-1RI and blocking the activation of downstream signaling pathways, including the MyD88-dependent and MyD88-independent cascades. This suppression reduces the production of inflammatory mediators such as prostaglandins, leukotrienes, and metalloproteinases, which are responsible for joint inflammation and degradation. Anakinra, a recombinant human IL-1 receptor antagonist, decreases inflammation and swelling in affected joints, reduces joint erosion, and enhances functional ability. It works by lowering the activation and proliferation of immune cells, such as T cells and macrophages. Anakinra is administered subcutaneously once a day and can be used either alone or in combination with MTX to treat RA. However, since TNF inhibitors exhibit far higher efficacy and have more convenient dose regimens, it is not utilized frequently.

Other IL-1 receptor inhibitors include canakinumab, rilonacept, and gevokizumab. Canakinumab is a monoclonal antibody that targets the beta subunit of IL-1 and is approved for the treatment of JIA, Still's disease, and gout. Rilonacept is a fusion protein that blocks IL-1α and IL-1β. Gevokizumab is a monoclonal antibody that targets IL-1β. These drugs are used to manage various inflammatory diseases by reducing the activity of IL-1, thereby mitigating inflammation and preventing joint damage.

2.15 PROBLEMS RELATED TO THE USE OF INTERLEUKIN-1 RECEPTOR INHIBITORS

IL-1 receptor inhibitors have been linked to several potential side effects and risks. Common side effects include headache, nausea, vomiting, and hepatotoxicity. These inhibitors also increase the risk of infections particularly when used in combination with TNF-α inhibitors. Additionally, they can reactivate latent infections such as tuberculosis. The safety and efficacy of these inhibitors during pregnancy and in the pediatric population are not well established, raising concerns about their use in these age groups. These inhibitors can also cause immunogenicity and hypersensitivity reactions. The need for high doses to achieve therapeutic effects underscores the importance of optimal dosing regimens to minimize adverse effects while maintaining efficacy.

2.16 INTERLEUKIN-6 RECEPTOR INHIBITORS

IL-6 inhibitors constitute a category of therapeutic agents that antagonize the biological activity of IL-6, a cytokine playing a pivotal role in the regulation of inflammation and immune response. IL-6 binds to its receptor, IL-6R, with high affinity, triggering the formation of a binary complex. This complex then recruits gp130, which acts as a signal-transducing subunit, to form a ternary complex. The subsequent activation of gp130 initiates a cascade of intracellular signaling events, involving the activation of Janus Kinases (JAKs), particularly JAK1 and JAK2, and the phosphorylation of signal transducer and activator of transcription (STAT) proteins, primarily STAT3. Phosphorylated STAT3 translocates to the nucleus, where it regulates the transcription of target genes, including those involved in inflammation, immune response, and cell survival. IL-6 also activates the MAPK/ERK and PI3K/AKT signaling pathways, thereby further amplifying its pro-inflammatory and antiapoptotic effects. The activation of IL-6 signaling leads to the production of acute-phase proteins, such as C-reactive protein and serum amyloid A, which play a crucial role in the systemic inflammatory response. Common IL-6 inhibitors include monoclonal antibodies such as tocilizumab, sarilumab, siltuximab, olokizumab, and clazakizumab. These drugs block IL-6 signaling and neutralize its activity, thereby mitigating inflammation and modulating the immune response.

The excessive production of IL-6 precipitates a plethora of deleterious consequences, culminating in a diverse array of adverse effects. The unrestrained activation of IL-6 signaling cascades triggers a pro-inflammatory response, fostering an environment conducive to tissue damage and dysfunction. This exacerbates the symptoms of RA, JIA, and Castleman disease, leading to enhanced articular erosion, pain, and diminished quality of life (Tanaka, Narazaki, & Kishimoto, 2018). The production of IL-6 contributes to the development of systemic manifestations, including fatigue, anemia, and hepatic dysfunction. It also increased the risk of cardiovascular events, such as myocardial infarction and stroke. Furthermore, chronic exposure to elevated IL-6 levels can lead to neurocognitive impairment, depression, and anxiety, underscoring the far-reaching consequences of uncontrolled IL-6 activity (Lee, 2020).

2.17 JAK INHIBITORS

JAK inhibitors are a class of drugs that block JAK enzymes, which are involved in intracellular signaling. JAK inhibitors target one or more of the four enzymes (TYK2, JAK-1, JAK-2, and JAK-3) to prevent their activation. By inhibiting JAK enzymes, these drugs also inhibit the activation of signaling pathways including the STAT pathway (McLornan, Pope, Gotlib, & Harrison, 2021). This inhibition decreases the signaling of pro-inflammatory cytokines such as IL-6, IL-12, and interferon-gamma (IFN-γ), leading to a decrease in inflammation and tissue damage. JAK inhibitors can also modulate the activity of immune cells, such as T cells and macrophages, which contribute to the pathogenesis of inflammatory diseases (Taylor, 2019).

Tofacitinib is an oral JAK inhibitor, used to treat conditions including RA, ankylosing spondylitis, and ulcerative colitis. Tofacitinib binds to the ATP-binding site of JAK enzymes, preventing the phosphorylation and activation of STAT proteins. The reduction in cytokine signaling leads to decreased inflammation (Dhillon, 2017).

Baricitinib is a potent and selective inhibitor of the JAK family of enzymes, with higher affinity for JAK1 and JAK2 compared to JAK3 and TYK2. Baricitinib inhibits JAK1 and JAK2, with an IC50 of 5.9 and 4.6 nM, respectively. It prevents the phosphorylation of STAT1, STAT3, and STAT5, thereby reducing the transcription of pro-inflammatory genes. Baricitinib decreases the signaling of IL-6, IL-12, IL-23, and IFN-γ (Al-Salama & Scott, 2018), which are key cytokines involved in autoimmune diseases. By affecting the activation, proliferation, and differentiation of immune cells, including T cells, B cells, and macrophages, baricitinib decreases inflammatory genes expression, including those involved in the NF-κB pathway. This leads to reduced inflammation, joint damage, and symptoms in RA and other autoimmune diseases (Kuriya, Cohen, & Keystone, 2017).

JAK inhibitors, while effective in treating various autoimmune conditions, are associated with several significant side effects and risks. One of the most serious risks is the increased susceptibility to infections, including tuberculosis, herpes zoster, and other serious infections. They may increase the risk of certain cancers. Gastrointestinal symptoms like nausea, indigestion, and diarrhea are common. Musculoskeletal problems, including joint pain, and hepatic adverse effects, including abnormal liver function tests and liver damage, can also occur. Additionally, JAK inhibitors can also increase cardiovascular risks, including heart attack, stroke, high cholesterol levels, and blood clots (Peng, Xiao, Ottaviani, Stebbing, & Wang, 2020). There have also been reports of more severe immunologic and hematological side effects, including lymphopenia, neutropenia, anemia, and an elevated risk of infection and malignancy. Nasopharyngitis and upper respiratory tract infections are also known to occur with the use of JAK inhibitors (Yamaoka, 2019).

2.18 NONSTEROIDAL ANTI-INFLAMMATORY DRUGS (NSAIDs)

NSAIDs inhibit the enzyme cyclooxygenase (COX), which is responsible for the transformation of arachidonic acid into prostaglandins and pro-inflammatory mediators. By decreasing prostaglandin production, NSAIDs reduce inflammation, swelling, and pain in RA. Some NSAIDs (e.g., celecoxib) selectively inhibit COX-2, which is more specific to inflammation, sparing COX-1, which protects the stomach lining. NSAIDs also reduce fever by blocking prostaglandin production in the hypothalamus. Common NSAIDs used for RA include ibuprofen, naproxen, meloxicam, and celecoxib (Panchal & Sabina, 2023).

2.19 CORTICOSTEROIDS

Corticosteroids are a class of potent anti-inflammatory agents that exert their effects through a multifaceted mechanism of action. They bind to glucocorticoid receptors and regulate gene transcription, leading to the suppression of pro-inflammatory cytokine production, including TNF-α, IL-1β, and IL-6. This transcriptional repression diminishes the expression of adhesion molecules, chemokines, and inflammatory enzymes, thereby curtailing leukocyte migration and activation. Corticosteroids also induce the expression of anti-inflammatory proteins, such as lipocortin-1, which inhibits phospholipase A2, thereby reducing eicosanoid production (Ethgen, de Lemos Esteves, Bruyere, & Reginster, 2013). In addition, corticosteroids modulate the immune response by suppressing T-cell activation, proliferation, and differentiation. They also inhibit the production of autoantibodies.

The net result of these molecular interventions is a reduction of synovial inflammation and joint damage. Some common corticosteroids include prednisolone, hydrocortisone, triamcinolone, betamethasone, fluticasone, and beclomethasone (Criswell et al., 1996).

While corticosteroids are effective in the treatment of many diseases, they can have a wide range of adverse effects. These include increased blood sugar levels, weight gain, fluid retention, mood changes (anxiety, depression, irritability), insomnia, and sleep disturbances. Long-term use of corticosteroids can also result in osteoporosis and bone fractures, adrenal insufficiency when stopped, and immunosuppression, which increases the risk of infections (Ethgen et al., 2013).

REFERENCES

Acharya, I., Smith, L. W., Banerjee, C., Camire, L. M., & Vij, R. (2024). Reactive arthritis after mpox vaccination. *Journal of Community Hospital Internal Medicine Perspectives*, *14*(1), 35.

Al-Salama, Z. T., & Scott, L. J. (2018). Baricitinib: a review in rheumatoid arthritis. *Drugs*, *78*, 761–772.

Aripova, N., Kremer, J. M., Pappas, D. A., Reed, G., England, B. R., Robinson, B. H.,… Mikuls, T. R. (2024). Anti-citrullinated protein antibody profiles predict changes in disease activity in patients with rheumatoid arthritis initiating biologics. *Rheumatology*, *63*(2), 542–550.

Babaahmadi, M., Tayebi, B., Gholipour, N. M., Kamardi, M. T., Heidari, S., Baharvand, H.,… Hassani, S.-N. (2023). Rheumatoid arthritis: the old issue, the new therapeutic approach. *Stem Cell Research & Therapy*, *14*(1), 268.

Bai, B., Li, T., Zhao, J., Zhao, Y., Zhang, X., Wang, T.,… Xu, J. (2023). The tyrosine phosphatase activity of PTPN22 is involved in T cell development via the regulation of TCR expression. *International Journal of Molecular Sciences*, *24*(19), 14505.

Bentaleb, I., Abdelghani, K. B., Rostom, S., Amine, B., Laatar, A., & Bahiri, R. (2020). Reactive arthritis: update. *Current Clinical Microbiology Reports*, *7*, 124–132.

Boer, C. G., Hatzikotoulas, K., Southam, L., Stefánsdóttir, L., Zhang, Y., de Almeida, R. C.,… Teder-Laving, M. (2021). Deciphering osteoarthritis genetics across 826,690 individuals from 9 populations. *Cell*, *184*(18), 4784–4818.

Boeren, A. M., Verstappen, M., Looijen, A. E., de Jong, P. H., & van der Helm-van Mil, A. H. (2024). Patients with rheumatoid arthritis presenting with mono-or oligo-arthritis and high VAS-ratings remain the most fatigued during 5 years of follow-up. *Rheumatology*, *63*(6), 1574–1581.

Chen, B., Ning, K., Sun, M.-L., & Zhang, X.-A. (2023). Regulation and therapy, the role of JAK2/STAT3 signaling pathway in OA: a systematic review. *Cell Communication and Signaling*, *21*(1), 67.

Crayne, C. B., & Beukelman, T. (2018). Juvenile idiopathic arthritis: oligoarthritis and polyarthritis. *Pediatric Clinics*, *65*(4), 657–674.

Criswell, L., Saag, K., Sems, K., Welch, V., Shea, B., Wells, G. A.,… Group, C. M. (1996). Moderate-term, low-dose corticosteroids for rheumatoid arthritis. *Cochrane Database of Systematic Reviews*, *2010*(1), 1–8.

Deng, G., Chen, X., Shao, L., Wu, Q., & Wang, S. (2023). Effectiveness and safety of 99Tc-methylene diphosphonate as a disease-modifying anti-rheumatic drug (DMARD) in combination with conventional synthetic (cs) DMARDs in the treatment of rheumatoid arthritis: a systematic review and meta-analysis of 34 randomized controlled trials. *Heliyon*, *9*(11), 1–13.

Dhillon, S. (2017). Tofacitinib: a review in rheumatoid arthritis. *Drugs*, *77*, 1987–2001.

Dombret, S., Skapenko, A., & Schulze-Koops, H. (2022). Reactive arthritis after SARS-CoV-2 infection. *Rmd Open*, *8*(2), e002519.

Ethgen, O., de Lemos Esteves, F., Bruyere, O., & Reginster, J.-Y. (2013). What do we know about the safety of corticosteroids in rheumatoid arthritis? *Current Medical Research and Opinion*, *29*(9), 1147–1160.

Ferreira, A., Marinho, A., Vita, P., Fonseca, T., & Vasconcelos, C. (2024). Post-infectious arthritis and reactive arthritis. In Y. Shoenfeld, N. Agmon-Levin, & N. R. Rose (Eds.), *Infection and Autoimmunity* (pp. 749–782). Elsevier.

Gaber, T., Strehl, C., & Buttgereit, F. (2017). Metabolic regulation of inflammation. *Nature Reviews Rheumatology*, *13*(5), 267–279.

Gravallese, E. M., & Firestein, G. S. (2023). Rheumatoid arthritis—common origins, divergent mechanisms. *New England Journal of Medicine*, *388*(6), 529–542.

Hermans, S. M., Lear, G., Case, B. S., & Buckley, H. L. (2023). The soil microbiome: an essential, but neglected, component of regenerative agroecosystems. *Iscience*, *26*(2), 1–10.

Ke, Y., Wang, C., Zhang, J., Zhong, X., Wang, R., Zeng, X., & Ba, X. (2019). The role of PARPs in inflammation—and metabolic—related diseases: molecular mechanisms and beyond. *Cells*, *8*(9), 1047.

Kerschbaumer, A., Sepriano, A., Bergstra, S. A., Smolen, J. S., Van Der Heijde, D., Caporali, R.,... Pope, J. E. (2023). Efficacy of synthetic and biological DMARDs: a systematic literature review informing the 2022 update of the EULAR recommendations for the management of rheumatoid arthritis. *Annals of the Rheumatic Diseases*, *82*(1), 95–106.
Khan, M. M., Ahmed, S., Hasan Sajib, M. K., Morshed, A. A., Mahbub-Uz-Zaman, K., & Haq, S. A. (2023). Tofacitinib versus methotrexate as the first-line disease-modifying antirheumatic drugs in the treatment of rheumatoid arthritis: an open-label randomized controlled trial. *International Journal of Rheumatic Diseases*, *26*(9), 1729–1736.
Khan, S., Mohan, K., Muzammil, S., Alam, M. A., & Khayyam, K. U. (2024). Current prospects in rheumatoid arthritis: pathophysiology, genetics, and treatments. *Recent Advances in Anti-Infective Drug Discovery Formerly Recent Patents on Anti-Infective Drug Discovery*, *19*(1), 36–55.
Kjørholt, K. E., Sundlisæter, N. P., Aga, A.-B., Sexton, J., Olsen, I. C., Fremstad, H.,... Bakland, G. (2024). Effects of tapering conventional synthetic disease-modifying antirheumatic drugs to drug-free remission versus stable treatment in rheumatoid arthritis (ARCTIC REWIND): 3-year results from an open-label, randomised controlled, non-inferiority trial. *The Lancet Rheumatology*, *6*(5), e268–e278.
Kuriya, B., Cohen, M. D., & Keystone, E. (2017). Baricitinib in rheumatoid arthritis: evidence-to-date and clinical potential. *Therapeutic Advances in Musculoskeletal Disease*, *9*(2), 37–44.
Lee, S. T. (2020). Inflammation, depression, and anxiety disorder: a population-based study examining the association between Interleukin-6 and the experiencing of depressive and anxiety symptoms. *Psychiatry Research*, *285*, 112809.
Levi, M. (2017). Inflammation and coagulation. In J.-M. Cavaillon & M. Singer (Eds.), *Inflammation: From Molecular and Cellular Mechanisms to the Clinic* (pp. 833–860). Wiley.
Mariani, F. M., Martelli, I., Pistone, F., Chericoni, E., Puxeddu, I., & Alunno, A. (2023). Pathogenesis of rheumatoid arthritis: one year in review 2023. *Clinical and Experimental Rheumatology*, *41*(9), 1725–1734.
Martini, A., Lovell, D. J., Albani, S., Brunner, H. I., Hyrich, K. L., Thompson, S. D., & Ruperto, N. (2022). Juvenile idiopathic arthritis. *Nature Reviews Disease Primers*, *8*(1), 5.
Martos, R., Tar, I., Nagy, A. C., Csősz, É., Kiss, C., & Márton, I. (2023). Hypercitrullination and anti-citrullinated protein antibodies in chronic apical periodontitis, a laboratory investigation. Does autoimmunity contribute to the pathogenesis? *International Endodontic Journal*, *56*(5), 584–592.
McLornan, D. P., Pope, J. E., Gotlib, J., & Harrison, C. N. (2021). Current and future status of JAK inhibitors. *The Lancet*, *398*(10302), 803–816.
Panchal, N. K., & Sabina, E. P. (2023). Non-steroidal anti-inflammatory drugs (NSAIDs): a current insight into its molecular mechanism eliciting organ toxicities. *Food and Chemical Toxicology*, *172*, 113598.
Patel, J. P., Srinivasa, N. K. K., Gande, A., Anusha, M., Dar, H., & Baji, D. B. (2023). The role of biologics in rheumatoid arthritis: a narrative review. *Cureus*, *15*(1), 1–10.
Peng, L., Xiao, K., Ottaviani, S., Stebbing, J., & Wang, Y.-J. (2020). A real-world disproportionality analysis of FDA Adverse Event Reporting System (FAERS) events for baricitinib. *Expert Opinion on Drug Safety*, *19*(11), 1505–1511.
Pisetsky, D. S. (2023). Pathogenesis of autoimmune disease. *Nature Reviews Nephrology*, *19*(8), 509–524.
Richard, M., Driban, J., & McAlindon, T. (2023). Pharmaceutical treatment of osteoarthritis. *Osteoarthritis and Cartilage*, *31*(4), 458–466.
Saricaoglu, E. M., Hasanoglu, I., & Guner, R. (2021). The first reactive arthritis case associated with COVID-19. *Journal of Medical Virology*, *93*(1), 192.
Sen, R., Riofrio, M., & Singh, J. A. (2024). A narrative review of the comparative safety of disease-modifying anti-rheumatic drugs used for the treatment of rheumatoid arthritis. *Expert Opinion on Drug Safety*, *23*(6), 687–714.
Sieghart, D., Konrad, C., Swiniarski, S., Haslacher, H., Aletaha, D., & Steiner, G. (2023). The diagnostic and prognostic value of IgG and IgA anti-citrullinated protein antibodies in patients with early rheumatoid arthritis. *Frontiers in Immunology*, *13*, 1096866.
Siegmund, D., & Wajant, H. (2023). TNF and TNF receptors as therapeutic targets for rheumatic diseases and beyond. *Nature Reviews Rheumatology*, *19*(9), 576–591.
Smolen, J. S., Landewé, R. B., Bergstra, S. A., Kerschbaumer, A., Sepriano, A., Aletaha, D.,... Pope, J. E. (2023). EULAR recommendations for the management of rheumatoid arthritis with synthetic and biological disease-modifying antirheumatic drugs: 2022 update. *Annals of the Rheumatic Diseases*, *82*(1), 3–18.

Studenic, P., Alunno, A., Sieghart, D., Bang, H., Aletaha, D., Blüml, S.,... Steiner, G. (2021). Presence of anti-acetylated peptide antibodies (AAPA) in inflammatory arthritis and other rheumatic diseases suggests discriminative diagnostic capacity towards early rheumatoid arthritis. *Therapeutic Advances in Musculoskeletal Disease*, *13*, 1759720X211022533.

Tanaka, T., Narazaki, M., & Kishimoto, T. (2018). Interleukin (IL-6) immunotherapy. *Cold Spring Harbor Perspectives in Biology*, *10*(8), a028456.

Tang, Y. (2024). Investigating the function of TRAF1 in NF-κB activation.

Tasneem, S., Liu, B., Li, B., Choudhary, M. I., & Wang, W. (2019). Molecular pharmacology of inflammation: medicinal plants as anti-inflammatory agents. *Pharmacological Research*, *139*, 126–140.

Taylor, P. C. (2019). Clinical efficacy of launched JAK inhibitors in rheumatoid arthritis. *Rheumatology*, *58*(Supplement_1), i17–i26.

Thompson, K. N., Bonham, K. S., Ilott, N. E., Britton, G. J., Colmenero, P., Bullers, S. J.,... Filer, A. (2023). Alterations in the gut microbiome implicate key taxa and metabolic pathways across inflammatory arthritis phenotypes. *Science Translational Medicine*, *15*(706), eabn4722.

Tracy, R. P. (2006). The five cardinal signs of inflammation: calor, dolor, rubor, tumor… and penuria (apologies to Aulus Cornelius Celsus, De medicina, c. AD 25). *The Journals of Gerontology Series A: Biological Sciences and Medical Sciences*, *61*(10), 1051–1052.

Tsetseri, M.-N., Silman, A. J., Keene, D. J., & Dakin, S. G. (2023). The role of the microbiome in rheumatoid arthritis: a review. *Rheumatology Advances in Practice*, *7*(2), rkad034.

Tsuge, K., Inazumi, T., Shimamoto, A., & Sugimoto, Y. (2019). Molecular mechanisms underlying prostaglandin E2-exacerbated inflammation and immune diseases. *International Immunology*, *31*(9), 597–606.

van den Bosch, M. H. (2021). Osteoarthritis year in review 2020: biology. *Osteoarthritis and Cartilage*, *29*(2), 143–150.

Van Hoovels, L., Jacobs, J., Vander Cruyssen, B., Van den Bremt, S., Verschueren, P., & Bossuyt, X. (2018). Performance characteristics of rheumatoid factor and anti-cyclic citrullinated peptide antibody assays may impact ACR/EULAR classification of rheumatoid arthritis. *Annals of the Rheumatic Diseases*, *77*(5), 667–677.

Vegas, L. P., Drouin, J., Dray-Spira, R., & Weill, A. (2023). Prevalence, mortality, and treatment of patients with rheumatoid arthritis: a cohort study of the French National Health Data System, 2010–2019. *Joint Bone Spine*, *90*(1), 105460.

Venetsanopoulou, A. I., Kalpourtzi, N., Alamanos, Y., Gavana, M., Vantarakis, A., Hadjichristodoulou, C.,... Drosos, A. A. (2023). Prevalence of rheumatoid arthritis in Greece: results from the national health examination survey EMENO. *Rheumatology International*, *43*(7), 1349–1355.

Weiss, P. F. (2013). Evaluation and treatment of enthesitis-related arthritis. *Current Medical Literature. Rheumatology/the Royal Society of Medicine*, *32*(2), 33.

Yamaoka, K. (2019). Tofacitinib for the treatment of rheumatoid arthritis: an update. *Expert Review of Clinical Immunology*, *15*(6), 577–588.

Zhang, F., Jonsson, A. H., Nathan, A., Millard, N., Curtis, M., Xiao, Q.,... Weisenfeld, D. (2023). Deconstruction of rheumatoid arthritis synovium defines inflammatory subtypes. *Nature*, *623*(7987), 616–624.

Zhang, X., Zhang, X., Yang, Y., Zhi, K., Chen, Y., Zhao, J.,... An, Y. (2023). Association between passive smoking and the risk of rheumatoid arthritis: a systematic review and meta-analysis. *Clinical Rheumatology*, *42*(3), 663–672.

3 Current Pharmacotherapy for Rheumatoid Arthritis

Mehak Fatima, Ammara Saleem, Aisha Mobashar, and Muhammad Furqan Akhtar

WHAT YOU WILL LEARN IN THIS CHAPTER?

This chapter will provide a comprehensive review of the concepts of available pharmaceutical drug classes for the treatment of rheumatoid arthritis, their mechanism of action, and special emphasis on their adverse drug reactions. This will also enable students and researchers to understand the drug targets and alternative treatments.

LIST OF ABBREVIATIONS

BVA	Bee venom acupuncture
COX	Cyclooxygenase
DHFR	Dihydrofolate reductase
JAK	Janus kinase
MTX	Methotrexate
STAT	Signal transducer and activator of transcription

3.1 ANALGESIC AND NSAIDs

Analgesics and nonsteroidal anti-inflammatory drugs (NSAIDs) are mostly utilized as a short-term treatment for rheumatoid arthritis (RA) in order to cope with the pain (Table 3.1) and inflammation of joints. The most widely used painkiller is acetaminophen, also known as paracetamol. It has a low risk of adverse effects. Over 20 distinct NSAIDs have been used, of which short-acting NSAIDs include ibuprofen, diclofenac, ketoprofen, and indomethacin; and long-acting NSAIDs include naproxen, celecoxib, meloxicam, nabumetone, and piroxicam (Kumar & Banik, 2013).

NSAIDs are widely used in RA for managing pain, inflammation, and stiffness, particularly in the early stages and while waiting for DMARDs to take effect. However, NSAIDs alone do not slow RA progression. Their efficacy stems from inhibiting cyclooxygenase (COX) enzymes, which are crucial in converting arachidonic acid into pro-inflammatory prostaglandins (PGs) (Gaffo, Saag, & Curtis, 2006). The opioid analgesics (codeine, dextropropoxyphene, and tramadol) are also used to treat chronic pain in RA patients (Cronstein, 1997).

3.1.1 Mechanism of Action of NSAIDs

There are two main COX isoforms: COX-1, which is constitutionally demonstrated and implicated in protecting the stomach lining and other functions, and COX-2, which is definite and plays a key role in inflammation. NSAIDs can be selective for COX-1 or COX-2, or nonselective, affecting both (Schuna, 1998). NSAIDs work primarily by blocking the activity of COX I and II. It stops arachidonic acid from being transformed into PG, prostacyclins, and thromboxanes. Dyspepsia, peptic ulcer disease, tinnitus, bleeding, and temporary increases in liver enzymes are among the common adverse effects of NSAIDs (Vane & Botting, 1998).

DOI: 10.1201/9781003535201-4

TABLE 3.1
Pharmaceutical Drug Classes Used for Rheumatoid Arthritis (RA) With Their Mechanisms of Action and Adverse Drug Reactions (ADRs)

Class	Mechanism of Action	Adverse Effect	References
Analgesics and NSAIDs Acetaminophen	To inhibit cyclooxygenase (COX) 1 and 2, by preventing the conversion of arachidonic acid to prostaglandins, prostacyclin, and thromboxanes	Hypertension, myocardial infarction, renal impairment	Kianian, Bansal, Lee, Zhang, and Bergese (2024)
Corticosteroids	Stimulate the expression of lipocortin-1, which inhibits the enzyme phospholipase A2 and the production of arachidonic acid	Osteoporosis, gastrointestinal bleeding, peptic ulcers, diabetes mellitus, cataracts, hypothalamic–pituitary–adrenal axis	De Bock and Sienaert (2024)
Nonbiological DMARDs			
Methotrexate	Reduce immune system activity and retain biological activity by converting into methotrexate-polyglutamates	Hepatotoxicity, hematological issues, skin rashes, nausea, dyspepsia	
Sulfasalazine	Effects on inflammation involve various mechanisms	Gastrointestinal symptoms, pulmonary complications, dermatologic complications	Ye et al. (2024)
Leflunomide	Inhibits dihydroorotate dehydrogenase (DHODH) and disrupts pyrimidine production, which is essential for DNA synthesis and cell proliferation	Nausea, diarrhea, skin rash, hair loss, hypertension, bone marrow suppression	Zewail (2024)
Hydroxychloroquine	Accumulates in lysosomes, increasing pH, which inhibits lysosomal enzymes and autophagy, impairing antigen presentation and lymphocyte function	Gastrointestinal issues, retinopathy, chronic kidney disease, exanthematous pustulosis	Huo, Wei, Yang, Lin, and Huang (2025)
Biologic DMARDs			
(TNF-α) inhibitors	Stimulates the expression of other inflammatory cytokines and adhesion molecules, contributing to synovitis	Malignancies, neurological complications, systemic lupus erythematosus	Riaz et al. (2025)
Anti-B-cell therapy Rituximab	Targets the CD20 molecule on certain B cells, depleting them extensively in peripheral blood and to varying degrees in synovial tissue and other sites	Infection, tuberculosis	de Carvalho and Skare (2025)
T-lymphocyte co-stimulation blocker Abatacept	Modulates T-cell activation by binding to CD80 and CD86 on antigen-presenting cells, blocking the required CD28 interaction between these cells and T cells	Musculoskeletal, infections, nervous system disorders, cardiac disorders, and neoplasms	Di Cola and Ruscitti (2024)
Interleukin-6 (IL-6) inhibitors	IL-6 initiates cell signaling by binding to its specific receptor, IL-6R (CD126)	Gastrointestinal perforation, neutropenia, thrombocytopenia	Di Cola and Ruscitti (2024)
JAK inhibitors Tofacitinib Ruxolitinib Baricitinib	Target cytokines involved in autoimmune diseases by inhibiting the JAK/STAT pathway, which play a pathogenic role in diseases	Nasopharyngitis, hypercholesterolemia, neutropenia, anemia, cytomegalovirus, herpes zoster, pulmonary tuberculosis	Virtanen et al. (2024)

Selective COX-2 inhibitors (e.g., celecoxib) were designed to lessen the gastrointestinal (GI) effects, but some COX-2 inhibitors, e.g., rofecoxib and valdecoxib, were withdrawn due to increased cardiovascular risks. Current data on COX-2 inhibitors' cardiovascular safety remain mixed, with some studies showing increased risks, particularly at higher doses, while others do not show a significant increase in cardiovascular mortality (van Laar et al., 2012).

Short-term NSAID use is generally well-tolerated, but long-term use can lead to GI complications such as ulcers and bleeding, especially in the elderly and those taking glucocorticoids or anticoagulants. Co-administration with GI-protective agents can reduce these risks. NSAIDs can also cause renal toxicity and increase cardiovascular risks, including myocardial infarction and heart failure (Kontny et al., 2000).

Aspirin, though historically significant, is less frequently used due to dosing difficulties and a higher incidence of gastritis compared to newer NSAIDs. PGs play a crucial role in protecting the gastric mucosa, and their inhibition by NSAIDs increases the risk of GI injury (Zerbini et al., 2017). The inhibition of PGs by NSAIDs also reduces renal blood flow, posing a risk of renal failure in high-risk populations such as the elderly and those with chronic renal disease (Graham, Agrawal, & Roth, 1988).

3.2 CORTICOSTEROIDS

RA has long been treated with glucocorticoids. While high-level doses of glucocorticoids pose significant risks. Daily doses of prednisone at ≤15 mg can effectively give analgesic and anti-inflammatory effects in many RA patients. Moreover, this dose may reverse the progression of disease (Caldwell & Furst, 1991). Glucocorticoids are often added to RA treatment temporarily to control disease activity. Combining them with DMARDs offers greater clinical benefits and slows radiographic progression. However, glucocorticoids are usually tapered off after the quarter or half of the year due to a longstanding negative impact (Criswell et al., 1996).

3.2.1 Mechanism of Action

Corticosteroids exert their effects through genomic and nongenomic pathways. Nongenomic effects, occurring rapidly, are not fully understood but may involve changes in cell membrane functions. These effects include pain relief and repression of adhesion molecules (Ferreira, Mohamed, & Emery, 2016). Innate effects are interfered with steroid receptors primarily through transactivation, DNA binding, and ligand binding domains. Alternative splicing of mRNA generates various receptor isoforms, of which gluocorticosteroid receptor α is the active receptor. When corticosteroids bind to CRα, the receptor undergoes conformational changes, dimerizes, and moves to the nucleus, where it binds to glucocorticosteroid response elements to regulate gene expression by either transactivation or transrepression. Additionally, corticosteroids hinder the activity of NF-κB and activator protein (AP-1) in gene transcription, inhibit histone acetylation, and modulate protein kinase cascades, ultimately leading to repression of pro-inflammatory genes and induction of anti-inflammatory effects. Additionally, corticosteroids control several elements of gene expression and cause the synthesis of the anti-inflammatory protein lipocortin (Berardicurti et al., 2020).

Glucocorticoids combat inflammation through several mechanisms. They enhance the expression of lipocortin-1, which inhibits the enzyme phospholipase A2 and reduces the synthesis of arachidonic acid (Choueiri, Chevalier, & Eymard, 2021). Additionally, the products of their transcription block the activity of nuclear factor κB (NF-κB) and the AP-1 protein, which are responsible for upregulating proinflammatory molecules like tumor necrosis factor alpha (TNF-α) and interleukin (IL)-1. The action of glucocorticoids, especially at larger doses, may be due to their rapid action via receptors on the membrane and second messengers (McWilliams et al., 2022).

3.2.2 Adverse Effects

Adverse consequences can arise with the prolonged usage of low-dose glucocorticoids in the management of RA. A higher incidence of osteoporosis and fissure, GI bleeding, ulcers (peptic), diabetes, infections, cataracts, and reduced response of the hypothalamic–pituitary–hormone axis are a few of these. Significant adverse cardiovascular events and infections are also possible. Rarely, complications associated with intra-articular injections can include systemic effects, acute synovitis, osteonecrosis, tendon rupture, and septic arthritis (Ruyssen-Witrand & Constantin, 2018).

Some common adverse effects of glucocorticoid treatment typically manifest early in treatment and are often unavoidable. These include emotional liability and increased appetite leading to weight gain, insomnia, and acne vulgaris. Prolonged use of supraphysiologic doses may result in a Cushing syndrome, suppression of the hypothalamic–pituitary–hormonal axis, delayed wound healing, myopathy, osteonecrosis, and a heightened risk of infections (Wilson et al., 2019). Depending on the cumulative dose, delayed side effects can include growth retardation, osteoporosis, cataracts, fatty liver, atherosclerosis, and skin shrinkage. There is also a chance of uncommon and unforeseen side effects including psychosis, pancreatitis, glaucoma, and pseudotumor cerebri (Mitchell et al., 2018).

3.3 TRADITIONAL (NONBIOLOGIC) DISEASE-MODIFYING ANTIRHEUMATIC DRUGS (DMARDs)

DMARDs such as methotrexate (MTX), hydroxychloroquine, sulfasalazine, and leflunomide are frequently used in the treatment of RA. Following the British Society for Rheumatology's guidelines, monitoring for these nonbiologic DMARDs is crucial. Due to their unfavorable risk-benefit profiles, older DMARDs like gold, penicillamine, cyclosporine, and azathioprine are no longer recommended. Moreover, antimicrobial DMARDs like minocycline and doxycycline are not recommended. For patients who do not achieve sufficient inflammation control with a single DMARD, combination therapy involving multiple DMARDs may be contemplated (Jurgens, Jacobs, & Bijlsma, 2011).

3.3.1 Methotrexate (MTX)

It is a common primary DMARD for RA treatment, and it is an anticancer drug as well. In RA, it is administered orally, intramuscularly, and subcutaneously. Its starting dose is 7.5–10 mg per week, with adjustments made to increase it to 20–25 mg per week every two weeks. Subcutaneous administration is preferred for patients with insufficient response regardless of dose escalation.

3.3.1.1 Mechanism of Action

It acts as a folic acid antagonist, inhibiting DNA synthesis and cell replication by binding to dihydrofolate reductase. In RA, its primary effect is to block enzymes involved in purine synthesis, which results in the accumulation of adenosine and the suppression of T-cell activation (Cronstein, 1997).

MTX's mechanism of action may be elucidated through its pharmacology. Following administration, MTX remains in circulation briefly before redistributing to tissues. Within cells and tissues, MTX is converted into long-lasting MTX-polyglutamates, which retain biological activity (Prasad et al., 2023). Studies indicate that in patients with RA receiving low-dose, intermittent MTX, these polyglutamates primarily accumulate in liver parenchyma. These metabolites persist in tissues for an extended period, up to two weeks in rats. Consequently, it is inferred that the therapeutic effects of MTX in RA result from the activity of these intracellular, polyglutamated derivatives, alike to the treatment of malignant cells with high doses of MTX (Ding et al., 2023).

3.3.1.2 Adverse Action

Adverse action of MTX includes GI symptoms that are predominant with nausea, vomiting, dyspepsia, and diarrhea. Folate supplementation, H2 blockers, and antacids typically manage these symptoms. Hematological effects such as primarily mild to moderate leukopenia and transient bone marrow depression, hepatic abnormalities with mild elevation of liver enzymes, recurrent skin rashes, accelerated nodulosis, and mild serum creatinine elevation (Gravallese & Firestein, 2023) also occur. RA patients taking MTX who have obesity, diabetes, alcohol use, and prior HBV or HCV infection are more prone to adverse effects such as stomatitis, nausea, vomiting, and diarrhea. Stomatitis can be managed with folic acid supplementation without compromising MTX's efficacy. GI symptoms may improve with dose adjustment. Rarely, MTX can cause pneumonia-like illness or pulmonary fibrosis, warranting evaluation in patients presenting with acute breathlessness or chronic cough (Chugh, Upadhyay, Dhami, & Nandave, 2024; Luo et al., 2024).

3.3.2 Sulfasalazine

Sulfasalazine, known as Azulfidine by Pharmacia, is utilized in the treatment of both inflammatory bowel disease (IBD) and inflammatory arthritis. This medication functions as a prodrug that is broken down by bacteria in the colon into two active compounds: sulfapyridine and 5-aminosalicylic acid. Notably, it may take several months after initiating treatment for sulfasalazine to elicit a therapeutic response (Zhao et al., 2024).

3.3.2.1 Mechanism of Action

Sulfasalazine exerts anti-inflammatory action by various mechanisms. Sulfasalazine and its metabolites weakly inhibit enzymes in the arachidonic acid cascade. It notably inhibits PG metabolism through 15-prostaglandin dehydrogenase activity. Additionally, it interferes with inflammatory cell functions, inhibiting chemotaxis, enzyme secretion, and superoxide production. These effects are mediated by interference with cellular signaling pathways, particularly phosphoinositide metabolism and heterotrimeric G-protein activity (Buch, Eyre, & McGonagle, 2021). In vitro studies demonstrate its ability to modulate the immune response by inhibiting T-cell growth, natural killer cell function, and B-cell activation. It also suppresses the release of various cytokines involved in immune activation. In clinical and animal studies, sulfasalazine showed immunomodulatory effects by reducing inflammatory responses in autoimmune diseases. These effects are evidenced by changes in lymphocyte subsets, immunoglobulin levels, and cytokine profiles in treated individuals (Cronstein & Aune, 2020).

3.3.2.2 Adverse Effects

Sulfasalazine can lead to a range of adverse effects, including GI symptoms like nausea, vomiting, anorexia, and abdominal discomfort, as well as headache, fever, and rash. Hematologic complications may include agranulocytosis, leukopenia, anemia, and hemolytic anemia. Pulmonary complications, though rare, can include dyspnea, cough, and fever. Liver damage, including hypersensitivity reactions, has been reported (Nunokawa et al., 2023). Dermatologic complications such as skin rash, exfoliative dermatitis, and Stevens–Johnson syndrome can occur. Male infertility has also been associated with sulfasalazine therapy. Other potential adverse effects include cardiac issues like tachycardia, cyanosis, hair loss, and neurotoxicity (Deshpande, Sonawale, Mulkalwar, Jain, & Goyal, 2023).

3.3.3 Leflunomide

Leflunomide, one of the newer DMARDs, is typically administered with an initial dose of 100 mg per day orally for three days, then a maintenance dose of 10–20 mg per day. However, the loading dose may be reduced or omitted in elderly patients or those with other health conditions to reduce the earliest side effects. It can be used alone or in combination with other drugs to treat/manage RA. Leflunomide is converted in the liver to its active form, teriflunomide, which is responsible for its therapeutic effects (M. B. Zewail, El-Gizawy, Asaad, Shabana, & El-Dakroury, 2023).

3.3.3.1 Mechanism of Action

Leflunomide is an isoxazole derivative used as an immunomodulatory drug, distinct from other known immunomodulators. It is a prodrug that converts rapidly to its active form, A77–1726, which inhibits dihydroorotate dehydrogenase (DHODH). This enzyme is pivotal for pyrimidine synthesis in activated lymphocytes. By inhibiting DHODH, A77–1726 disrupts pyrimidine production, which is essential for DNA synthesis and cell proliferation. A77–1726 is over 99.5% protein-bound in plasma and exhibits a long half-life due to extensive protein binding and enterohepatic recirculation (Zheng et al., 2023). It does not interact significantly with common medications like warfarin or oral contraceptives. However, it may weakly displace NSAIDs, potentially raising blood pressure in some patients. In activated lymphocytes, pyrimidine ribonucleotide synthesis is critical during the cell cycle's G1 phase, when DNA replication and cell proliferation are initiated. A77–1726 causes G1 phase arrest by depleting pyrimidine levels, which activates the cell-cycle regulator p53. This mechanism helps control autoimmune lymphocytes without severely affecting normal immune functions, minimizing the risk of opportunistic infections and toxicity (Zheng et al., 2023).

3.3.3.2 Adverse Effects

Adverse effects of leflunomide include nausea and diarrhea, which affect 20%–30% of patients but may lessen with continued treatment. Skin rash and reversible hair loss occur in 5%–10% of patients. Less common side effects include severe bone marrow suppression, infections, persistent abnormal liver function tests despite dose adjustments, and hypertension (Erten, 2023). Leflunomide should not be used in patients with severe immunodeficiency, compromised bone marrow function, or severe uncontrolled infections. Avoid consuming excessive amounts of alcohol due to the potential for liver impairment. Leflunomide also inhibits cytochrome P450 2C9, which can lead to interactions with medications such as phenytoin and warfarin. When administered to pregnant women, it can increase the risk of fetal death or teratogenic effects (Hoque, Akram, & Saha, 2023).

3.3.4 Hydroxychloroquine

Hydroxychloroquine and chloroquine work through various mechanisms for the management of RA, which are discussed below.

These drugs act either through lysosomal activity or autophagy inhibition. These drugs accumulate in lysosomes and increase the pH, which inhibits lysosomal enzymes and autophagy, impairing antigen presentation and lymphocyte function.

Hydroxycholoquine and chloroquine act by inhibition of signaling pathways. They disrupt Toll-like receptor (TLR) signaling by altering endosomal pH and inhibiting TLR7 and TLR9. They also inhibit the cGAS–STING pathway, reducing proinflammatory cytokine production. They also act by affecting cytokine production and immune system activation (Yusuf, Charbel Issa, & Ahn, 2023).

Hydroxychloroquine and chloroquine reduce pro-inflammatory cytokines (IL-1, IL-6, TNF-α, and IFNγ) by inhibiting TLR pathways and other immune processes. These drugs are believed to lower the possibility of thrombotic complications in RA cases with systemic inflammation and antiphospholipid syndrome. They may lower fasting glucose and lipid levels, inhibit anti-phospholipid antibody binding, and improve endothelial function. However, further large-scale studies are needed to confirm these therapeutic benefits (Melles et al., 2023).

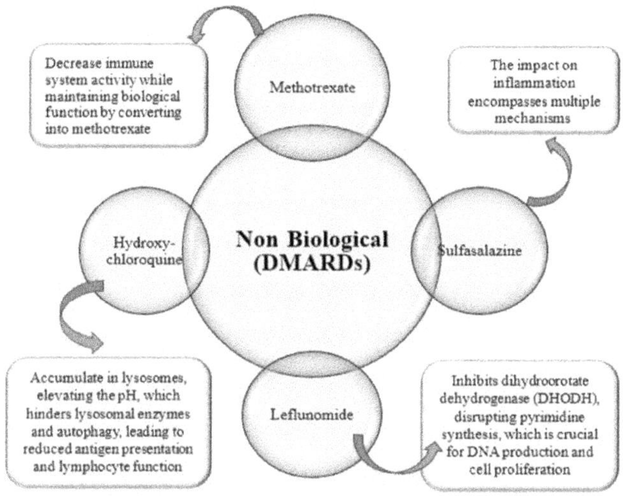

FIGURE 3.1 Mechanism of action of nonbiological DMARDs.

3.3.4.1 Adverse Effects

Hydroxychloroquine and chloroquine have a good safety profile and are considered immunomodulatory rather than immunosuppressant drugs. Common adverse effects include GI issues. Rarely, these drugs are associated with myopathy and cardiotoxic effects, including rhythm disorders and cardiomyopathy, although conclusive evidence is lacking. The most serious drawback is retinopathy, which is more common with chloroquine, causing retinal damage and vision deficiency. Adverse conditions for retinopathy include high drug dosage, long-lasting use, high advancing dose, chronic kidney disease, and concurrent tamoxifen use (Talarico et al., 2023). Hyperpigmentation, pruritus, photosensitivity, urticaria, psoriasis, systemic symptoms, and drug reactions involving eosinophilia, erythroderma, blistering, erythema multiforme, porphyria, hair loss, hyperpigmentation, stomatitis, and melanonychia were less-frequent side effects (Fierro et al., 2024).

Figure 3.1 illustrates the mechanism of action of biological demands, which involve complex physiological processes. These demands trigger signaling pathways, leading to cellular responses. The classification includes various biological processes, such as metabolic regulation, immune response, and hormonal signaling. This mechanism enables the body to adapt and respond to internal and external stimuli.

3.4 BIOLOGIC DMARDs

Biologic DMARDs include various therapies, for instance, TNF-α inhibitors, anti-B cell therapy, T-cell co-stimulation blockers, anti-IL-6, and anti-IL-1β. Biologic DMARDs, commonly referred to as biologics, quickly slow down the advancement of joint damage resulting from RA (Takase-Minegishi et al., 2024). Some DMARDs are discussed in detail here and described in Figure 3.2.

3.4.1 TNF-α Inhibitors (e.g., Adalimumab, Infliximab, Etanercept)

TNF, primarily produced by monocytes and macrophages, is a key cytokine involved in inflammation, particularly in RA. It stimulates the expression of other inflammatory cytokines and adhesion molecules, contributing to synovitis. TNF receptors, TNF-R1 and TNF-R2, are present on

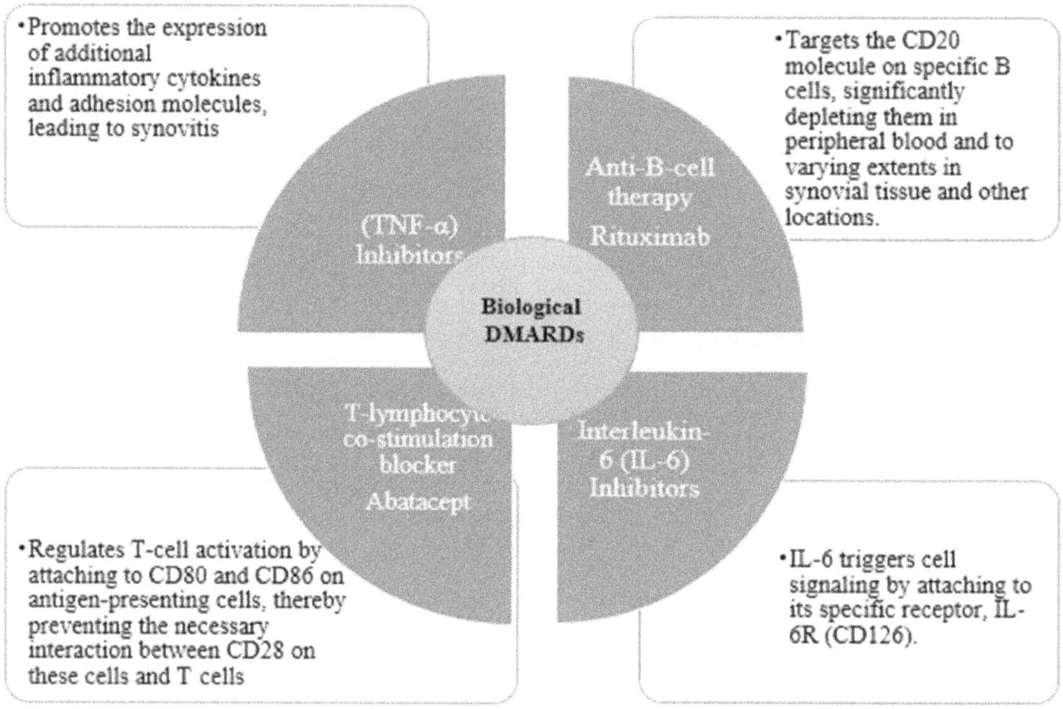

FIGURE 3.2 Mechanism of action of biological DMARDs.

most nucleated cells, with TNF-R1 primarily mediating inflammatory responses. Excess TNF-α in RA exacerbates joint inflammation, making it a therapeutic target. Anti-TNF-α therapy, including drugs like etanercept, infliximab, and adalimumab, aims to neutralize TNF-α, leading to improved synovial inflammation and reduced joint destruction in RA patients (Leone, Mangano, Petralia, Nicoletti, & Fagone, 2023).

TNF-α, produced mainly by immune cells, interacts with TNFR1 and TNFR2 receptors to trigger cellular responses. Activation of TNFR1 forms specific signaling complexes, namely, complexes I, IIa, IIb, and IIc, leading to inflammation, cell survival, proliferation, apoptosis, or necroptosis. Complex I activation at the plasma membrane induces inflammation and cell survival via NF-κB and mitogen-activated protein kinase pathways. Complexes IIa and IIb in the cytoplasm trigger apoptosis, while complex IIc induces necroptosis and inflammation (Souza, Caetano, Magalhães, & Castelucci, 2023).

3.4.1.1 Adverse Effect

Inhibiting TNF-α using agents such as etanercept and infliximab is generally well tolerated, with adverse effects primarily involving serious infections, malignancies, and some neurological complications. Long-term use of these blockers has not shown a significant increase in severe adverse events, with rates of extreme infections and malignancies comparable to those observed in control groups. Notably, there is a concern regarding the unmasking of latent tuberculosis with TNF-α inhibitory therapy, particularly with antibody therapies (Steinhart, 2024). Neurological disorders, such as aseptic meningitis and worsening of demyelinating diseases, have been reported. Syndromes resembling systemic lupus erythematosus have also been observed, along with rare cases of serious blood dyscrasias. Despite these concerns, TNF-α inhibitors are generally tolerated due to their potential to normalize TNF-α activity while preserving essential protective functions in RA patients (Caporali, Conti, & Iannone, 2024; Steinhart, 2024).

3.4.1.2 Anti-B-Cell Therapy

In RA, B cells are significant contributors to the disease process. Targeting these cells involves the use of antibodies against the CD-20 surface marker, which is present on all B cells. Ongoing evaluation includes exploring other targets like anti-CD19. Rituximab stands as the sole licensed therapy for specifically targeting B cells in the treatment of RA patients (Martin et al., 2023).

3.4.2 Rituximab

Rituximab, a monoclonal antibody that targets CD20 on B-lymphocytes, induces complement-dependent B-cell cytotoxicity and antibody-dependent cellular toxicity. It is administered intravenously in 1 g dose on days 1 and 15, in conjunction with MTX. Subsequent doses can be given every 24–52 weeks depending on clinical evaluation, with a minimum interval of 16 weeks between treatments. B-cell retrieval typically starts about 6 months post-therapy, reaching normal levels by 12 months post-treatment (Chan et al., 2023). This therapy, combined with MTX, has shown efficacy in patients resistant to MTX alone or TNF-α inhibitors. Long-term safety data indicate a profile similar to other biologics, with rare reports of progressive multifocal leukoencephalopathy prompting cautious use, especially after TNF-α inhibitor failure. Rituximab is particularly effective in seropositive patients for rheumatoid factor (RF) or anti-CCP antibodies. Infusion-related reactions are managed with premedication, including intravenous methylprednisolone and anti-histamines (Vesperinas-Castro & Cortés-Vicente, 2023).

3.4.2.1 Mechanism of Action

Rituximab targets the CD20 molecule on certain B cells, depleting them extensively in peripheral blood and to varying degrees in synovial stratum and other sites. Impersonal response in RA appears linked to the reduction of synovial tissue B cells and possibly peripheral B cells. Post-treatment B-cell reconstitution typically involves immature, naive B cells, but relapse of disease occurs in many patients (Khandelwal et al., 2023). Rituximab depletes B cells through several mechanisms, including antibody-dependent cellular cytotoxicity, complement-dependent cytotoxicity, and B-cell apoptosis. While the exact way B-cell depletion contributes to clinical improvement of RA is not completely understood, it may involve impacts on B-cell antigen presentation, cytokine production, and the generation of autoantibodies, such as RF (Bastari, Indrajaya, & Indra, 2024).

3.4.2.2 Adverse Effects

Long-term studies show sustained efficacy and safety, with repeated treatments maintaining clinical benefits over several years. B-cell depletion may increase susceptibility to infections, especially in patients with prior immunosuppressive treatment. However, overall adverse event rates are low, with no increased risk of tuberculosis or malignancy observed. Predictors of response to rituximab include serological status and baseline disease activity. However, more research is needed to identify reliable predictors in everyday clinical practice (Lemoine et al., 2023).

3.4.3 T-Lymphocyte Co-Stimulation Blocker

T cells require two signals for full activation. The first signal is provided by antigen-presenting cells (APCs) when the T-cell receptor binds to the major histocompatibility complex. The second signal comes from the co-stimulation of CD80/CD86 on APCs interacting with CD28 on T cells. Although T cells also have an additional co-stimulatory receptor called the inducible co-stimulator, it has not yet been explored as a potential therapeutic target (Wang et al., 2024).

3.4.4 Abatacept

Abatacept is a medication that selectively modulates co-stimulation by inhibiting T-cell activation through its binding to CD80 and CD86 on APCs. This action disrupts the essential interaction between APCs and T cells mediated by CD28. It can be administered either intravenously or subcutaneously. For patients weighing under 60 kg, the intravenous dose is 500 mg; for those between 60 and 100 kg, it is 750 mg; and for individuals over 100 kg, it is 1,000 mg. After the initial infusion, this dose is repeated at two and four weeks and then every four weeks thereafter (Rech et al., 2024).

3.4.4.1 Mechanism of Action

Abatacept, a treatment for RA, modulates T-cell activation by binding to CD80 and CD86 on APCs, thereby blocking the necessary interaction with CD28 on T cells. This action leads to a decrease in serum levels of various inflammatory markers, including IL-2 receptor, IL-6, RF, C-reactive protein (CRP), and TNF-α. Clinical studies have demonstrated that abatacept results in dose-dependent reductions of these markers. Its pharmacokinetics was studied in healthy adults and RA patients, revealing comparable results (da Rosa et al., 2024). Multiple intravenous infusions resulted in proportional increases in serum concentration, reaching steady state by day 60. Continued monthly dosing did not lead to systemic accumulation. Subcutaneous administration demonstrated linear pharmacokinetics with comparable systemic clearance and half-life to intravenous administration. Population analyses indicated that abatacept clearance increases with body weight, while age and gender (after adjusting for body weight) do not affect clearance. Furthermore, concomitant use of medications such as MTX, NSAIDs, corticosteroids, or TNF-blocking agents does not influence abatacept clearance. Additional studies are needed to evaluate the drug's effects in patients with renal or hepatic impairment (Adelakun, Haddad, Mirza, Dover, & Golombek, 2024).

3.4.4.2 Adverse Effect

Adverse events reported in the study included headache, nasopharyngitis, nausea, diarrhea, upper respiratory tract infection, dizziness, pain in back, influenza, cough, dyspepsia, pharyngitis, high blood pressure, fatigue, upper abdominal pain, sinusitis, bronchitis, and infusion reactions (both acute and peri-infusional). Life-threatening adverse events, primarily musculoskeletal and connective tissue disorders, infections, nervous system disorders, cardiac disorders, and neoplasms, were observed (Yamada et al., 2024).

Abatacept administration in patients with chronic obstructive pulmonary disease (COPD) may lead to an enhanced occurrence of COPD-related adverse reactions. Additionally, the formation of antibodies was observed in 2%–41% of patients receiving abatacept (Rech et al., 2024).

3.4.5 Interleukin-6 (IL-6) Inhibitors

3.4.5.1 Mechanism of Action

IL-6 triggers cell signaling by attaching to its specific receptor, IL-6R (CD126). For effective signal transduction, the IL-6/IL-6R complex must also interact with the widely expressed transmembrane protein glycoprotein 130 (gp130 or CD130). This interaction results in the homodimerization of gp130, which activates signaling through the Janus-activated kinase (JAK) and signal transducer and activator of transcription (STAT) pathways. IL-6R is present on various cell types. IL-6 can also bind to soluble forms of IL-6R (Swaroop et al., 2024). The soluble IL-6/IL-6R complex can bind to gp130 on a wider array of cells, allowing activation through gp130 in tissues that typically do not express IL-6R, a phenomenon known as trans-signaling. This mechanism may account for the extensive role of IL-6 in systemic inflammatory conditions such as RA. Tocilizumab (TCZ), formerly known as myeloma receptor antibody, is a humanized IgG1 monoclonal antibody that targets the 80 kDa subunit of IL-6R. By binding with high affinity to IL-6R, TCZ inhibits the dimerization of the IL-6/IL-6R complex with gp130, thereby blocking downstream signaling (Harmalkar et al., 2024).

3.4.5.2 Adverse Effect

Patients who have a history of serious or opportunistic infections, have been exposed to tuberculosis, or have active infections should not use TCZ. Patients should visit a doctor if they develop new stomach pain because using TCZ increases the risk of GI perforation (Vilbert et al., 2024). Moreover, side effects from TCZ include thrombocytopenia, neutropenia, raised liver enzymes, and elevated cholesterol levels. Monitoring laboratory results is advised every 4–6 weeks. A higher risk of cancer, hypersensitivity reactions, demyelinating illnesses, hepatic impairment, and the need to use caution while administering live immunizations are other possible side effects (Chertow et al., 2024).

Figure 3.2 illustrates the mechanism of action of biological disease-modifying antirheumatic drugs (bDMARDs). These agents target specific molecules involved in inflammation and immune response, such as TNF-α, IL-6, or CD20. By inhibiting these targets, bDMARDs modulate the immune system, reducing inflammation and slowing disease progression in conditions like RA. This targeted approach provides therapeutic benefits and improves patient outcomes.

3.4.6 Janus Kinase (JAK) Inhibitors

The JAK family consists of four members: JAK1, JAK2, JAK3, and TYK2. Research has demonstrated that various JAK isoforms and STAT proteins are present in synovial tissue and cells (Song, Lee, Jo, & Kwon, 2024). A specific group of receptors, categorized as Type I and Type II cytokine receptors, is targeted by numerous pro-inflammatory cytokines associated with the pathophysiology of RA. These receptors have distinct structural features compared to other receptors, such as those that interact with TNF-α and IL-1β. The JAK–STAT pathway is necessary for cytokines that attach to Type I and Type II receptors in order to send signals. Because of this, different JAK inhibitors, or jakinibs, have different levels of specificity and selectivity for the JAK enzymes (Ciccia & Caporali, 2024).

In Japan, there are now four approved JAK inhibitors and one is under approval (filgotinib). While Upadacitinib was approved in 2020, tofacitinib, baricitinib, and peficitinib were already introduced for the treatment of RA (Bredemeier et al., 2024).

Figure 3.3 illustrates the JAK/STAT signaling pathway, where cytokine binding activates JAKs, leading to STAT phosphorylation, dimerization, and nuclear translocation, regulating gene expression, and influencing immune response, cell growth, and inflammation.

3.4.6.1 Mechanism of Action

Jakinibs, such as tofacitinib, ruxolitinib, and baricitinib, target cytokines involved in autoimmune diseases by inhibiting the JAK/STAT pathway. These drugs block key cytokines like IL-6, IL-12, IL-23, interferons (IFNs), and GM-CSF, which play a pathogenic role in diseases like RA, psoriatic arthritis, IBD, and eczema. Tofacitinib specifically inhibits cytokines that signal through JAK3 (e.g., IL-2, IL-4, and IL-7), JAK1 (e.g., IL-6, IL-11, and IFNs), and JAK2 (e.g., IL-3, IL-5, and GM-CSF) (Tsuchiya et al., 2024). This inhibition affects both adaptive and innate immune responses by interfering with Th1 and Th17 cell differentiation, B-cell function, and chemokine production. Ruxolitinib and baricitinib, which also inhibit JAK1 and JAK2, block many of the same cytokines and share similar mechanisms of action with tofacitinib. These drugs effectively reduce inflammation and autoimmune responses by targeting multiple cytokine pathways (Chandrashekara, 2024; Kwon et al., 2024).

The JAK–STAT pathway includes type I and II cytokine receptors, as well as receptors for IFNs and growth factors, all of which depend on JAKs to mediate downstream responses. Upon cytokine binding, receptor-associated JAKs trans phosphorylate and activate each other, then phosphorylate

FIGURE 3.3 JAK/STAT signaling pathway.

the receptor tail, creating a docking site for STATs. Phosphorylated STATs separate, form dimers, and move to the nucleus to modulate gene expression. This pathway's importance is highlighted by studies on knockout mice and primary immunodeficiencies like severe combined immunodeficiency. Overexpression of JAK–STAT pathways is linked to autoimmune diseases and malignancies, and their blockade is considered a therapeutic approach for autoimmune diseases like RA (Dreo et al., 2024).

3.4.7 Tofacitinib

Several clinical trials have tested a twice-daily dose of tofacitinib; first-generation, selective oral JAK1,3 inhibitors with reduced efficacy against JAK2 and TYK2 have undergone several clinical trials conducted in patients with RA. More than 80 nations have granted approval for its clinical usage (Ruscitti et al., 2024).

3.4.8 Baricitinib

Baricitinib, a novel, potent, and selective first-generation oral JAK1,2 inhibitors, was created by structurally altering tofacitinib. It exhibited modest activity against TYK2 and significantly lower activity against JAK3. Baricitinib was recently granted a license in Europe to treat RA, following the successful completion of a phase III clinical study programme, in which it was administered once daily (Calvo-Garcia et al., 2024).

3.4.9 Filgotinib

Filgotinib performs its action by blocking specifically the activity of JAK1, an enzyme that is involved in intracellular signaling. It also suppresses the signaling of cytokines, such as IL-6, IL-12, and IFN-γ. It majorly contributes to decreasing inflammation and modulates immune response. It reduces the symptoms of RA such as stiffness, swelling, and joint pain (Skouvaklidou et al., 2024).

3.5 ADVERSE EFFECTS OF JAK INHIBITORS

The most frequently reported adverse effects include nasopharyngitis, hypercholesterolemia, neutropenia, and anemia. Reports have indicated the presence of significant infections such as bacterial pneumonia, disseminated herpes zoster, pulmonary tuberculosis, cryptococcal pneumonia, and pneumocystis pneumonia. The main risk factors for these infections include diabetes, concurrent glucocorticoid usage, and advanced age. While a direct correlation between lower lymphocyte count and infection is unknown, there may be one between severe lymphopenia and life-threatening illnesses (Kesavulu, Yashmeen Nikhat, Kumar, & Pooja, 2024). Clinical trials suggest a potential link between hypercholesterolemia and cardiovascular events in patients treated with tofacitinib. However, long-term studies have not shown an elevated incidence of cardiovascular events, despite the rapid reduction in CRP levels observed with tofacitinib, similar to the effects seen with the anti-IL-6 receptor antibody TCZ. Tofacitinib modifies the composition of high-density lipoprotein, shifting it from proatherogenic to less atherogenic forms. Additionally, JAK inhibition has been associated with a brown-like metabolic activity in white adipocytes, which could potentially make the hypercholesterolemia associated with tofacitinib less atherogenic. Further investigation is needed to fully understand it (Wlassits, Müller, Fenzl, Lamprecht, & Erlacher, 2024).

JAK2 inhibition affects erythropoietin signaling and colony stimulating factors, which can lead to anemia, thrombocytopenia, and neutropenia when taking jakinibs. They are also linked to slight elevations in creatinine and hypercholesterolemia, which may be caused by blocking IL-6 signaling (Bonelli et al., 2024).

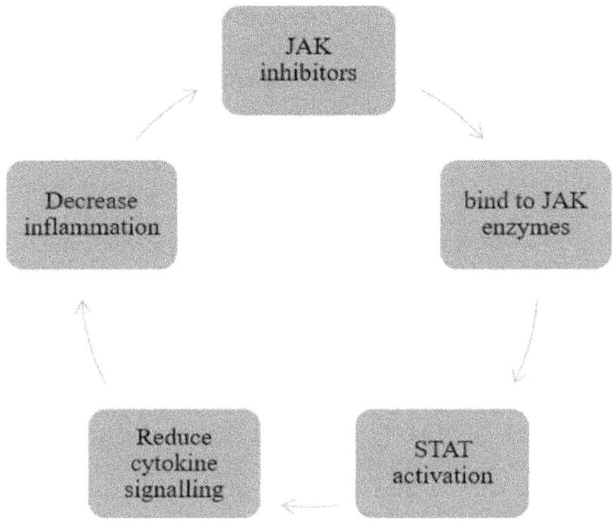

FIGURE 3.4 Mechanism of action of JAK inhibitors.

Figure 3.4 illustrates the mechanism of action of JAK inhibitors, which block the activity of JAKs involved in cytokine signaling. By inhibiting JAKs, these agents prevent the phosphorylation of STATs, thereby reducing the downstream signaling and gene expression. This leads to a decrease in inflammatory responses and immune system activation, making JAK inhibitors effective in treating autoimmune diseases and inflammatory conditions.

3.6 ALTERNATIVE TREATMENT

Complementary and alternative medicine modalities offer additional treatment options for RA. Acupuncture, both in its traditional Chinese model and in the Western scientific model, aims to alleviate pain. The traditional approach focuses on restoring the flow of energy (Qi) through meridians, while the Western approach is based on the "gate control" theory, where acupuncture releases neurotransmitters that block pain transmission (Pruthviraj, Mukherjee, Abilash Krishnan, & Sundaram, 2024). Techniques include manual needle stimulation, electroacupuncture, heat stimulation (moxibustion), and injection acupuncture. Clinical studies indicate trends toward pain reduction and improved disease activity, although some results are not statistically significant.

Bee venom acupuncture (BVA) integrates the pharmacological properties of bee venom with mechanical stimulation from acupuncture. It has shown anti-inflammatory and pain-relieving effects in both animal models and patients with RA. Clinical studies have shown significant decreases in joint tenderness, swelling, morning stiffness, erythrocyte sedimentation rate, CRP levels, and health assessment questionnaire scores after BVA treatment. While generally safe with proper practice, BVA can cause serious adverse effects like anaphylaxis. Both acupuncture and BVA can be considered useful adjunct treatments for managing pain and inflammation in RA.

Herbal remedies have been utilized for their anti-inflammatory properties in treating RA and other inflammatory conditions. Despite limited evidence and potential side effects, the American College of Rheumatology advises cautious use of these treatments early in RA management to prevent joint damage (Ghosh, Pathak, Malsawmdawngkimi, Kumar, & Gull, 2024). Key herbs include green tea, which contains epigallocatechin gallate, shown to inhibit inflammatory pathways; Celastur known for its anti-inflammatory and anti-oxidant effects (Ning et al., 2024); Cat's Claw, which exhibits anti-inflammatory and immunomodulating properties (Arado et al., 2024);

Lepidium meyenii (RNI 249), which promotes joint restoration (Ulloa del Carpio et al., 2024); and Thunder God Vine (*Tripterygium wilfordii* Hook F), used for various inflammatory diseases. These herbal options offer alternatives for managing RA symptoms, but require close monitoring due to potential side effects (Ghosh et al., 2024).

REFERENCES

Adelakun, A. A., Haddad, A. W., Mirza, N., Dover, M., & Golombek, S. (2024). Transverse myelitis: an adverse reaction to Abatacept. *Cureus*, *16*(4), 1–3.

Arado, G. M., Amatto, P. D. P. G., Marins, M., Rizzi, E. S., França, S. D. C., Coppede, J. D. S., ... Pereira, A. M. S. (2024). Anti-inflammatory and/or immunomodulatory activities of Uncaria tomentosa (cat's claw) extracts: a systematic review and meta-analysis of in vivo studies. *Frontiers in Pharmacology*, *15*, 1378408.

Bastari, R., Indrajaya, T., & Indra, S. (2024). The role of Rituximab in the management of heart failure. *Bioscientia Medicina: Journal of Biomedicine and Translational Research*, *8*(7), 4594–4599.

Berardicurti, O., Ruscitti, P., Pavlych, V., Conforti, A., Giacomelli, R., & Cipriani, P. (2020). Glucocorticoids in rheumatoid arthritis: the silent companion in the therapeutic strategy. *Expert Review of Clinical Pharmacology*, *13*(6), 593–604.

Bonelli, M., Kerschbaumer, A., Kastrati, K., Ghoreschi, K., Gadina, M., Heinz, L. X., ... Laurence, A. (2024). Selectivity, efficacy and safety of JAKinibs: new evidence for a still evolving story. *Annals of the Rheumatic Diseases*, *83*(2), 139–160.

Bredemeier, M., Duarte, Â. L., Pinheiro, M. M., Kahlow, B. S., Macieira, J. C., Ranza, R., ... Bértolo, M. B. (2024). The effect of antimalarials on the safety and persistence of treatment with biologic agents or Janus kinase inhibitors in rheumatoid arthritis. *Rheumatology*, *63*(2), 456–465.

Buch, M. H., Eyre, S., & McGonagle, D. (2021). Persistent inflammatory and non-inflammatory mechanisms in refractory rheumatoid arthritis. *Nature Reviews Rheumatology*, *17*(1), 17–33.

Calvo-Garcia, A., Ramírez Herráiz, E., Llorente Cubas, I. M., Varas De Dios, B., Benedí González, J., Morell Baladrón, A., & García-Vicuña, R. (2024). The real-world effectiveness, persistence, adherence, and safety of Janus Kinase inhibitor Baricitinib in rheumatoid arthritis: a long-term study. *Journal of Clinical Medicine*, *13*(9), 2517.

Caporali, R., Conti, F., & Iannone, F. (2024). Management of patients with inflammatory rheumatic diseases after treatment failure with a first tumour necrosis factor inhibitor: a narrative review. *Modern Rheumatology*, *34*(1), 11–26.

Chan, E. Y.-H., Yap, D. Y.-H., Colucci, M., Ma, A. L.-T., Parekh, R. S., & Tullus, K. (2023). Use of rituximab in childhood idiopathic nephrotic syndrome. *Clinical Journal of the American Society of Nephrology*, *18*(4), 533–548.

Chandrashekara, S. (2024). Pharmacokinetic review of janus kinase inhibitors and its clinical implications for the management of rheumatoid arthritis. *Expert Opinion on Drug Metabolism & Toxicology* (just-accepted), *20*(8), 741–748.

Chertow, G. M., Chang, A. M., Felker, G. M., Heise, M., Velkoska, E., Fellström, B., ... Goodman, S. G. (2024). IL-6 inhibition with clazakizumab in patients receiving maintenance dialysis: a randomized phase 2b trial. *Nature Medicine*, *30*(8), 2328–2336.

Choueiri, M., Chevalier, X., & Eymard, F. (2021). Intraarticular corticosteroids for hip osteoarthritis: a review. *Cartilage*, *13*(1_suppl), 122S–131S.

Chugh, D., Upadhyay, J., Dhami, P., & Nandave, M. (2024). Unraveling the enigma of rheumatoid arthritis: exploring etiology, pathophysiology and its treatment. In T. Behl, S. Singh, & N. Sharma (Eds.), *Targeting Angiogenesis, Inflammation, and Oxidative Stress in Chronic Diseases* (pp. 75–91). Elsevier.

Ciccia, F., & Caporali, R. (2024). Proposals for the rheumatological use of JAK inhibitors. *Nature Reviews Rheumatology*, *20*(2), 65–66.

Criswell, L., Saag, K., Sems, K., Welch, V., Shea, B., Wells, G. A., ... Group, C. M. (1996). Moderate-term, low-dose corticosteroids for rheumatoid arthritis. *Cochrane Database of Systematic Reviews*, *2010*(1), 1–14.

Cronstein, B. N. (1997). The mechanism of action of methotrexate. *Rheumatic Disease Clinics of North America*, *23*(4), 739–755.

Cronstein, B. N., & Aune, T. M. (2020). Methotrexate and its mechanisms of action in inflammatory arthritis. *Nature Reviews Rheumatology*, *16*(3), 145–154.

da Rosa, L. C., Scales, H. E., Benson, R. A., Brewer, J. M., McInnes, I. B., & Garside, P. (2024). The effect of abatacept on T-cell activation is not long-lived in vivo. *Discovery Immunology*, *3*(1), kyad029.

De Bock, M., & Sienaert, P. (2024). Corticosteroids and mania: a systematic review. *The World Journal of Biological Psychiatry*, *25*(3), 161–174.

de Carvalho, J. F., & Skare, T. L. (2025). Rituximab combined with intravenous immunoglobulin in autoimmune diseases: a systematic review. *Advances in Rheumatology*, *65*(1), 1–10.

Deshpande, G., Sonawale, A., Mulkalwar, A., Jain, H., & Goyal, A. (2023). Short-term efficacy and adverse effects of sulfasalazine in the management of axial spondyloarthropathy. *Cureus*, *15*(12), 1–11.

Di Cola, I., & Ruscitti, P. (2024). The latest advances in the use of biological DMARDs to treat Still's disease. *Expert Opinion on Biological Therapy*, *24*(1–2), 63–75.

Ding, Q., Hu, W., Wang, R., Yang, Q., Zhu, M., Li, M., … Zhu, Y. Z. (2023). Signaling pathways in rheumatoid arthritis: implications for targeted therapy. *Signal Transduction and Targeted Therapy*, *8*(1), 68.

Dreo, B., Muralikrishnan, A. S., Husic, R., Lackner, A., Brügmann, T., Haudum, P., … Stradner, M. (2024). JAK/STAT signaling in rheumatoid arthritis leukocytes is uncoupled from serum cytokines in a subset of patients. *Clinical Immunology*, *264*, 110238.

Erten, P. T. (2023). Evaluation of the efficacy, side effects, and drug survival data with methotrexate-leflunomide combination therapy in patients with inflammatory arthritis. *Meandros Medical & Dental Journal*, *24*(3), 1–13.

Ferreira, J. F., Mohamed, A. A. A., & Emery, P. (2016). Glucocorticoids and rheumatoid arthritis. *Rheumatic Disease Clinics*, *42*(1), 33–46.

Fierro, J. J., Velásquez-Berrío, M., Ospina, A., Henning, S., de Leeuw, K., & Cadavid J, Á. P. (2024). The effects of hydroxychloroquine and its promising use in refractory obstetric antiphospholipid syndrome. *Rheumatology International*, *44*(2), 223–234.

Gaffo, A., Saag, K. G., & Curtis, J. R. (2006). Treatment of rheumatoid arthritis. *American Journal of Health-System Pharmacy*, *63*(24), 2451–2465.

Ghosh, N., Pathak, S., Malsawmdawngkimi, Kumar, G., & Gull, A. (2024). Natural products and traditional herbal medicines as managerial therapies to combat rheumatoid arthritis. *Clinical & Translational Metabolism*, *22*(1), 2.

Graham, D., Agrawal, N., & Roth, S. (1988). Prevention of NSAID-induced gastric ulcer with misoprostol: multicentre, double-blind, placebo-controlled trial. *The Lancet*, *332*(8623), 1277–1280.

Gravallese, E. M., & Firestein, G. S. (2023). Rheumatoid arthritis—common origins, divergent mechanisms. *New England Journal of Medicine*, *388*(6), 529–542.

Harmalkar, D. S., Sivaraman, A., Nada, H., Lee, J., Kang, H., Choi, Y., & Lee, K. (2024). Natural products as IL-6 inhibitors for inflammatory diseases: synthetic and SAR perspective. *Medicinal Research Reviews*, *44*(4), 1683–1726..

Hoque, M., Akram, T., & Saha, S. N. (2023). A review on Methotrexate used in rheumatoid arthritis. *International Journal of Research*, *10*(9), 321–341.

Huo, R., Wei, C., Yang, Y., Lin, J., & Huang, X. (2025). Hydroxychloroquine: a double-edged sword. *Molecular Medicine Reports*, *31*(4), 1–11.

Jurgens, M., Jacobs, J., & Bijlsma, J. (2011). The use of conventional disease-modifying anti-rheumatic drugs in established RA. *Best Practice & Research Clinical Rheumatology*, *25*(4), 523–533.

Kesavulu, C. C., Yashmeen Nikhat, P., Kumar, J. V., & Pooja, M. (2024). A review on potential adverse effects associated with Janus-kinase inhibitors. *International Journal of Basic & Clinical Pharmacology*, *13*(3), 418.

Khandelwal, K., Jajoo, V., Bajpai, K., Madke, B., Prasad, R., Wanjari, M. B., … Taksande, A. B. (2023). Rituximab in pemphigus vulgaris: a review of monoclonal antibody therapy in dermatology. *Cureus*, *15*(6), 1–12.

Kianian, S., Bansal, J., Lee, C., Zhang, K., & Bergese, S. D. (2024). Perioperative multimodal analgesia: a review of efficacy and safety of the treatment options. *Anesthesiology and Perioperative Science*, *2*(1), 9.

Kontny, E., Szczepańska, K., Kowalczewski, J., Kurowska, M., Janicka, I., Marcinkiewicz, J., & Maśliński, W. (2000). The mechanism of taurine chloramine inhibition of cytokine (interleukin-6, interleukin-8) production by rheumatoid arthritis fibroblast-like synoviocytes. *Arthritis & Rheumatism: Official Journal of the American College of Rheumatology*, *43*(10), 2169–2177.

Kumar, P., & Banik, S. (2013). Pharmacotherapy options in rheumatoid arthritis. *Clinical Medicine Insights: Arthritis and Musculoskeletal Disorders*, *6*, CMAMD. S5558.

Kwon, O. C., Choi, W., Ahn, S. M., Oh, J. S., Hong, S., Lee, C.-K., … Kim, Y.-G. (2024). Drug survival and change of disease activity using a second janus kinase inhibitor in patients with difficult-to-treat rheumatoid arthritis who failed to a janus kinase inhibitor and subsequent biologics. *Advances in Rheumatology*, *64*, 26.

Lemoine, L., Agier, M. S., Audemard-Verger, A., Chouchana, L., Michot, J., Jonville-Béra, A. P., & Maillot, F. (2023). Adverse effect of *Pneumocystis Jirovecii* infection associated with rituximab therapy for autoimmune disease are more frequently reported in older vs. younger patient. *Therapies, 78*(4), 409–417.

Leone, G. M., Mangano, K., Petralia, M. C., Nicoletti, F., & Fagone, P. (2023). Past, present and (foreseeable) future of biological anti-TNF alpha therapy. *Journal of Clinical Medicine, 12*(4), 1630.

Luo, F., Yuan, X.-M., Xiong, H., Huang, C., Chen, C.-M., Ma, W.-K., & Yao, X.-M. (2024). Exploring the mechanism of action of *Tripterygium wilfordii* Hook F in the treatment of rheumatoid arthritis based on network pharmacology and molecular docking. *Tropical Journal of Pharmaceutical Research, 23*(2), 327–337.

Martin, T., Usmani, S. Z., Berdeja, J. G., Agha, M., Cohen, A. D., Hari, P., ... Lesokhin, A. (2023). Ciltacabtagene autoleucel, an anti–B-cell maturation antigen chimeric antigen receptor T-cell therapy, for relapsed/refractory multiple myeloma: CARTITUDE-1 2-year follow-up. *Journal of Clinical Oncology, 41*(6), 1265–1274.

McWilliams, D. F., Thankaraj, D., Jones-Diette, J., Morgan, R., Ifesemen, O. S., Shenker, N. G., & Walsh, D. A. (2022). The efficacy of systemic glucocorticosteroids for pain in rheumatoid arthritis: a systematic literature review and meta-analysis. *Rheumatology, 61*(1), 76–89.

Melles, R. B., Jorge, A. M., Marmor, M. F., Zhou, B., Conell, C., Niu, J., ... Choi, H. K. (2023). Hydroxychloroquine dose and risk for incident retinopathy: a cohort study. *Annals of Internal Medicine, 176*(2), 166–173.

Mitchell, E. L., Lau, P. K. H., Khoo, C., Liew, D., Leung, J., Liu, B., ... Smith, K. (2018). Rheumatic immune-related adverse events secondary to anti–programmed death-1 antibodies and preliminary analysis on the impact of corticosteroids on anti-tumour response: a case series. *European Journal of Cancer, 105*, 88–102.

Ning, R., Zhao, J., Chen, L., Mu, H., Chen, Z., Yang, K., ... He, S. (2024). Macrolide sesquiterpene pyridine alkaloids from Celastrus monospermus and evaluation of their immunosuppressive and anti-osteoclastogenesis activities. *Bioorganic Chemistry, 145*, 107246.

Nunokawa, T., Chinen, N., Shimada, K., Kimura, M., Tateishi, M., Chen, F. Y., ... Sugihara, M. (2023). Efficacy of sulfasalazine for the prevention of Pneumocystis pneumonia in patients with rheumatoid arthritis: a multicentric self-controlled case series study. *Journal of Infection and Chemotherapy, 29*(2), 193–197.

Prasad, P., Verma, S., Surbhi, Ganguly, N. K., Chaturvedi, V., & Mittal, S. A. (2023). Rheumatoid arthritis: advances in treatment strategies. *Molecular and Cellular Biochemistry, 478*(1), 69–88.

Pruthviraj, S., Mukherjee, S., Abilash Krishnan, V., & Sundaram, T. (2024). Chronic arsenic neurotoxicity following complementary and alternative medicine use in rheumatoid arthritis. *Indian Journal of Rheumatology, 19*(3), 1-10.

Rech, J., Tascilar, K., Hagen, M., Kleyer, A., Manger, B., Schoenau, V., ... Braun, J. (2024). Abatacept inhibits inflammation and onset of rheumatoid arthritis in individuals at high risk (ARIAA): a randomised, international, multicentre, double-blind, placebo-controlled trial. *The Lancet, 403*(10429), 850–859.

Riaz, M., Rasool, G., Yousaf, R., Fatima, H., Munir, N., & Ejaz, H. (2025). Anti-rheumatic potential of biological DMARDS and protagonistic role of bio-markers in early detection and management of rheumatoid arthritis. *Innate Immunity, 31*, 17534259251324820.

Ruscitti, P., Di Muzio, C., Di Cola, I., Panahi, A. S., Sensini, F., Ursini, F., ... Cipriani, P. (2024). *AB0562 The Effects of Tofacitinib in Simultaneously Improving Glycaemic Parameters and Inflammatory Markers in Rheumatoid Arthritis Patients with Comorbid Type 2 Diabetes. A Proof-of-Concept,* Open, Prospective, Clinical Study. BMJ Publishing Group Ltd.

Ruyssen-Witrand, A., & Constantin, A. (2018). Controversies in rheumatoid arthritis glucocorticoid therapy. *Joint Bone Spine, 85*(4), 417–422.

Schuna, A. A. (1998). Update on treatment of rheumatoid arthritis. *Journal of the American Pharmaceutical Association (1996), 38*(6), 728–737.

Skouvaklidou, E., Deligeorgakis, D., Skalkou, A., Skepastianos, V., Tsafis, K., Papadimitriou, E., ... Tzitiridou-Chatzopoulou, M. (2024). Focus on filgotinib in rheumatoid arthritis: a trial-based review. *Mediterranean Journal of Rheumatology, 35*(Suppl 1), 1.

Song, Y.-K., Lee, J., Jo, J., & Kwon, J.-W. (2024). Comparison of active tuberculosis occurrence associated with Janus kinase inhibitors and biological DMARDs in rheumatoid arthritis. *Rmd Open, 10*(2), e003946.

Souza, R. F., Caetano, M. A. F., Magalhães, H. I. R., & Castelucci, P. (2023). Study of tumor necrosis factor receptor in the inflammatory bowel disease. *World Journal of Gastroenterology, 29*(18), 2733.

Steinhart, A. H. (2024). Anti-tumor necrosis factor therapy: safety and toxicity. In Miguel Regueiro & Jason Swoger (Eds.), *Clinical Challenges and Complications of IBD* (pp. 303–326). CRC Press.

Swaroop, A. K., Negi, P., Kar, A., Mariappan, E., Natarajan, J., & Selvaraj, J. (2024). Navigating IL-6: from molecular mechanisms to therapeutic breakthroughs. *Cytokine & Growth Factor Reviews, 76*, 48–76.

Takase-Minegishi, K., Böhringer, S., Nam, J. L., Kaneko, Y., Behrens, F., Saevarsdottir, S., … Landewé, R. (2024). The impact of autoantibodies on the efficacy of biological disease-modifying anti-rheumatic drugs in rheumatoid arthritis: meta-analysis of randomized controlled trials. *Rheumatology*, *64*(2), 548-560.

Talarico, F., Chakravarty, S., Liu, Y. S., Greenshaw, A. J., Passos, I. C., & Cao, B. (2023). Systematic review of psychiatric adverse effects induced by chloroquine and hydroxychloroquine: case reports and population studies. *Annals of Pharmacotherapy*, *57*(4), 463–479.

Tsuchiya, H., Ota, M., Takahashi, H., Hatano, H., Ogawa, M., Nakajima, S., … Fujio, K. (2024). Epigenetic targets of Janus kinase inhibitors are linked to genetic risks of rheumatoid arthritis. *Inflammation and Regeneration*, *44*(1), 29.

Ulloa del Carpio, N., Alvarado-Corella, D., Quiñones-Laveriano, D. M., Araya-Sibaja, A., Vega-Baudrit, J., Monagas-Juan, M., … Villar-López, M. (2024). Exploring the chemical and pharmacological variability of Lepidium meyenii: a comprehensive review of the effects of maca. *Frontiers in Pharmacology*, *15*, 1360422.

Vane, J. R., & Botting, R. M. (1998). Anti-inflammatory drugs and their mechanism of action. *Inflammation Research*, *47*, 78–87.

van Laar, M., Pergolizzi Jr, J. V., Mellinghoff, H.-U., Merchante, I. M., Nalamachu, S., O'Brien, J., … Raffa, R. B. (2012). Pain treatment in arthritis-related pain: beyond NSAIDs. *The Open Rheumatology Journal*, 6, 320.

Vesperinas-Castro, A., & Cortés-Vicente, E. (2023). Rituximab treatment in myasthenia gravis. *Frontiers in Neurology*, *14*, 1275533.

Vilbert, M., Koch Hein, E. C., Madeira, T., Michelon, I., N. Priantti, J., Lobo, M., … Wu, C.-Y. (2024). Safety and efficacy of interleukin-6 blockade for immune-related adverse events: a systematic review and meta-analysis. *Journal of Clinical Oncology*, *42*(16_suppl), e24149–e24149.

Virtanen, A., Spinelli, F. R., Telliez, J. B., O'Shea, J. J., Silvennoinen, O., & Gadina, M. (2024). JAK inhibitor selectivity: new opportunities, better drugs? *Nature Reviews Rheumatology*, *20*(10), 649–665.

Wang, C.-M., Jan Wu, Y.-J., Zheng, J.-W., Huang, L. Y., Tan, K. P., & Chen, J.-Y. (2024). T cell expressions of aberrant gene signatures and co-inhibitory receptors (Co-IRs) as predictors of renal damage and lupus disease activity. *Journal of Biomedical Science*, *31*(1), 41.

Wilson, J. C., Sarsour, K., Gale, S., Pethö-Schramm, A., Jick, S. S., & Meier, C. R. (2019). Incidence and risk of glucocorticoid-associated adverse effects in patients with rheumatoid arthritis. *Arthritis Care & Research*, *71*(4), 498–511.

Wlassits, R., Müller, M., Fenzl, K. H., Lamprecht, T., & Erlacher, L. (2024). JAK-inhibitors–a story of success and adverse events. *Open Access Rheumatology: Research and Reviews*, *16*, 43–53.

Yamada, Z., Muraoka, S., Kawazoe, M., Hirose, W., Kono, H., Yasuda, S., … Nanki, T. (2024). Long-term effects of abatacept on atherosclerosis and arthritis in older vs. younger patients with rheumatoid arthritis: 3-year results of a prospective, multicenter, observational study. *Arthritis Research & Therapy*, *26*(1), 87.

Ye, W., Ding, Y., Li, M., Tian, Z., Wang, S., & Liu, Z. (2024). Safety assessment of sulfasalazine: a pharmacovigilance study based on FAERS database. *Frontiers in Pharmacology*, *15*, 1452300.

Yusuf, I. H., Charbel Issa, P., & Ahn, S. J. (2023). Hydroxychloroquine-induced retinal toxicity. *Frontiers in Pharmacology*, *14*, 1196783.

Zerbini, C., Clark, P., Mendez-Sanchez, L., Pereira, R., Messina, O., Uña, C., … Lane, N. (2017). Biologic therapies and bone loss in rheumatoid arthritis. *Osteoporosis International*, *28*, 429–446.

Zewail, M. (2024). Leflunomide nanocarriers: a new prospect of therapeutic applications. *Journal of Microencapsulation*, *41*(8), 715–738.

Zewail, M. B., El-Gizawy, S. A., Asaad, G. F., Shabana, M. E., & El-Dakroury, W. A. (2023). Chitosan coated clove oil-based nanoemulsion: an attractive option for oral delivery of leflunomide in rheumatoid arthritis. *International Journal of Pharmaceutics*, *643*, 123224.

Zhao, C., Yu, Y., Yin, G., Xu, C., Wang, J., Wang, L., … Zhou, B. (2024). Sulfasalazine promotes ferroptosis through AKT-ERK1/2 and P53-SLC7A11 in rheumatoid arthritis. *Inflammopharmacology*, *32*(2), 1277–1294.

Zheng, K., Chen, Y., Liu, S., He, C., Yang, Y., Wu, D., … Zhang, F. (2023). Leflunomide: traditional immunosuppressant with concurrent antiviral effects. *International Journal of Rheumatic Diseases*, *26*(2), 195–209.

4 Combination Therapy and Dual Inhibition of Signaling Pathways in Rheumatoid Arthritis

Tanzeela Awan, Uzma Saher, Justyna Wozniak, Ayesha Fatima, and Ammara Saleem

WHAT YOU WILL LEARN IN THIS CHAPTER?

This chapter provides a comprehensive overview of available treatments and their combinations used for rheumatoid arthritis (RA). It also provides an insight into the potential of phytochemical monotherapy and its combinations with disease-modifying antirheumatic drugs (DMARDs) in the treatment of RA. It highlights the significance of using evidence-based integrative methods in clinical practice to improve patient outcomes.

LIST OF ABBREVIATIONS

AIA	Adjuvant-induced arthritis
bDMARDs	Biologic DMARDs
DMARD	Disease-modifying antirheumatic drug
EGCG	Epigallocatechin gallate
ESR	Erythrocyte sedimentation rate
EULAR	European League Against Rheumatoid Arthritis
JAK	Janus-activated kinase
MDA	Malondialdehyde
MHC	Major histocompatibility complex
MTX	Methotrexate
NSAIDs	Nonsteroidal anti-inflammatory drugs
RA	Rheumatoid arthritis
RSV	Resveratrol
TNFi	TNF inhibitor
tsDMARDs	Targeted synthetic DMARDs

4.1 INTRODUCTION

Rheumatoid arthritis (RA) is a chronic inflammatory autoimmune disorder, which mainly affects multiple joints of the body (Guo et al., 2018). RA first affects small joints before moving on to larger joints and eventually to other organs of the body like kidneys, heart, lungs, and skin. Joint bones and cartilage are frequently damaged, and ligaments and tendons deteriorate (Lee, Kim, Cho, & Lee, 2017). It is more prevalent in women than in men and mainly affects the elderly. Geographic variations in the 2002 prevalence rate were observed, ranging from 0.5% to 1% of the population (Silman & Pearson, 2002).

DOI: 10.1201/9781003535201-5

An immune response against the lining membrane of joints, the synovium, is initiated by both genetic and environmental factors, including smoking, periodontitis, alteration in gut microbiota, and exposure to silica. However, much remains unclear regarding the specific processes and triggers of tissue destruction in RA. Major histocompatibility complex (MHC) II genetic variants have been associated with an increased risk of RA in several different ethnic groups. MHC II is encoded by the human leukocyte antigen system and many of its receptors are linked to an elevated risk of RA (Nepom, 1998). The disease can result in poor quality of life for the affected patients and has been proven to be a great socioeconomic burden. Clinical signs of symmetric joint involvement (affects the same joints on both sides of the body) can include redness, edema, and possibly limited range of motion in addition to arthralgia (Van Der Linden et al., 2010). The treatment of RA is still challenging because of the involvement of multiple factors in the pathogenesis.

4.2 THERAPEUTIC INTERVENTIONS

The aim of RA therapy is to minimize joint inflammation and pain, maximize joint function, and prevent joint deformation and destruction. Treatment plans include medicine, exercise, heightened patient awareness, and rest. Individualized treatment regimens are frequently developed in response to the patient's needs and overall health status. This encompasses things including how the disease is progressing, which joints are afflicted, age, overall health, occupation, therapy adherence, and awareness of the issue (Staheli, 1998). RA is considered incurable; however, symptomatic treatment may help to reduce pain and swelling in joints. Certain treatments may lessen the severity of the illness and enhance the patient's quality of life. Moreover, recent advancements in targeted drug therapy have revolutionized the treatment strategies of RA.

Treatment options for RA include monotherapy and combination therapy. Strikingly, several randomized controlled trials have demonstrated that combining a biologic disease-modifying antirheumatic drug (DMARD) with a traditional DMARD, like methotrexate (MTX), is more effective than using a single drug. Treatment initiated at the first in the illness was described to halt radiographic development, which usually occurs in the initial few months. The two main pillars of modern RA treatment are disease modification and symptom management. Patients receiving delayed medication show an increased risk of radiographic joint space narrowing and bone erosions. Bony erosions appear on radiographs in individuals with poorly managed RA within two years of the condition's onset, and these erosive alterations are associated with worse functional outcomes. A quick referral to a rheumatologist is therefore required for a patient with otherwise unexplained new-onset polyarthritis. This allows for the establishment of the diagnosis of RA and the start of a medication-based treatment strategy that aims to prevent deformity while achieving disease remission.

The treatment strategies used for RA are discussed below:

4.2.1 Monotherapy

4.2.1.1 Modulation of Oral and Gut Microbiota

Modulation of gut microbiota is recommended for the treatment of inflammation in RA. The use of probiotics in the diet helps improve the beneficial bacterial species in microbiota, which ultimately reduces the body's inflammatory response and escalates the disease outcome. Dysbiosis of oral microbiota may result in periodontitis and gingivitis, increasing the risk of developing RA, as the inflamed gingiva contains citrullinated proteins. These citrullinated proteins are identified to contribute to the progression of RA (Bennike et al., 2017). Furthermore, even before the onset of symptomatic clinical arthritis, changes in the gut microbiota are related to RA. Examining the gut microbiota has become a practical approach for anticipating and mitigating the emergence of RA, as it has revealed new information about factors that either increase or decrease susceptibility to the disease. The enzymatic products produced by the human gut microbiota may also have a direct or indirect impact on the toxicity, clinical effectiveness, and bioavailability of drugs.

On the other hand, by altering the composition of the gut microbiota, some drugs and active substances can affect the immune system and bolster the host's defenses. Even if notable anomalies in particular microbial communities are linked to the advancement of RA, many crucial issues need to be resolved to promote the progress of gut microbiota-targeted therapies. Future research needs to identify the sources of dysbiosis and pinpoint the precise mechanism and timing by which gut dysbiosis affects the development of RA. Second, to obtain useful targets for diagnosis, prognosis, and treatment, it is necessary to define the connection between the disease and the changed microbial composition as well as the molecular pathways controlling RA development. Third, to maximize therapy options that restore microbial symbiosis in patients with RA, it is crucial to understand the mechanisms underlying the effects of DMARDs on the gut microbiota and the consequences of gut dysbiosis on the modulation of treatment response. All things considered, more investigation into these unexplored areas might enhance therapy effectiveness, lower the risk of toxicity, and enhance RA clinical results (Fan, Jiang, & He, 2023).

4.2.1.2 NSAIDs

Nonsteroidal anti-inflammatory drugs (NSAIDs) are a group of drugs with anti-inflammatory actions. Since RA is an inflammatory disorder, NSAIDs are used as a first-line therapy (Gaffo, Saag, & Curtis, 2006). However, these drugs alone can only minimize the disease-associated pain and stiffness. First-line treatment aims to reduce inflammation and relieve pain. Among them, acetyl salicylic acid, naproxen, ibuprofen, and etodolac are more common. Aspirin is one of the oldest NSAIDs used for joint pain and is an effective anti-inflammatory when used at high doses (Ong, Lirk, Tan, & Seymour, 2007). However, adverse drug reactions need to be considered, as tinnitus, hearing loss, and gastric intolerance can be caused if NSAIDs are administered in high doses. Other, more recently released NSAIDs, such as COX-2 inhibitors (coxib), are equally effective as aspirin and require fewer daily doses. NSAIDs function by inhibiting cyclooxygenase (COX) to prevent the synthesis of prostaglandins and prostacyclin. In patients with osteoarthritis, diclofenac, ibuprofen, tolmetin, and naproxen are the most effective oral NSAIDs for pain and functional restoration, while etoricoxib and diclofenac are the least effective for joint symptoms (Sohail et al., 2023). NSAIDs alone are not able to reduce or slow down the progression of the disease. Therefore, for the treatment of RA, NSAIDs must be used in combination with DMARDs.

NSAIDs possess a relatively low-risk profile of side effects when used for a short period. However, long-term utilization of these medications can cause some serious adverse effects, the most prominent of which are gastrointestinal (GI) problems including nausea, vomiting, stomach discomfort, ulcers, GI bleeding, and dyspepsia; these can be mitigated by administering the medication with food, antacids, proton pump inhibitors, or misoprostol. Other organs, such as the kidneys, may also be affected by the drug's nephrotoxicity (Ong et al., 2007). A more recent NSAID, celecoxib (Celebrex), is a selective COX-2 inhibitor with less GI side effects. For long-term use, selective COX-2 inhibitors such as coxibs are recommended, resulting in a 50% decline in the adverse effects generally observed with conventional NSAIDs. Though these drugs are implicated successfully, cardiovascular side effects have also been observed with selective COX-2 inhibitors (Alhammadi et al., 2022) and need to be taken into account for individual patient treatment.

4.2.1.3 Corticosteroids

Corticosteroids are more effective in anti-inflammatory therapy; however, they possess more adverse drug reactions in comparison to NSAIDs, and should only be used for short term, at modest doses, when RA symptoms are aggravated. Corticosteroid injections intra-articularly are effective in treating localized inflammatory symptoms. Glucocorticoids, a class of corticosteroids that bind to the glucocorticoid receptor, have been a pillar in the treatment of RA for a considerable amount of time. Their strong anti-inflammatory and immunosuppressive qualities help diminish the symptoms of the disease and improve patient outcomes. Even though excessive amounts of glucocorticoids result in intolerable toxicity, many individuals may experience less pain and swelling at appropriate daily dosages of prednisone (or its equivalent) of ≤15 mg. In RA patients, this dosage of glucocorticoids may also have a minor beneficial effect on disease modification (van Everdingen, Jacobs, Siewertsz Van Reesema, & Bijlsma, 2002).

These synthetic analogs of endogenous cortisol exert their effects primarily through binding to the glucocorticoid receptor, leading to the transcriptional repression of pro-inflammatory cytokines such as tumor necrosis factor-α (TNF-α), interleukin-1 (IL-1), and interleukin-6 (IL-6), which play a pivotal role in the pathophysiology of RA. Additionally, glucocorticoids induce the production of anti-inflammatory proteins, including lipocortin-1, which further mitigates inflammatory responses. The induced expression of lipocortin-1 results in the inhibition of the enzyme phospholipase A2, which is responsible for the production of arachidonic acid, a precursor of prostaglandins. In addition, corticosteroids inhibit the action of the nuclear factor kappa B (NF-κB) and AP-1 protein and upregulate TNF-α and IL-1 (proinflammatory molecules). They also exert their effect through membrane-associated receptors and second messengers (Rothhut & Russo-Marie, 1984). As a result, they unfortunately also cause some severe side effects including immunosuppression (unwanted excessive immune suppression may lead to other disorders), bone thinning, diabetes, and even an increase in body weight. Supplements like calcium and vitamin D can help patients reduce the thinning of the bone. In addition to using supplements, reducing the dose can also help improve the health status of RA patients. Of note, the administration of corticosteroids must not be stopped suddenly, as this may result in RA flare-ups by suppression of the hypothalamic–pituitary–adrenal axis (Liu et al., 2013).

Clinical studies have demonstrated that short-term glucocorticoid therapy can significantly reduce disease activity and joint damage, thus providing rapid symptom relief. However, their long-term use is often limited by a spectrum of adverse drug effects such as osteoporosis, hyperglycemia, and increased susceptibility to infections.

Therefore, current therapeutic strategies advocate for the use of glucocorticoids in conjunction with DMARDs as a bridge therapy to maximize efficacy while minimizing potential risks. To improve therapeutic outcomes and reduce side effects in RA patients, research on developing selective glucocorticoid receptor modulators, improving glucocorticoid regimens, and investigating innovative delivery methods is conducted. Glucocorticoids alone are not recommended due to dosage-dependent severe adverse drug effects (Saag et al., 1994).

4.2.1.4 Opioid Analgesics

Opioid analgesics can also be used for pain relief in cases of RA. They can be used for the short term, as their benefits outweigh their adverse effects. Among them, the most popular weak opioids are codeine, dextropropoxyphene, and tramadol. These options are preferred over other alternatives (Richards, Whittle, van der Heijde, & Buchbinder, 2012).

4.2.1.5 Disease-Modifying Antirheumatic Drugs (DMARDs)

Analgesics, NSAIDs, corticosteroids, and finally DMARDs have been the pharmaceuticals used to treat RA over the years. The main therapeutic drugs that serve as the cornerstone of RA treatment are DMARDs. By focusing on the fundamental processes of RA, they comprise a broad family of pharmacological drugs used to stop or delay the disease's progression ("Systematic Review: Comparative Effectiveness and Harms of Disease-Modifying Medications for Rheumatoid Arthritis," 2008). The main pillar of RA treatment is the use of biologic DMARDs (bDMARDs), targeted synthetic DMARDs (tsDMARDs), conventional DMARDs, and conventional synthetic DMARDs (csDMARDs).

4.2.1.5.1 Conventional Synthetic DMARDs (csDMARDs)

csDMARDs, such as MTX, sulfasalazine, and hydroxychloroquine, are the traditional first-line therapies for RA. These drugs are well known for their anti-inflammatory and immunosuppressive properties, which minimize joint damage and enhance long-term results ("Systematic Review: Comparative Effectiveness and Harms of Disease-Modifying Medications for Rheumatoid Arthritis," 2008). Especially, MTX is considered the basis of RA treatment due to its efficacy and safety profile. MTX has been the mainstay of RA treatment in the USA for nearly 20 years, as it has proven to be low cost, highly tolerable, reliable, and effective over a prolonged period. When compared to other DMARDs, MTX may lower mortality, primarily from cardiovascular causes, according to current research. Up to four different mechanisms of action have been proposed regarding MTX. It first functions as an antifolate agent by preventing the division of fast-dividing cells, such as lymphocytes and other mediators of inflammation. Second, MTX may reduce the build-up of harmful substances, including polyamines, which are tetrahydrofolate-dependent and implicated in RA tissue damage. Third, MTX may lower glutathione levels inside cells, thereby reducing harm (Gaffo et al., 2006). However, major adverse effects found with the use of MTX are GI toxicity and hepatotoxicity, including hepatitis and cirrhosis. Further mouth ulcers, cytopenia, and interstitial pneumonitis are observed. With renal excretion being the main pathway of MTX elimination (80%), renal-impaired patients are more prone to adverse drug reactions, including paradoxical worsening of RA and an increase in rheumatoid nodules (Rheumatoid Arthritis Clinical Trial Archive Group, 1995). MTX is the gold standard to which newer RA treatments are compared. Its effectiveness is on par with other conventional DMARDs, if not better for many patients when compared to the more recent biological drugs (Bathon et al., 2000).

Hydroxychloroquine, another conventional DMARD, has the highest safety profile. Hydroxychloroquine does not result in hepatotoxicity or renal failure when compared to other traditional DMARDs. Rash and diarrhea are two of the most frequent side effects of hydroxychloroquine. Retinopathy/maculopathy is an uncommon but potentially harmful adverse effect of hydroxychloroquine that is more common with larger cumulative dosages. Risk factors for hydroxychloroquine maculopathy include older age, chronic kidney disease, more than five years of continuous therapy, and dosages greater than 5 mg/kg/day. Patients using hydroxychloroquine should be evaluated by an ophthalmologist utilizing ocular coherence tomography regularly. This test can identify maculopathy well in advance of symptoms related to visual field impairments because of its great sensitivity. Hydroxychloroquine seldom exhibits cardiomyopathy, leukopenia, myopathy, and anemia.

The use of combination therapy with conventional DMARDs has nowadays drawn more attention. The benefits of combining three DMARDs (MTX, hydroxychloroquine, and sulfasalazine) over DMARD monotherapy have been amply proven in many clinical trials. Additionally, there is proof that the three DMARDs together are more effective than the separate combinations of MTX and hydroxychloroquine or MTX and sulfasalazine (Möttönen et al., 1999).

4.2.1.5.2 Biologic DMARDs (bDMARDs)

bDMARDs are monoclonal antibodies and proteins that target specific immune system components such as cytokines. These consist of TNFi (e.g., infliximab, etanercept, and adalimumab), IL-6 receptor antagonists (e.g., tocilizumab), and B- and T-cell-depleting agents (e.g., rituximab and abatacept). bDMARDs can significantly decrease disease activity by specifically targeting these inflammatory mediators. Their main actions consist of (i) interfering with the synthesis or function of cytokines, (ii) blocking the "second signal" necessary for T-cell activation, (iii) suppressing activated T cells, or (iv) reducing the number of B cells or preventing the substances that stimulate B cells. A small chemical inhibitor called tofacitinib inhibits the protein tyrosine kinase JAK, which is important in modulating cytokine signaling (Aletaha & Smolen, 2018).

Aside from being quite expensive, biologic DMARDs are, however, also associated with a higher risk for serious infections, which include fungal, bacterial, and viral infections. Anti-TNF-α therapy is found to be associated with bacterial infections such as *Salmonella* and *Mycobacterium tuberculosis* (Park et al., 2020). Hepatitis B/C, herpes zoster, and TB can all reactivate. Further, other severe side effects such as sporadic events of hepatotoxicity and bone marrow suppression have been reported. Anti-TNF medications have the potential to exacerbate conditions such as drug-induced lupus, demyelinating central nervous system disorders, and severe congestive heart failure. They may also be associated with lymphomas and skin malignancies other than melanoma. The administration of JAK and IL-6 inhibitors can result in pancytopenia, hyperlipidemia, and increased liver function tests. Further, chronic obstructive pulmonary disease has been documented to exacerbate because of abatacept, which blocks the immune activity of T cells. Inflammatory bowel disease can be brought on by or made worse by IL-17 inhibitors. In patients receiving rituximab, progressive multifocal leukoencephalopathy has been documented.

4.2.1.5.3 Targeted Synthetic DMARDs (tsDMARDs)

tsDMARDs, such as Janus kinase (JAK) inhibitors (e.g., tofacitinib and baricitinib), are the latest advancement in the treatment of RA. These small molecules inhibit intracellular signaling pathways fundamental to the inflammatory process (Fletcher et al., 2022). This approach offers a targeted therapy for modulating the immune response in RA. tsDMARDs are an oral substitute to biologics and are useful for patients who do not respond effectively to csDMARDs or bDMARDs.

4.3 PATHWAYS INVOLVED IN RA

tsDMARDs function by inhibiting one of the signaling pathways involved in the progression of RA. Targeting these pathways limits the development of disease. Some major pathways are described below.

4.3.1 JAK–STAT Signaling Pathway

The Janus-activated kinase (JAK)–signal transducer and activator of transcription (STAT) pathway is one of the critical signaling pathways activated by cytokine-induced molecules. Blocking JAK/STAT-mediated NF-κB activation can be used as a therapeutic strategy, as some cytokines that activate the JAK–STAT pathway can also lead to NF-κB activation. While TNF-α itself does not primarily signal through the JAK–STAT pathway, TNF-α-induced gene expression is able to trigger JAK–STAT signaling (Bolli, Dawn, & Xuan, 2003). The activation of this pathway contributes to cell differentiation, apoptosis, inflammation, and immune responses and is found to be abnormally activated during the development of RA (Angelini et al., 2020). Pathway inhibition may result in suppression of the overactive immune system and reduction of pain and swelling in RA (Oike et al., 2017; Tripathi et al., 2017). Tofacitinib (Dhillon, 2017; Van Vollenhoven et al., 2012), filgotinib, peficitinib (Takeuchi et al., 2019; Tanaka & Izutsu, 2020), baricitinib (Al-Salama & Scott, 2018; Taylor et al., 2017), and upadacitinib (Serhal & Edwards, 2019) are examples of JAK–STAT pathway inhibitors that have been approved and used for the treatment of RA.

4.3.2 MAPK Signaling Pathway

Mitogen-activated protein kinase (MAPK) signaling pathway regulates various cell functions such as gene expression, migration, survival, apoptosis, and cell differentiation (Ralph & Morand, 2008). These processes have a vital role in the progression of RA. Upregulated activation of this pathway results in an increased immune response and inflammatory hyperplasia of synovial cells (Cargnello & Roux, 2011). P38, JNK, and ERK are three major molecules of downstream pathways, which further regulate many crucial cellular processes involved in inflammatory reactions (Whitaker & Cook, 2021). Hence, the MAPK pathway, p38-mediated signaling in particular, is a potential therapeutic target for treating immune-mediated diseases, including RA (Sujitha & Rasool, 2017).

4.3.3 PI3K Signaling Pathway

The phosphatidylinositol 3 kinase (PI3K)–AKT pathway also critically regulates various cellular functions including cell survival, proliferation, migration, and angiogenesis. Research has shown that this pathway is involved in the occurrence and progression of RA. The activation of this pathway results in the abnormal growth of fibroblast-like synoviocytes (FLSs) and causes inflammation through the participation of several cytokines such as IL-1, IL-6, IL-17, and TNF-α (Kwok et al., 2012; Tsai et al., 2017). These events significantly contribute to the pathogenesis of RA, causing joint stiffness and deformity (Dinesh & Rasool, 2018; Xuan et al., 2017). Studies have shown that PI3K/AKT/mammalian target of rapamycin (mTOR) signaling contributes to the development of RA, providing evidence that inhibiting this pathway could also alleviate disease progression (Ba et al., 2021; Li et al., 2013).

4.3.4 SYK Signaling Pathway

Phosphorylated spleen tyrosine kinase (SYK) was found to be upregulated in the B cells of RA patients. SYK activates Bruton's tyrosine kinase (BTK), which regulates B cells and related signaling (Corneth, Klein Wolterink, & Hendriks, 2015). Since RA involves dysregulation of B- and T-lymphocyte proliferation and B cell-regulated signaling, it ultimately produces autoantibodies and cytokines, causing disease progression (Arneson, Carroll, & Ruderman, 2021; Shinohara et al., 2008). Thus, BTK is an attractive candidate target for anti-RA treatment (Gaballa & Pinilla-Ibarz, 2021; Gabizon & London, 2020).

4.3.5 Wnt Signaling Pathway

The Wnt signaling pathway is typically active during embryonic development and cancer. However, it also plays a crucial role in the progression of RA by regulating synovial inflammation and bone metabolism (Miao et al., 2013). Dysregulation of Wnt signaling in RA can lead to increased inflammation and abnormal bone remodeling, contributing to joint damage and disease progression. The molecules that activate this pathway, such as NAV2, are considered attractive therapeutic targets for the treatment of RA. Targeting NAV2 and other components of the Wnt pathway may help modulate these processes, potentially reducing inflammation and preventing bone erosion in RA patients (Chakrabarti et al., 2020).

4.3.6 Notch Signaling Pathway

Notch signaling is crucial for normal cell functioning including cell differentiation and proliferation. It also plays a key role in the regulation of inflammatory processes. In RA, the Notch signaling pathway is activated and modulates inflammation by inducing the secretion of proinflammatory cytokines from synoviocytes. Inhibition of Notch-3 signaling has been shown to reduce inflammation and joint damage (Wei et al., 2020).

4.4 PHYTOCHEMICALS AS THERAPY FOR RA

Many natural substances originating from plants, known as phytochemicals, have been identified to possess pharmacological and biological properties. Alkaloids, polyphenols, flavonoids, and terpenoids are examples of secondary metabolites that have proven to be a helpful source of medications for treating conditions like diabetes, cancer, RA, Alzheimer's, Parkinson's, and cardiovascular problems.

In the past, people with RA diagnoses have experimented with dietary modifications in the hope of reducing their symptoms. Individuals often seek guidance to incorporate dietary adjustments as an additional therapeutic approach. Due to the unfavorable side effects, the patients are hesitant to rely solely on pharmaceutical treatments for managing RA (Smedslund, Byfuglien, Olsen, & Hagen, 2010).

There is growing interest in using phytochemicals as an adjunct or alternative therapy for RA. While monotherapy with phytochemicals may not replace DMARDs in moderate to severe cases, they can serve as valuable adjuncts in mild cases or as preventive strategies.

There is a wide variety of foods from both plant and animal sources in the human diet. Numerous phytochemicals, such as polyphenols, flavonoids, alkaloids, glycosides, terpenoids, phytosterols, and sterols, are found in plant-based foods. Below is a brief description of how certain phytochemical elements affect RA.

4.4.1 Apigenin

Among the flavone class of dietary flavonoids, apigenin is present in fruits, vegetables, spices, and herbs. Synoviocytes, which resemble activated RA fibroblasts, are essential for both the onset and progression of RA. These cells are connected to collagenase activity, which participates in the breakdown of collagen in the joints, contributing to joint damage and RA progression. Apigenin has been shown to reduce the collagenase activity linked to RA. In human RA FLSs (RA-FLSs), apigenin amplifies the lethal effects of TNF-related apoptosis-inducing ligand (TRAIL), most likely mediated through inactivation of the PI3-K/Akt signaling pathway (Lee et al., 2007; Sun et al., 2012). Increased amounts of IL-6, IL-1β, and TNF-α cytokines were observed in synoviocytes after an inflammatory response triggered by TNF-α. It was found that TNF-α activates the NF-κB pathways, which increases the differential expression of pro-inflammatory cytokines, transthyretin, and the receptor for advanced glycation end-products. Apigenin's anti-inflammatory properties prevent TNF-α-mediated deregulation or expression linked to the pathophysiology of RA (Monu, Agnihotri, Saquib, & Biswas, 2024).

4.4.2 Epigallocatechin-3-Gallate (EGCG)

EGCG is a polyphenol obtained from the plants *Camellia assamica* and *Camellia sinensis*. EGCG has shown anti-inflammatory effects through the modulation of cytokines and immune cells. It is a catechin monomer that is isolated from tea leaves and possesses significant anti-inflammatory and antioxidant properties. According to studies (Fechtner, Singh, Chourasia, & Ahmed, 2017), EGCG is the most effective catechin in green tea extract for reducing inflammation in human RA synovial fibroblasts. Its potential antiarthritic effects include increased production of phosphorylated extracellular signal-regulated kinase (pERK), nuclear factor erythroid 2-related factor 2 (Nrf2), and heme oxygenase-1 (HO-1), as well as decreased STAT3 activation. Subsequent research has demonstrated that EGCG suppresses Th17 cell development by blocking STAT3 activation. It also suppresses the mTOR and the subsequent activation of hypoxia-inducible factor 1-alpha, which may reduce Th17 and increase regulatory T cells (Treg). These investigations point to EGCG's possible benefit in the management of RA (Ahmed, Pakozdi, & Koch, 2006).

4.4.3 Resveratrol

Resveratrol (RES), a natural antioxidant (3,5,4-trihydroxystilbene), is found in many plants, including *Polygonum cuspidatum*, *Cassia obtusifolia*, *Veratrum nigrum*, grapes, soybeans, and nuts. It exhibits a range of pharmacological activities and can be used as a treatment for autoimmune diseases, including RA (Fernández-Rodríguez et al., 2021). RES decreases the levels of RA-associated biochemical markers in the serum, such as IL-6, MMP-3, carboxylated osteocalcin, TNF-α, and C-reactive protein. Silent information regulator 1 (SIRT1) expression in RA patients' synovial tissue and FLSs is notably lower than in the group of healthy controls. SIRT1 is a crucial regulator of the pathophysiology of the disease and has been found to have anti-inflammatory and anticytokine properties. RES activates SIRT1, which in turn can inhibit the inflammatory response and reduce the production of proinflammatory cytokines and matrix metalloproteinases. This activation helps mitigate the inflammatory processes and tissue degradation associated with RA, making RES a promising therapeutic agent in the management of the disease (Oliveira et al., 2017).

4.4.4 Quercetin

The natural flavonoid quercetin is found in a wide range of fruits, vegetables, and plants like apples, onions, and broccoli. It is also present in beverages like tea. Quercetin possesses a diverse range of pharmacological actions including anti-inflammatory and antioxidant properties. It contributes to the amelioration of RA's clinical manifestations by reducing inflammation and inhibiting the development of pannus, enabling its use as an herbal medication to help treat RA (Javadi et al., 2017; Ji et al., 2013).

According to a study (Kim et al., 2019), quercetin can block Th17 differentiation and IL17 secretion, reduce RANKL synthesis, and inhibit monocyte-to-osteoclast formation in IL-17-stimulated RA-FLS. It has immunomodulatory effects on RA-FLS via the mTOR, ERK, and NF-κB pathways. Another study conducted in collagen-induced arthritis (CIA) rats showed that quercetin (Yang et al., 2018) was able to alleviate arthritic symptoms. The possible mechanism of action was predicted to involve controlling the Th17/Treg balance, preventing NLRP3 inflammasome activation, and triggering the HO-1-mediated anti-inflammatory response.

Additional (Endale et al., 2013) research on quercetin's anti-RA molecular mechanism demonstrated that its anti-inflammatory potential on lipopolysaccharide (LPS)-induced RAW264.7 cells was attained by blocking the synthesis of Toll like receptor-4/myeloid differentiation factor-88 (MyD88)/PI3K complex and inhibiting inflammatory mediators (TNF-α, IL-1β, IL-6, and macrophage colony-stimulating factor). According to these studies, quercetin primarily inhibits the development of osteoclasts, regulates the immune system, and reduces inflammation to produce its anti-RA effects.

4.4.5 Baicalin

Baicalin is a flavonoid active ingredient isolated from *Scutellaria radix* roots. Previous research (Chen et al., 2014) has demonstrated that baicalin possesses anti-inflammatory activity on RA-FLS and suppresses the proliferation of RA-FLS induced by IL-1β. These effects are attributed to baicalin's suppression of the MAPK/ERK/p38 signaling pathway mediated by macrophage migration inhibitory factor and NF-κB transcriptional activity. A study showed that baicalin (Yang, Yang, & Zou, 2013) downregulates the production of adhesion molecules (intracellular adhesion molecule 1 and vascular cell adhesion molecule 1) and inflammatory agents (IL-6 and TNF-α) in synovial cells triggered by IL-17, reduces joint inflammation in CIA mice, and blocks the body's Th17 cell proliferation. Baicalin administration might lessen joint inflammation in CIA rats (Wang et al., 2014). This effect was associated with baicalin's ability to suppress NF-κB p65 acetylation by raising SIRT1 and decreasing NF-κB p65 protein expression in synovial tissue and FLSs. Baicalin's anti-RA action is linked to the downregulation of inflammatory markers (MMP-2, MMP-9, iNOS, and COX-2), proinflammatory factors (TNF-α, IL-1β, and IL6), and the induction of monocyte apoptosis in the synovial fluid of CIA mice (Bai et al., 2020). Additionally, baicalin inhibits JAK1/STAT3 and TLR2/MYD88/NF-κB p65 signal transduction. The research mentioned above demonstrates a significant anti-RA action of baicalin.

4.4.6 Curcumin

Curcumin, known for its potent anti-inflammatory and antioxidant properties, is the main curcuminoid in turmeric (*Curcuma longa*). It inhibits NF-κB and other proinflammatory pathways. An intraperitoneal injection of an extract containing curcuminoids considerably reduced both acute (75%) and chronic (68%) joint inflammation in the streptococcal cell wall-induced arthritis in female Lewis rats (Funk et al., 2006). The safety and efficacy of curcumin therapy in individuals with active RA were initially proposed by Chandran and Goel (2012). Curcumin mediates anti-inflammatory effects by downregulating proinflammatory cytokine expressions such as TNF-α, interleukins (IL-1, IL-2, IL-6, IL-8, IL-12), and chemokines, as well as proinflammatory enzymes such as nitric oxide synthase (iNOS), COX-2, and Arachidonate 5-lipoxygenase. These anti-inflammatory actions are facilitated by the deactivation of NF-κB, which is a key molecular correlate of inflammatory reactions in patients with arthritis (Roshak, Callahan, & Blake, 2002). According to a recent investigation, curcumin functions as an alternative TNF-α blocker both *in vivo* and *in vitro*. Interestingly, low doses of curcumin have also been shown to effectively manage and treat osteoarthritis, highlighting its potential as a versatile anti-inflammatory agent.

4.4.7 Arctigenin

Arctigenin is a naturally occurring phenylpropanoid dibenzylbutyrolactone lignin. Although there is no concrete evidence of arctigenin's anti-RA effect, several encouraging findings suggest its potential efficacy against RA. It has been found that by downregulating iNOS, arctigenin inhibits the production of NO in the presence of LPS. Further investigations discovered that arctigenin inhibits NFκB activation, PI3K, AKT, and IKKβ phosphorylation, along with the inhibition of IL-1β, TNF-α, and IL-6 expression in inflammatory conditions (Hyam et al., 2013). Additionally, arctigenin suppresses the activation of MAP kinases (ERK1/2, p38 kinase, and JNK) by reducing MKK activities, leading to AP-1 inactivation and a decrease in TNF-α production. Arctigenin's antioxidant activity also prevents reactive oxygen species-dependent STAT activation, contributing to its anti-inflammatory effects (Kou, Qi, Dai, Luo, & Yin, 2011).

4.4.8 Ferulic Acid

Ferulic acid belongs to the polyphenol class of natural chemical compounds. Based on available data, ferulic acid might significantly reduce inflammation linked to arthritis in various trials. Ferulic acid-rich phenolic compounds have been shown to considerably reduce adjuvant-induced arthritis (AIA), and dimerization of ferulic acid prevents COX-2 production in LPS-stimulated mouse macrophages (Hirata et al., 2005).

When chondrocytes are co-incubated with sodium ferulate, a sodium salt of ferulic acid, the rate of apoptosis in the chondrocytes is significantly reduced by IL-1β. This improvement occurs alongside a dose-dependent decrease in the expressions of TNF-α, TNFR-1, TNF receptor-associated death domain, caspase-8, and caspase-3 (Qin et al., 2012).

4.4.9 Gallic Acid

Gallic acid (3, 4, 5-trihydroxybenzoic acid) is a natural polyphenolic compound found in grapes, wine, sumac, oak bark, tea leaves, and gall nuts. Emerging evidence suggests that gallic acid possesses anti-RA properties. TNF-treated fibroblast-like synoviocytes isolated from RA (RA-FLS) patients showed dose- and time-dependent inhibition of many proinflammatory genes upon treatment with gallic acid. Moreover, gallic acid causes RA FLS to undergo caspase-3-dependent apoptosis (Chung et al., 2010).

4.5 COMBINATION OF DMARDS FOR THE TREATMENT OF RHEUMATOID ARTHRITIS

The use of combination DMARD therapy in RA is well established and often recommended in clinical practice. This approach involves using two or more DMARDs concurrently to achieve better disease control, reduce symptoms, and slow disease progression compared to monotherapy.

4.5.1 Rationale for Combination DMARD Therapy

4.5.1.1 Enhanced Efficacy

When DMARDs with distinct modes of action are combined, they can work in concert to target various pathways that are implicated in the inflammation and autoimmune response that characterize RA. This strategy aims to inhibit disease activity more thoroughly than is typically possible with a monotherapy.

4.5.1.2 Reduced Dosage and Toxicity

Clinicians may be able to use lower dosages of each drug by combining DMARDs, which may lessen the possibility of side effects linked to greater doses of individual drugs. This is especially critical given that RA treatment is a long-term process.

4.5.1.3 Delay in Disease Progression

Research has indicated that, in contrast to sequential monotherapy, early aggressive treatment with combination DMARD therapy can significantly postpone joint deterioration and preserve joint function.

4.5.2 Frequently Used Combinations of DMARDs

Several combinations of DMARDs are commonly used in clinical practice, based on their efficacy, safety profiles, and synergistic mechanisms as described in Table 4.1.

4.5.2.1 Combination of MTX and Sulfasalazine

This combination is frequently used, particularly in patients who have not responded adequately to MTX alone. Sulfasalazine, an anti-inflammatory DMARD, complements MTX's immunosuppressive effects.

TABLE 4.1
Combinations Used for Antiarthritis Therapy

Combination of Anti-RA Drugs	Outcome	Reference
Combination of csDMARDs		
MTX+Salfasalazine	Improved disease activity with less toxicity	Capell et al. (2007), Haagsma (1994), Dougados et al. (1999), Haagsma et al. (1997)
MTX+hydroxychloroquine	Increased bioavailability of MTX and decreased the disease activity	Ferraz et al. (1994)
MTX+Ciclosporin	More effective than monotherapy	Stein et al. (1997), Tugwell et al. (1995)
MTX+Leflunomide	Decreased disease activity at 24 weeks, however, withdrawals were common due to adverse effects	Kremer et al. (2004)
MTX+Gold	Combination is more effective but adverse effects were also higher	(Williams, 1992)
MTX+Azathioprine	Equal effectiveness either monotherapy or combination but adverse effects were observed more in the case of combination	(Willkens and Stablein (1996)
Salfasalazine+Leflunomide	Slightly effective combination	Dougados et al. (2005)
Salfasalazine+hydroxychloroquine	Similar effect of combination and monotherapy	Faarvang et al. (1993)
Combination of csDMARDs and b/tsDMARDs		
MTX+Abatacept	High disease activity score compared to Abatacept monotherapy	Emery et al. (2015)
MTX+Adalimumab	ACR20/50/70 response rates and DAS28(CRP) remission rates were significantly higher compared to monotherapy	Breedveld et al. (2006)
MTX+Baricitinib	High ACR response rate and disease remission with the combination	Fleischmann et al. (2017)
csDMARD+Certolizumab	High ACR20/50/70 response rate compared to monotherapy	Weinblatt et al. (2012)
MTX+Etanercept	High ACR20/50/70 response rate with the combination	Kameda (2011)

(Continued)

TABLE 4.1 (*Continued*)
Combinations Used for Antiarthritis Therapy

Combination of Anti-RA Drugs	Outcome	Reference
MTX+Tocilizumab	High disease remission compared to the monotherapy	Burmester et al. (2017)
Combination of csDMARDs and phytochemicals		
MTX+Epigallocatechin gallate	Decreased disease activity	Roy et al. (2012)
Dexamethasone+Epigallocatechin gallate	Enhanced antiarthritic activity	Roy et al. (2013)
MTX+Curcumin	Synergistic antiarthritic activity with reduced hepatocellulartoxicity	Banji, Pinnapureddy, Banji, Saidulu, and Hayath (2011)
MTX+Quercetin	Improved disease state with the combination	Haleagrahara et al. (2018), Natarajan, Krithica, Madhan, and Sehgal (2011)
MTX+Resveratrol	Enhanced MTX bioavailability and reduced side effects	Jia et al. (2016)

Out of the four trials found, two examined the combination of sulfasalazine and MTX-like monotherapy with either drug alone (Dougados et al., 1999; Haagsma, Van Riel, De Jong, & Van de Putte, 1997). Randomized controlled trials with double blinding were used in both investigations. When compared to monotherapy of either medication, Haagsma et al. first demonstrated that MTX–sulfasalazine (15 mg/week and 3 g/day, respectively) produced a slight additive effect, which, however, failed to achieve statistical significance (Haagsma et al., 1997). On the other hand, Dougadas et al. demonstrated using a large sample size involving 209 patients that the combination therapy of MTX and sulfasalazine resulted in a statistically significant ($P=0.019$) improvement in the 1-year disease activity score in early RA (Dougados et al., 1999).

The two remaining studies examined a step-up strategy where patients who did not respond well to sulfasalazine were subsequently given MTX. One study was small and not blinded, using a median dose of sulfasalazine at 2.5 g/day and MTX at 15 mg/week, both in combination therapy and monotherapy (86). The other study was conducted on a large sample size; it was blinded and used higher doses of MTX, with median doses for combination therapy being 2.5 g/day for sulfasalazine and 12.5 mg/week for MTX. The two studies demonstrated that combining MTX with sulfasalazine was better than MTX monotherapy. However, the latter MASCOT study, which evaluated combination therapy against each drug alone, did not demonstrate a significant difference in radiological outcomes or functional disability. Nonetheless, it did reveal a notable therapeutic benefit in overall disease management (87).

4.5.2.2 Combination of MTX and Hydroxychloroquine

Hydroxychloroquine, an antimalarial agent with immunomodulatory properties, is often added to MTX to improve control of RA symptoms and reduce the risk of joint damage.

In their study, Ferraz et al. contrasted the effectiveness of MTX (7.5 mg/week) combined with 250 mg/week of chloroquine versus MTX alone (7.5 mg/week) (Ferraz et al., 1994). Although there was no discernible change in overall effectiveness between the groups, the combination group experienced more adverse effects. However, the MTX–chloroquine group performed better in terms of grip strength, painful joint count, and functional ability, compared to the MTX–placebo group ($P=0.034$, 0.038, and 0.036, respectively).

Investigations have also been conducted into the effectiveness of hydroxychloroquine as a down-titration therapy from MTX–hydroxychloroquine combination therapy (Clegg et al., 1997). Following 24 weeks of open treatment, 121 patients who responded to the combination therapy were randomly allocated to either continue hydroxychloroquine or to switch to a placebo. The second round was double-blind. The combined therapy's initial benefits were extended with hydroxychloroquine. Moreover, it has been demonstrated that the combined administration of hydroxychloroquine and MTX increases the drug's bioavailability. While this combination enhances therapeutic benefits, it also increases the potential for toxicity (Carmichael, Beal, Day, & Tett, 2002).

4.5.2.3 Combination of MTX and Ciclosporin

A study involving 148 patients who responded suboptimally to MTX (≤15 mg/week) randomly assigned them to receive either ciclosporin at an average dose of 2.97 mg/kg/day or a placebo (Tugwell et al., 1995). After 24 weeks, 47% of patients given the step-up MTX–cyclosporin combination showed a 20% betterment in the American College of Rheumatology criteria (ACR20), compared to 16% of patients receiving a placebo. Except for hypertrichosis and a rise in serum creatinine levels from baseline, the incidence of side effects was not increased by adding ciclosporin.

For an additional 24 weeks, all patients received the combination as part of the initial double-blind therapy phase (Stein et al., 1997). Almost all outcome markers improved for those who adhered to the MTX–ciclosporin combination regimen. There was a noteworthy improvement in the joint inflammation for individuals who transitioned from MTX–placebo to MTX–ciclosporin. Another trial found that in early RA, MTX–ciclosporin combination therapy was more effective at decreasing disease progression than ciclosporin monotherapy (Gerards et al., 2003). Initially, 120 individuals with active, early-stage RA were randomly allocated to receive either MTX+ciclosporin or ciclosporin alone. Only a few individuals in the combination group met the criteria for remission after 48 weeks, compared to those receiving monotherapy. ACR20, ACR50, and ACR70 responses occurred more frequently in patients administering combination therapy; however, only the difference in ACR50 responses between the two groups was statistically significant. Patients in the monotherapy group had noticeably faster disease progression on serial radiographs. Rates of adverse events were comparable in both groups, with the majority related to ciclosporin (Marchesoni et al., 2002).

Finally, in a 12-month placebo-controlled study, where 61 patients with early RA were randomized to receive either MTX alone or MTX with ciclosporin, the MTX–ciclosporin combination therapy has been shown to reduce radiographic progression compared to MTX monotherapy (Marchesoni et al., 2003).

4.5.2.4 Combination of MTX and Leflunomide

Leflunomide inhibits pyrimidine synthesis and has immunomodulatory effects similar to MTX. This combination is effective for patients who do not achieve adequate disease control with MTX alone.

When leflunomide was used in combination with MTX, combination-treated patients improved more compared to MTX-treated patients in terms of disease activity at 24 weeks. This was evidenced by a considerably higher number of patients reaching ACR20/50/70 responses. Additionally, except for an alteration in the erythrocyte sedimentation rate (ESR), all individual indicators of disease activity indicated a larger improvement in the MTX–leflunomide group (Kremer et al., 2002).

In the MTX-leflunomide group, adverse events were more common, causing drug withdrawal, while withdrawals due to diminished efficacy were more common in the MTX–placebo group. Combination of MTX and leflunomide was more likely to cause mild leukopenia and neutropenia, as well as elevated aminotransferase levels (Kremer et al., 2004).

4.5.2.5 Combination of MTX and Gold

The clinical benefit of MTX–gold combination therapy is comparable to that of both monotherapies. There were no statistical differences among the groups in clinical and laboratory test results, and no patient in any group attained clinical remission. While GI problems were more common in patients getting combination treatment, the frequency of adverse effects was similar across all treatment groups. Despite this, individuals receiving combination therapy experienced a higher number of adverse events that required treatment withdrawal. Conversely, patients receiving combination therapy had fewer withdrawals due to a decrease in treatment efficacy (Williams, 1992).

In a different, single-blind trial, the rate of radiographic progression showed that auranofin (oral gold)-treated patients had significantly worsening radiographic erosion and joint space narrowing scores (Méndez, Daniel, Reading, Ward, & Alarcón, 1993). The MTX and gold combination therapy was evaluated in individuals with established RA who did not respond well to MTX alone (Méndez et al., 1993). ACR20/50/70 responses were significantly more common in patients receiving MTX–gold combination therapy than in patients receiving MTX alone. These benefits remained significant even when concurrent steroid use was considered (with a substantial increase in the placebo group). In both groups, adverse event rates were comparable (Méndez et al., 1993).

4.5.2.6 Combination of MTX and Azathioprine

The same patient cohort was documented in two trials at 24 and 48 weeks (Willkens & Stablein, 1996; Willkens et al., 1992). The RA patient received MTX and azathioprine combination after not responding to other drugs alone. The between-group comparison revealed that, following a 24-week course of treatment, combination treatment produced noticeably greater improvements in patients.

However, high toxicity rates were observed in the combination therapy group at both 24 and 48 weeks. When compared to either monotherapy group, the combination treatment group's high toxicity rates (89% and 78%, respectively) at 24 and 48 weeks were most likely to be responsible for therapy withdrawal (Willkens & Stablein, 1996).

4.5.2.7 Sulfasalazine Combinations

Trials examining the impact of sulfasalazine in conjunction with other DMARDs on RA are comparatively rare. The conclusions drawn from the randomized, placebo-controlled, blinded trials are constrained by their small sample sizes. Patients who had an inadequate response to leflunomide monotherapy showed slight improvement when sulfasalazine was added to their treatment regimen. However, the combination therapy of hydroxychloroquine and sulfasalazine did not show a significantly different response compared to sulfasalazine monotherapy in RA patients (Faarvang et al., 1993). Additionally, sulfasalazine has been safely used with ciprofloxacin and penicillamine, both of which have been shown to improve the disease outcome (Faarvang et al., 1993; Rojkovich, 1999).

4.5.2.8 Triple Therapy (Methotrexate + Sulfasalazine + Hydroxychloroquine)

The management of RA typically involves DMARDs as the cornerstone of treatment. However, recent research has explored the potential benefits of combining DMARDs with phytochemicals derived from plants to enhance therapeutic outcomes and mitigate side effects as described in Table 4.1.

In the sequential monotherapy group, all patients were started on sulfasalazine, then switched to MTX, azathioprine, and finally, any combination of injectable gold, hydroxychloroquine, podophyllotoxin, auranofin, or penicillamine. Compared to sequential monotherapy, triple therapy produced a markedly higher proportion of ACR50 responses. Moreover, a considerably higher proportion of patients using combination therapy met the requirements for ACR remission (Möttönen et al., 1999).

Triple therapy is used as a "step-up" for those patients who do not respond well to monotherapy with any of the DMARDs. Patients showed improved disease status with a higher ACR50 response rate (Ferraccioli et al., 2002). However, it is challenging to conduct accurate between-group comparisons due to volatility and variable dose modifications.

4.6 COMBINATION OF DMARDS AND PHYTOCHEMICALS

Phytochemicals extracted from plants have been used as a treatment for inflammatory diseases such as RA for centuries. The medicinal potential of phytochemicals is frequently underestimated, despite the abundance of scientific data to support it. DMARDs are classified as both biological and synthetic and are used to treat RA. Pneumococcus influenza, herpes zoster, hepatitis, latent tuberculosis, and other serious infections are common adverse effects of long-term DMARD therapy. Therapy discontinuation is occasionally the result of these side effects. A relatively recent concept in the treatment of RA suggests that combining DMARDs with pharmacologically active phytochemicals may offer a more effective option.

Phytochemicals, bioactive compounds derived from plants, have gained attention due to their anti-inflammatory, antioxidant, and immunomodulatory properties. Phytochemicals like curcumin, RES, and EGCG are derived from turmeric, grapes, and green tea, respectively. These compounds have shown promising effects in preclinical studies and early clinical trials for their potential to complement DMARD therapy.

4.6.1 Combination of EGCG and MTX

EGCG is a potent antioxidant and the main catechin found in green tea as described above and has been linked to several treatment advantages in the treatment of cancer, neurological disorders, hypertension, and RA (Clement, 2009; Singh, Shankar, & Srivastava, 2011). Intra-articular injections of EGCG were reported to significantly reduce cartilage degradation in a CIA model, hence improving the physical integrity (Natarajan, Madhan, & Tiku, 2015). Given the research supporting EGCG's antioxidant, anti-inflammatory, and immunosuppressive qualities, it may be a suitable addition to an antirheumatic treatment plan.

To assess the enhanced antiarthritic effects of adding EGCG to dexamethasone, a study was carried out using AIA rats. The immunohistochemistry results showed a reduction in proinflammatory cytokine levels and the edema of the rat paws with combination therapy. Additionally, there was a notable increase in antioxidant effects, bone mineral density, and serum plasma protein levels in the group receiving combination therapy. Histopathological and radiographic investigations further highlighted the superior antiarthritic activity of the combination of dexamethasone and EGCG. Overall, the study indicated that by lowering cytokines and raising antioxidant markers, EGCG enhanced the antiarthritic and antioxidant effects of dexamethasone (Roy, Sannigrahi, Ghosh, Pusp, & Roy, 2013).

AIA rats were used to study the combinatorial effect of EGCG and MTX in the development of RA. For four weeks, AIA rats were administered a weekly dose of MTX plus EGCG. The ESR and hind paw edema showed a considerable decrease in the rats who received the combination therapy. Additionally, analysis of the tibiotarsal joints showed clear joint space, devoid of any symptoms of inflammatory cell infiltration or bone disintegration. While suppressing lipid peroxidation, the combination treatment considerably raised antioxidant activity when compared to monotherapy. Furthermore, the combined therapy showed a significant reduction in proinflammatory cytokines as indicated by immunohistochemical analysis. On the whole, the research suggests that the antiarthritic and antioxidant efficacy of MTX and EGCG combination therapy is enhanced, indicating that EGCG is beneficial when used in combination with other drugs (Roy, Sannigrahi, Vaddepalli, Ghosh, & Pusp, 2012).

4.6.2 Combination of Curcumin and MTX

Curcumin, extracted from the rhizomes of *Curcuma longa*, is renowned for its various pharmacological activities as described above, including anti-inflammatory and antiarthritic properties. Curcumin was shown to be safe and effective (Saif et al., 2024), well-tolerated, and free of adverse effects in a clinical trial that looked into its effectiveness and safety in RA patients. Therefore, assessing this polyphenol's potential as an auxiliary in treating RA is intriguing. In a rat model, the impact of curcumin in conjunction with MTX was assessed for treatment against arthritis.

The combination of MTX and curcumin significantly reduced paw swelling. Additionally, using curcumin with MTX mitigated the adverse effects of MTX on the liver, as evidenced by improved liver function tests. Curcumin also inhibited the increase in lipid peroxides, such as malondialdehyde (MDA), induced by MTX. Curcumin's inhibition of lipid peroxides like MDA induced by MTX reduces oxidative stress and prevents cellular damage, thereby protecting against hepatotoxicity and inflammation. This enhances the overall therapeutic efficacy of the combination while minimizing side effects. The histopathological examination further demonstrated that the combination therapy significantly reduced hepatocellular toxicity, including fibrosis and inflammation.

Consequently, the study supports curcumin's capacity to work synergistically with MTX to prevent hepatocellular injury while exerting an antiarthritic effect. The combination therapy's synergistic impact may allow for a decrease in MTX dosage without sacrificing treatment efficacy for RA.

4.6.3 Combination of Quercetin and MTX

Several studies have shown quercetin to be an anti-inflammatory and antioxidant agent, making it a promising candidate for the treatment of RA. Thus, combining quercetin with synthetic DMARDs like MTX may provide enhanced therapeutic benefits. In an inflammatory arthritis mouse model, the combination therapy of quercetin and MTX reduced disease-related weight loss, minimized paw inflammation, and significantly decreased the mean arthritis score. Proinflammatory cytokine levels were notably lowered by the combination treatment, along with IL-6 levels. Additionally, this combined therapy exhibited a significant reduction in the breakdown of bone and cartilage, indicating a synergistic antirheumatic effect.

Moreover, quercetin improved the adverse effects of MTX on the intestine by inhibiting apoptosis and promoting enterocyte proliferation, preventing intestinal damage, and improving intestinal healing. The intestinal p-ERK protein expression was significantly elevated in the combination therapy group, leading to increased cell proliferation (Smolen et al., 2017). The combination also decreased caspase-3 expression, resulting in the inhibition of apoptosis. Overall, quercetin can protect against intestinal damage caused by MTX.

4.6.4 Combination of Resveratrol and MTX

Resveratrol (RES) possesses antioxidant properties and has been demonstrated to protect against metabolic, neurological, and cardiovascular diseases due to its impact on arachidonic metabolism. Studies on RSV demonstrated that it has anti-inflammatory and antioxidant activities in synoviocytes through *in vitro* research, and immunomodulatory properties on T and B cells are particularly relevant to rheumatic illnesses. RSV treatment has been shown to reduce autoantibody synthesis, Th17 cells in draining lymph nodes, and proinflammatory cytokine levels, thereby modulating humoral and cellular responses (Xuzhu et al., 2012). Combining RSV with current RA therapy, such as MTX, may offer enhanced therapeutic benefits. Research revealed that RSV inhibits P-gp and MRP-2, increasing the intestinal absorption of MTX. Furthermore, combination therapy increased MTX plasma exposure, extended its half-life, and reduced clearance. The study also demonstrated reduced nephrotoxicity when the combination was administered compared to MTX alone (Jia et al., 2016).

When comparing RSV+MTX and RSV+adalimumab combinations to MTX or adalimumab alone, the levels of MCP-1 were considerably reduced. RSV plus MTX has a significant lowering effect on MCP-1 levels in individuals with moderate disease activity (Lomholt, Mellemkjær, Iversen, Pedersen, & Kragstrup, 2018). Overall, the research supports the idea that RSV would be a useful supplementary option to MTX, enhancing MTX absorption, decreasing side effects, and minimizing the risks linked to prolonged treatment of RA.

4.7 COMBINATION OF CSDMARDS WITH BDMARDS

According to European League Against Rheumatoid Arthritis (EULAR) recommendations, the first-line treatment combination for RA is MTX together with glucocorticoids. If this combination effectively reduces RA symptoms, the treatment decision is based on the disease prognosis. If there is no production of autoantibodies and the disease activity is slow, patients are either switched to another csDMARD or a combination of two csDMARDs is recommended. If autoantibodies are present and high disease activity is observed, the addition of bDMARD (TNFi) or tsDMARD (JAK inhibitor) to csDMARD is advised. If this combination also fails, switching to another bDMARD or tsDMARD is an option. However, better treatment outcomes are observed with the combination therapy of csDMARDs with bDMARD or tsDMARD. The use of MTX in combination enhances the bioavailability of bDMARDs, assisting in the reduction of inflammation and radiographic progression of the disease.

Efficacy analysis of the combination of MTX and adalimumab demonstrated that after 104 weeks of treatment, the combination therapy showed significantly improved ACR20/50/70 response rates and DAS28 remission rates compared to MTX and adalimumab monotherapy (Breedveld et al., 2006). Similarly, other studies have shown that the combination of MTX and baricitinib is more effective in achieving ACR20/50/70 response and DAS28 disease remission compared to MTX alone (Fleischmann et al., 2017). The combinations of MTX with etanercept (Kameda, 2011) or MTX with tocilizumab (Burmester et al., 2017) have also proven to be effective in achieving higher rates of ACR20/50/70 responses and disease remission compared to their respective monotherapies.

Curtis et al. studied two combination strategies including triple therapy with three different csDMARDs (MTX, sulfasalazine, and hydroxychloroquine) and combination therapy with MTX and a TNF inhibitor (TNFi). Although the sample size for triple therapy was very low because few people initiated this therapy, dual therapy of MTX and TNFi was found to be effective, showing persistent and clinically effective outcomes compared with patients receiving triple therapy (Curtis et al., 2020).

4.8 CONCLUSION: A TREAT-TO-TARGET APPROACH

Historically, the Western world has undervalued the medicinal potential of natural substances derived from plants, such as flavonoids, polyphenols, and alkaloids. Despite abundant scientific evidence supporting the anti-inflammatory, antiarthritic, and immunomodulatory properties of these plant-based chemicals, their capacity for treating autoimmune disorders like RA is often overlooked. Numerous *in vitro*, *in vivo*, and clinical studies have verified the usefulness of combining phytochemicals with DMARDs, which are known to enhance the efficacy of conventional anti-rheumatic treatments, both synthetic and biological, unquestionably offering promise as therapeutic agents for RA. Nonetheless, the side effects of long-term RA medication, including hepatotoxicity, nephrotoxicity, as well as pulmonary and GI toxicities, cannot be ignored. Incorporating phytochemicals into RA treatment regimens could reduce these adverse effects and improve overall therapeutic outcomes.

Additionally, bDMARDs are expensive and delivered through intravenous or subcutaneous injections, which pose risks of viral or bacterial infections such as tuberculosis, herpes zoster, and pneumonitis. bDMARD therapy is recommended only when MTX alone or in combination with the first-line standard DMARDs fails to yield a satisfactory response. Phytochemicals present promising options for combination therapy with standard csDMARDs due to their ease of administration, better acceptance, strong safety profile, and cost-effectiveness. This therapeutic intervention seeks to improve the efficacy, safety, and tolerability of long-term RA treatment while minimizing side effects.

To fully understand the potential of these combinations in enhancing RA therapy, further research through more comprehensive clinical and experimental trials is necessary. Additionally, the use of small-molecule DMARDs, such as tofacitinib and baricitinib (JAK/STAT inhibitors), represents another emerging trend in RA treatment. Nevertheless, these DMARDs are linked to thrombosis, liver and renal damage, and severe infections. This creates a basis for exploring other phytochemicals as potential JAK/STAT pathway inhibitors in conjunction with conventional DMARDs to improve tolerance. To enhance existing RA therapy, it is important to investigate the use of phytochemicals as JAK/STAT inhibitors in combination with DMARDs, particularly in light of the unclear future of these agents. This chapter also provides an overview of addressing drug delivery challenges related to phytochemicals, recommending the progress in the drug development of phytochemicals and their effectiveness when used with a single or combination of DMARDs for patients who do not respond to monotherapies.

As csDMARD and bDMARDs can be used as a combination therapy and their doses can be adjusted according to clinical outcomes, the concept of "treat-to-target" (TTT) has developed in recent years. This strategy includes (i) choosing a goal for the therapy which may be complete disease remission or lowering disease activity, (ii) assessment of disease activity (DAS28) every 2–3 months, (iii) using b/tsDMARDs along with csDMARDs to reach the goal of disease remission, and (iv) co-administration of glucocorticoids for a short time, if needed.

While the TTT strategy is effective for managing RA, combining DMARDs with phytochemicals offers a promising alternative. TTT involves setting specific therapeutic goals and using combinations of DMARDs, often with bSMARDs or tsDMARDs, but it can be costly and carry significant side effects.

Phytochemicals, with their anti-inflammatory, antioxidant, and immunomodulatory properties, can complement conventional DMARDs, potentially reducing the required dosage and associated side effects. This combined approach aims to enhance disease control and improve patient quality of life by integrating the benefits of both conventional and natural therapies. Further research is needed to optimize these combinations and fully realize their potential in RA treatment.

REFERENCES

Ahmed, S., Pakozdi, A., & Koch, A. E. (2006). Regulation of interleukin-1β–induced chemokine production and matrix metalloproteinase 2 activation by epigallocatechin-3-gallate in rheumatoid arthritis synovial fibroblasts. *Arthritis & Rheumatism: Official Journal of the American College of Rheumatology, 54*(8), 2393–2401.

Al-Salama, Z. T., & Scott, L. J. (2018). Baricitinib: a review in rheumatoid arthritis. *Drugs, 78*, 761–772.

Aletaha, D., & Smolen, J. S. (2018). Diagnosis and management of rheumatoid arthritis: a review. *JAMA, 320*(13), 1360–1372. https://doi.org/10.1001/jama.2018.13103

Alhammadi, N., Asiri, A. H., Alshahrani, F. M., Alqahtani, A. Y., Al Qout, M. M., Alnami, R. A., … Alqahtani, A. (2022). Gastrointestinal complications associated with non-steroidal anti-inflammatory drug use among adults: a retrospective, single-center study. *Cureus, 14*(6), 1–12.

Angelini, J., Talotta, R., Roncato, R., Fornasier, G., Barbiero, G., Dal Cin, L., … Scaglione, F. (2020). JAK-inhibitors for the treatment of rheumatoid arthritis: a focus on the present and an outlook on the future. *Biomolecules, 10*(7), 1002.

Arneson, L. C., Carroll, K. J., & Ruderman, E. M. (2021). Bruton's tyrosine kinase inhibition for the treatment of rheumatoid arthritis. *ImmunoTargets and Therapy, 28*(10), 333–342.

Ba, X., Huang, Y., Shen, P., Huang, Y., Wang, H., Han, L., … Qin, K. (2021). WTD attenuating rheumatoid arthritis via suppressing angiogenesis and modulating the PI3K/AKT/mTOR/HIF-1α pathway. *Frontiers in Pharmacology, 12*, 696802.

Bai, L., Bai, Y., Yang, Y., Zhang, W., Huang, L., Ma, R., … Wan, Q. (2020). Baicalin alleviates collagen-induced arthritis and suppresses TLR2/MYD88/NF-κB p65 signaling in rats and HFLS-RAs. *Molecular Medicine Reports, 22*(4), 2833–2841.

Banji, D., Pinnapureddy, J., Banji, O. J. F., Saidulu, A., & Hayath, M. S. (2011). Synergistic activity of curcumin with methotrexate in ameliorating Freund's Complete Adjuvant induced arthritis with reduced hepatotoxicity in experimental animals. *European Journal of Pharmacology, 668*(1), 293–298. https://doi.org/10.1016/j.ejphar.2011.06.006

Bathon, J. M., Martin, R. W., Fleischmann, R. M., Tesser, J. R., Schiff, M. H., Keystone, E. C., … Weaver, A. L. (2000). A comparison of etanercept and methotrexate in patients with early rheumatoid arthritis. *New England Journal of Medicine, 343*(22), 1586–1593.

Benjamin, O., Goyal, A., & Lappin, S. L. (2024). Disease-modifying antirheumatic drugs (DMARD). In B. K. Nimmana & Preeti Patel (Eds.), *StatPearls*. StatPearls Publishing.

Bennike, T. B., Ellingsen, T., Glerup, H., Bonderup, O. K., Carlsen, T. G., Meyer, M. K., … Andersen, V. (2017). Proteome analysis of rheumatoid arthritis gut mucosa. *Journal of Proteome Research, 16*(1), 346–354.

Bolli, R., Dawn, B., & Xuan, Y.-T. (2003). Role of the JAK–STAT pathway in protection against myocardial ischemia/reperfusion injury. *Trends in Cardiovascular Medicine, 13*(2), 72–79.

Breedveld, F. C., Weisman, M. H., Kavanaugh, A. F., Cohen, S. B., Pavelka, K., van Vollenhoven, R., … Spencer-Green, G. T. (2006). The PREMIER study: a multicenter, randomized, double-blind clinical trial of combination therapy with adalimumab plus methotrexate versus methotrexate alone or adalimumab alone in patients with early, aggressive rheumatoid arthritis who had not had previous methotrexate treatment. *Arthritis Rheumatology, 54*(1), 26–37. https://doi.org/10.1002/art.21519

Burmester, G. R., Rigby, W. F., Van Vollenhoven, R. F., Kay, J., Rubbert-Roth, A., Blanco, R., ... Dimonaco, S. (2017). Tocilizumab combination therapy or monotherapy or methotrexate monotherapy in methotrexate-naive patients with early rheumatoid arthritis: 2-year clinical and radiographic results from the randomised, placebo-controlled FUNCTION trial. *Annals of the Rheumatic Diseases, 76*(7), 1279–1284.

Capell, H. A., Madhok, R., Porter, D. R., Munro, R. A., McInnes, I. B., Hunter, J. A., ... Sambrook, M. (2007). Combination therapy with sulfasalazine and methotrexate is more effective than either drug alone in patients with rheumatoid arthritis with a suboptimal response to sulfasalazine: results from the double-blind placebo-controlled MASCOT study. *Annals of the Rheumatic Diseases, 66*(2), 235–241.

Cargnello, M., & Roux, P. P. (2011). Activation and function of the MAPKs and their substrates, the MAPK-activated protein kinases. *Microbiology and Molecular Biology Reviews, 75*(1), 50–83.

Carmichael, S. J., Beal, J., Day, R. O., & Tett, S. E. (2002). Combination therapy with methotrexate and hydroxychloroquine for rheumatoid arthritis increases exposure to methotrexate. *The Journal of Rheumatology, 29*(10), 2077–2083.

Chakrabarti, S., Hore, Z., Pattison, L. A., Lalnunhlimi, S., Bhebhe, C. N., Callejo, G., ... Smith, E. S. J. (2020). Sensitization of knee-innervating sensory neurons by tumor necrosis factor-α-activated fibroblast-like synoviocytes: an in vitro, coculture model of inflammatory pain. *Pain, 161*(9), 2129–2141. https://doi.org/10.1097/j.pain.0000000000001890

Chandran, B., & Goel, A. (2012). A randomized, pilot study to assess the efficacy and safety of curcumin in patients with active rheumatoid arthritis. *Phytotherapy Research, 26*(11), 1719–1725.

Chen, S., Yang, Y., Feng, H., Wang, H., Zhao, R., & Liu, H. (2014). Baicalein inhibits interleukin-1β-induced proliferation of human rheumatoid arthritis fibroblast-like synoviocytes. *Inflammation, 37*(1), 163–169.

Chung, S.-J., Kim, T.-Y., Kwon, Y., Park, Y., Lee, S., & Park, M. (2010). Gallic acid diminishes cellular proliferation and pro-inflammatory gene expressions in fibroblast like synoviocytes from patients with rheumatoid arthritis. *Arthritis Rheumatology, 62*, 27.

Clegg, D., Dietz, F., Duffy, J., Willkens, R., Hurd, E., Germain, B., ... Sleckman, J. (1997). Safety and efficacy of hydroxychloroquine as maintenance therapy for rheumatoid arthritis after combination therapy with methotrexate and hydroxychloroquine. *The Journal of Rheumatology, 24*(10), 1896–1902.

Clement, Y. (2009). Can green tea do that? A literature review of the clinical evidence. *Preventive Medicine, 49*(2), 83–87. https://doi.org/10.1016/j.ypmed.2009.05.005

Corneth, O. B., Klein Wolterink, R. G., & Hendriks, R. W. (2015). BTK signaling in B cell differentiation and autoimmunity. *B Cell Receptor Signaling, 393*, 67–105.

Curtis, J., Palmer, J., Reed, G., Greenberg, J., Pappas, D., & Harrold, L. (2020). Real-world outcomes associated with triple therapy versus TNFi/MTX therapy. *Arthritis Care Research, 73*(8), 1114–1124.

Dhillon, S. (2017). Tofacitinib: a review in rheumatoid arthritis. *Drugs, 77*, 1987–2001.

Dinesh, P., & Rasool, M. (2018). Berberine inhibits IL-21/IL-21R mediated inflammatory proliferation of fibroblast-like synoviocytes through the attenuation of PI3K/Akt signaling pathway and ameliorates IL-21 mediated osteoclastogenesis. *Cytokine, 106*, 54–66.

Dougados, M., Combe, B., Cantagrel, A., Goupille, P., Olive, P., Schattenkirchner, M., ... Zeidler, H. (1999). Combination therapy in early rheumatoid arthritis: a randomised, controlled, double blind 52 week clinical trial of sulphasalazine and methotrexate compared with the single components. *Annals of the Rheumatic Diseases, 58*(4), 220–225.

Dougados, M., Emery, P., Lemmel, E., Zerbini, C., Brin, S., & van Riel, P. (2005). When a DMARD fails, should patients switch to sulfasalazine or add sulfasalazine to continuing leflunomide? *Annals of the Rheumatic Diseases, 64*(1), 44–51.

Emery, P., Burmester, G. R., Bykerk, V. P., Combe, B. G., Furst, D. E., Barré, E., ... Huizinga, T. W. (2015). Evaluating drug-free remission with abatacept in early rheumatoid arthritis: results from the phase 3b, multicentre, randomised, active-controlled AVERT study of 24 months, with a 12-month, double-blind treatment period. *Annals of the Rheumatic Diseases, 74*(1), 19–26.

Endale, M., Park, S.-C., Kim, S., Kim, S.-H., Yang, Y., Cho, J. Y., & Rhee, M. H. (2013). Quercetin disrupts tyrosine-phosphorylated phosphatidylinositol 3-kinase and myeloid differentiation factor-88 association, and inhibits MAPK/AP-1 and IKK/NF-κB-induced inflammatory mediators production in RAW 264.7 cells. *Immunobiology, 218*(12), 1452–1467.

Faarvang, K., Egsmose, C., Kryger, P., Pødenphant, J., Ingeman-Nielsen, M., & Hansen, T. (1993). Hydroxychloroquine and sulphasalazine alone and in combination in rheumatoid arthritis: a randomised double blind trial. *Annals of the Rheumatic Diseases, 52*(10), 711–715.

Fan, J., Jiang, T., & He, D. (2023). Advances in the implications of the gut microbiota on the treatment efficacy of disease-modifying anti-rheumatic drugs in rheumatoid arthritis. *Frontiers in Immunology, 14*, 1189036. https://doi.org/10.3389/fimmu.2023.1189036

Fechtner, S., Singh, A., Chourasia, M., & Ahmed, S. (2017). Molecular insights into the differences in anti-inflammatory activities of green tea catechins on IL-1β signaling in rheumatoid arthritis synovial fibroblasts. *Toxicology and Applied Pharmacology, 329*, 112–120.

Fernández-Rodríguez, J. A., Almonte-Becerril, M., Ramil-Gómez, O., Hermida-Carballo, L., Viñas-Diz, S., Vela-Anero, Á., … López-Armada, M. J. (2021). Autophagy activation by resveratrol reduces severity of experimental rheumatoid arthritis. *Molecular Nutrition & Food Research, 65*(2), 2000377.

Ferraccioli, G., Gremese, E., Tomietto, P., Favret, G., Damato, R., & Di Poi, E. (2002). Analysis of improvements, full responses, remission and toxicity in rheumatoid patients treated with step-up combination therapy (methotrexate, cyclosporin A, sulphasalazine) or monotherapy for three years. *Rheumatology, 41*(8), 892–898.

Ferraz, M., Pinheiro, G., Helfenstein, M., Albuquerque, E., Rezende, C., Roimicher, L., … Atra, E. (1994). Combination therapy with methotrexate and chloroquine in rheumatoid arthritis: a multicenter randomized placebo-controlled trial. *Scandinavian Journal of Rheumatology, 23*(5), 231–236.

Fleischmann, R., Schiff, M., van der Heijde, D., Ramos-Remus, C., Spindler, A., Stanislav, M., … Takeuchi, T. (2017). Baricitinib, methotrexate, or combination in patients with rheumatoid arthritis and no or limited prior disease-modifying antirheumatic drug treatment. *Arthritis and Rheumatism, 69*(3), 506–517. https://doi.org/10.1002/art.39953

Fletcher, A., Lassere, M., March, L., Hill, C., Barrett, C., Carroll, G., & Buchbinder, R. (2022). Patterns of biologic and targeted-synthetic disease-modifying antirheumatic drug use in rheumatoid arthritis in Australia. *Rheumatology, 61*(10), 3939–3951. https://doi.org/10.1093/rheumatology/keac048

Funk, J. L., Oyarzo, J. N., Frye, J. B., Chen, G., Lantz, R. C., Jolad, S. D., … Timmermann, B. N. (2006). Turmeric extracts containing curcuminoids prevent experimental rheumatoid arthritis. *Journal of Natural Products, 69*(3), 351–355.

Gaballa, S., & Pinilla-Ibarz, J. (2021). BTK inhibitors in chronic lymphocytic leukemia. *Current Hematologic Malignancy Reports, 16*(5), 422–432.

Gabizon, R., & London, N. (2020). A fast and clean BTK inhibitor. *Journal of Medicinal Chemistry, 63*(10), 5100–5101.

Gaffo, A., Saag, K. G., & Curtis, J. R. (2006). Treatment of rheumatoid arthritis. *American Journal of Health-System Pharmacy, 63*(24), 2451–2465. https://doi.org/10.2146/ajhp050514

Gerards, A., Landewe, R., Prins, A., Bruijn, G., Laan, R., & Dijkmans, B. (2003). Cyclosporin A monotherapy versus cyclosporin A and methotrexate combination therapy in patients with early rheumatoid arthritis: a double blind randomised placebo controlled trial. *Annals of the Rheumatic Diseases, 62*(4), 291–296.

Guo, Q., Wang, Y., Xu, D., Nossent, J., Pavlos, N. J., & Xu, J. (2018). Rheumatoid arthritis: pathological mechanisms and modern pharmacologic therapies. *Bone Research, 6*, 15. https://doi.org/10.1038/s41413-018-0016-9

Haagsma, C. J. (1994). Combination of methotrexate and sulphasalazine vs methotrexate alone: a randomized open clinical trial in rheumatoid arthritis patients persistent to sulphasalazine therapy. *British Journal of Rheumatology, 33*, 1049–1055.

Haagsma, C. J., Van Riel, P., De Jong, A., & Van de Putte, L. (1997). Combination of sulphasalazine and methotrexate versus the single components in early rheumatoid arthritis: a randomized, controlled, double-blind, 52 week clinical trial. *British Journal of Rheumatology, 36*(10), 1082–1088.

Haleagrahara, N., Hodgson, K., Miranda-Hernandez, S., Hughes, S., Kulur, A. B., & Ketheesan, N. (2018). Flavonoid quercetin-methotrexate combination inhibits inflammatory mediators and matrix metalloproteinase expression, providing protection to joints in collagen-induced arthritis. *Inflammopharmacology, 26*(5), 1219–1232. https://doi.org/10.1007/s10787-018-0464-2

Hirata, A., Murakami, Y., Atsumi, T., Shoji, M., Ogiwara, T., Shibuya, K., … Fujisawa, S. (2005). Ferulic acid dimer inhibits lipopolysaccharide-stimulated cyclooxygenase-2 expression in macrophages. *In Vivo, 19*(5), 849–853.

Hyam, S. R., Lee, I.-A., Gu, W., Kim, K.-A., Jeong, J.-J., Jang, S.-E., … Kim, D.-H. (2013). Arctigenin ameliorates inflammation in vitro and in vivo by inhibiting the PI3K/AKT pathway and polarizing M1 macrophages to M2-like macrophages. *European Journal of Pharmacology, 708*(1–3), 21–29.

Javadi, F., Ahmadzadeh, A., Eghtesadi, S., Aryaeian, N., Zabihiyeganeh, M., Rahimi Foroushani, A., & Jazayeri, S. (2017). The effect of quercetin on inflammatory factors and clinical symptoms in women with rheumatoid arthritis: a double-blind, randomized controlled trial. *Journal of the American College of Nutrition, 36*(1), 9–15.

Ji, J.-J., Lin, Y., Huang, S.-S., Zhang, H.-L., Diao, Y.-P., & Li, K. (2013). Quercetin: a potential natural drug for adjuvant treatment of rheumatoid arthritis. *African Journal of Traditional, Complementary, and Alternative Medicines, 10*(3), 418.

Jia, Y., Liu, Z., Wang, C., Meng, Q., Huo, X., Liu, Q., … Liu, K. (2016). P-gp, MRP2 and OAT1/OAT3 mediate the drug-drug interaction between resveratrol and methotrexate. *Toxicology and Applied Pharmacology, 306*, 27–35. https://doi.org/10.1016/j.taap.2016.06.030

Kameda, H. (2011). Etanercept (ETN) plus methotrexate (MTX) combination therapy resulted in a better outcome in joint damage and physical function than ETN monotherapy even in patients with active rheumatoid arthritis despite MTX treatment: 52-week results from the JESMR study. *Journal of Rheumatology, 20*(6), 531–538.

Kim, H.-R., Kim, B.-M., Won, J.-Y., Lee, K.-A., Ko, H. M., Kang, Y. S., … Kim, K.-W. (2019). Quercetin, a plant polyphenol, has potential for the prevention of bone destruction in rheumatoid arthritis. *Journal of Medicinal Food, 22*(2), 152–161.

Kou, X., Qi, S., Dai, W., Luo, L., & Yin, Z. (2011). Arctigenin inhibits lipopolysaccharide-induced iNOS expression in RAW264. 7 cells through suppressing JAK-STAT signal pathway. *International Immunopharmacology, 11*(8), 1095–1102.

Kremer, J., Genovese, M., Cannon, G. W., Caldwell, J., Cush, J., Furst, D. E., … Kavanaugh, A. (2004). Combination leflunomide and methotrexate (MTX) therapy for patients with active rheumatoid arthritis failing MTX monotherapy: open-label extension of a randomized, double-blind, placebo controlled trial. *The Journal of Rheumatology, 31*(8), 1521–1531.

Kremer, J. M., Genovese, M. C., Cannon, G. W., Caldwell, J. R., Cush, J. J., Furst, D. E., … Bensen, W. M. (2002). Concomitant leflunomide therapy in patients with active rheumatoid arthritis despite stable doses of methotrexate: a randomized, double-blind, placebo-controlled trial. *Annals of Internal Medicine, 137*(9), 726–733.

Kwok, S. K., Cho, M. L., Park, M. K., Oh, H. J., Park, J. S., Her, Y. M., … Park, K. S. (2012). Interleukin-21 promotes osteoclastogenesis in humans with rheumatoid arthritis and in mice with collagen-induced arthritis. *Arthritis & Rheumatism, 64*(3), 740–751.

Lee, J.-E., Kim, I. J., Cho, M.-S., & Lee, J. (2017). A case of rheumatoid vasculitis involving hepatic artery in early rheumatoid arthritis. *Journal of Korean Medical Science, 32*(7), 1207.

Lee, J.-H., Zhou, H. Y., Cho, S. Y., Kim, Y. S., Lee, Y. S., & Jeong, C. S. (2007). Anti-inflammatory mechanisms of apigenin: inhibition of cyclooxygenase-2 expression, adhesion of monocytes to human umbilical vein endothelial cells, and expression of cellular adhesion molecules. *Archives of Pharmacal Research, 30*, 1318–1327.

Li, G.-Q., Zhang, Y., Liu, D., Qian, Y.-Y., Zhang, H., Guo, S.-Y., … Liu, Y.-Q. (2013). PI3 kinase/Akt/HIF-1α pathway is associated with hypoxia-induced epithelial–mesenchymal transition in fibroblast-like synoviocytes of rheumatoid arthritis. *Molecular and Cellular Biochemistry, 372*, 221–231.

Liu, D., Ahmet, A., Ward, L., Krishnamoorthy, P., Mandelcorn, E. D., Leigh, R., … Kim, H. (2013). A practical guide to the monitoring and management of the complications of systemic corticosteroid therapy. *Allergy, Asthma & Clinical Immunology, 9*, 1–25.

Lomholt, S., Mellemkjær, A., Iversen, M. B., Pedersen, S. B., & Kragstrup, T. W. (2018). Resveratrol displays anti-inflammatory properties in an ex vivo model of immune mediated inflammatory arthritis. *BMC Rheumatology, 2*, 1–10.

Marchesoni, A., Battafarano, N., Arreghini, M., Panni, B., Gallazzi, M., & Tosi, S. (2003). Radiographic progression in early rheumatoid arthritis: a 12-month randomized controlled study comparing the combination of cyclosporin and methotrexate with methotrexate alone. *Rheumatology, 42*(12), 1545–1549.

Marchesoni, A., Battafarano, N., Arreghini, M., Pellerito, R., Cagnoli, M., Prudente, P., … Tosi, S. (2002). Step-down approach using either cyclosporin A or methotrexate as maintenance therapy in early rheumatoid arthritis. *Arthritis Care & Research: Official Journal of the American College of Rheumatology, 47*(1), 59–66.

Méndez, A. L., Daniel, W. W., Reading, J. C., Ward, J. R., & Alarcón, G. S. (1993). Radiographic assessment of disease progression in rheumatoid arthritis patients enrolled in the cooperative systematic studies of the rheumatic diseases program randomized clinical trial of methotrexate, auranofin, or a combination of the two. *Arthritis & Rheumatism: Official Journal of the American College of Rheumatology, 36*(10), 1364–1369.

Miao, C. G., Yang, Y. Y., He, X., Li, X. F., Huang, C., Huang, Y., … Li, J. (2013). Wnt signaling pathway in rheumatoid arthritis, with special emphasis on the different roles in synovial inflammation and bone remodeling. *Cell Signal, 25*(10), 2069–2078. https://doi.org/10.1016/j.cellsig.2013.04.002

Monu, Agnihotri, P., Saquib, M., & Biswas, S. (2024). Targeting TNF-α-induced expression of TTR and RAGE in rheumatoid arthritis: apigenin's mediated therapeutic approach. *Cytokine, 179*, 156616. https://doi.org/10.1016/j.cyto.2024.156616

Möttönen, T., Hannonen, P., Leirisalo-Repo, M., Nissilä, M., Kautiainen, H., Korpela, M., ... Vuori, K. (1999). Comparison of combination therapy with single-drug therapy in early rheumatoid arthritis: a randomised trial. *The Lancet, 353*(9164), 1568–1573.

Natarajan, V., Krithica, N., Madhan, B., & Sehgal, P. K. (2011). Formulation and evaluation of quercetin polycaprolactone microspheres for the treatment of rheumatoid arthritis. *Journal of Pharmaceutical Sciences, 100*(1), 195–205. https://doi.org/10.1002/jps.22266

Natarajan, V., Madhan, B., & Tiku, M. L. (2015). Intra-articular injections of polyphenols protect articular cartilage from inflammation-induced degradation: suggesting a potential role in cartilage therapeutics. *PloS One, 10*(6), e0127165.

Nepom, G. T. (1998). Major histocompatibility complex-directed susceptibility to rheumatoid arthritis. *Advances in Immunology, 68*, 315–332. https://doi.org/10.1016/s0065-2776(08)60563-5

Oike, T., Sato, Y., Kobayashi, T., Miyamoto, K., Nakamura, S., Kaneko, Y., ... Matsumoto, M. (2017). Stat3 as a potential therapeutic target for rheumatoid arthritis. *Scientific Reports, 7*(1), 10965.

Oliveira, A. L. D. B., Monteiro, V. V. S., Navegantes-Lima, K. C., Reis, J. F., Gomes, R. D. S., Rodrigues, D. V. S., ... Monteiro, M. C. (2017). Resveratrol role in autoimmune disease—a mini-review. *Nutrients, 9*(12), 1306.

Ong, C., Lirk, P., Tan, C., & Seymour, R. (2007). An evidence-based update on nonsteroidal anti-inflammatory drugs. *Clinical Medicine & Research, 5*(1), 19–34.

Park, E. J., Kim, H., Jung, S. M., Sung, Y. K., Baek, H. J., & Lee, J. (2020). The use of biological disease-modifying antirheumatic drugs for inflammatory arthritis in Korea: results of a Korean Expert Consensus. *Korean Journal of Internal Medicine, 35*(1), 41–59. https://doi.org/10.3904/kjim.2019.411

Qin, J., Shang, L., Ping, A.-S., Li, J., Li, X.-j., Yu, H., ... Wang, H. (2012). TNF/TNFR signal transduction pathway-mediated anti-apoptosis and anti-inflammatory effects of sodium ferulate on IL-1β-induced rat osteoarthritis chondrocytes in vitro. *Arthritis Research & Therapy, 14*, 1–12.

Ralph, J. A., & Morand, E. F. (2008). MAPK phosphatases as novel targets for rheumatoid arthritis. *Expert Opinion on Therapeutic Targets, 12*(7), 795–808.

Rheumatoid Arthritis Clinical Trial Archive Group. (1995). The effect of age and renal function on the efficacy and toxicity of methotrexate in rheumatoid arthritis. *Journal of Rheumatology, 22*(2), 218–223.

Richards, B. L., Whittle, S. L., van der Heijde, D. M., & Buchbinder, R. (2012). The efficacy and safety of antidepressants in inflammatory arthritis: a cochrane systematic review. *The Journal of Rheumatology Supplement, 90*, 21–27.

Rojkovich, L. H., Bálint, G., Szegedi, G., Varjú, T., Tamási, L., Molnár, É., Szilágyi, M., Szocsik, K. B. (1999). Cyclosporin and sulfasalazine combination in the treatment of early rheumatoid arthritis. *Scandinavian Journal of Rheumatology, 28*(4), 216–221.

Roshak, A. K., Callahan, J. F., & Blake, S. M. (2002). Small-molecule inhibitors of NF-κB for the treatment of inflammatory joint disease. *Current Opinion in Pharmacology, 2*(3), 316–321.

Rothhut, B., & Russo-Marie, F. (1984). Novel concepts in the mode of action of anti-inflammatory steroids. *Agents and Actions Supplements, 14*, 171–180.

Roy, S., Sannigrahi, S., Ghosh, B., Pusp, P., & Roy, T. (2013). Combination therapy of dexamethasone with epigallocatechin enhances tibiotarsal bone articulation and modulates oxidative status correlates with cartilage cytokines expression in the early phase of experimental arthritis. *European Journal of Pharmacology, 698*(1–3), 444–454.

Roy, S., Sannigrahi, S., Vaddepalli, R. P., Ghosh, B., & Pusp, P. (2012). A novel combination of methotrexate and epigallocatechin attenuates the overexpression of pro-inflammatory cartilage cytokines and modulates antioxidant status in adjuvant arthritic rats. *Inflammation, 35*(4), 1435–1447. https://doi.org/10.1007/s10753-012-9457-2

Saag, K. G., Koehnke, R., Caldwell, J. R., Brasington, R., Burmeister, L. F., Zimmerman, B., ... Furst, D. E. (1994). Low dose long-term corticosteroid therapy in rheumatoid arthritis: an analysis of serious adverse events. *American Journal of Medicine, 96*(2), 115–123. https://doi.org/10.1016/0002-9343(94)90131-7

Saif, A., Omer, M. O., Sattar, A., Tipu, Y., Alharbi, H. M., Saher, U., & Awan, T. (2024). Comprehensive analysis of curcumin zinc oxide nanoparticles, synthesis, characterization, and cytogenotoxic profiling. *ACS Omega, 9*(26), 28186–28193.

Serhal, L., & Edwards, C. J. (2019). Upadacitinib for the treatment of rheumatoid arthritis. *Expert Review of Clinical Immunology, 15*(1), 13–25.

Shinohara, M., Koga, T., Okamoto, K., Sakaguchi, S., Arai, K., Yasuda, H., ... Geha, R. S. (2008). Tyrosine kinases Btk and Tec regulate osteoclast differentiation by linking RANK and ITAM signals. *Cell, 132*(5), 794–806.

Silman, A. J., & Pearson, J. E. (2002). Epidemiology and genetics of rheumatoid arthritis. *Arthritis Research & Therapy, 4*, 1–8.

Singh, B. N., Shankar, S., & Srivastava, R. K. (2011). Green tea catechin, epigallocatechin-3-gallate (EGCG): mechanisms, perspectives and clinical applications. *Biochemical Pharmacology, 82*(12), 1807–1821. https://doi.org/10.1016/j.bcp.2011.07.093

Smedslund, G., Byfuglien, M. G., Olsen, S. U., & Hagen, K. B. (2010). Effectiveness and safety of dietary interventions for rheumatoid arthritis: a systematic review of randomized controlled trials. *Journal of the American Dietetic Association, 110*(5), 727–735.

Smolen, J. S., Landewé, R., Bijlsma, J., Burmester, G., Chatzidionysiou, K., Dougados, M., … Van Vollenhoven, R. (2017). EULAR recommendations for the management of rheumatoid arthritis with synthetic and biological disease-modifying antirheumatic drugs: 2016 update. *Annals of the Rheumatic Diseases, 76*(6), 960–977.

Sohail, R., Mathew, M., Patel, K. K., Reddy, S. A., Haider, Z., Naria, M., … Akbar, A. (2023). Effects of non-steroidal anti-inflammatory drugs (NSAIDs) and gastroprotective NSAIDs on the gastrointestinal tract: a narrative review. *Cureus, 15*(4), e37080. https://doi.org/10.7759/cureus.37080

Staheli, L. T. (1998). *Arthrogryposis: A Text Atlas*. Cambridge University Press.

Stein, C. M., Pincus, T., Yocum, D., Tugwell, P., Wells, G., Gluck, O., … McKendry, R. (1997). Combination treatment of severe rheumatoid arthritis with cyclosporine and methotrexate for forty-eight weeks. An open-label extension study. *Arthritis & Rheumatism, 40*(10), 1843–1851.

Sujitha, S., & Rasool, M. (2017). MicroRNAs and bioactive compounds on TLR/MAPK signaling in rheumatoid arthritis. *Clinica Chimica Acta, 473*, 106–115.

Sun, Q.-W., Jiang, S.-M., Yang, K., Zheng, J.-M., Zhang, L., & Xu, W.-D. (2012). Apigenin enhances the cytotoxic effects of tumor necrosis factor-related apoptosis-inducing ligand in human rheumatoid arthritis fibroblast-like synoviocytes. *Molecular Biology Reports, 39*, 5529–5535.

Takeuchi, T., Tanaka, Y., Tanaka, S., Kawakami, A., Iwasaki, M., Katayama, K., … Kaneko, Y. (2019). Efficacy and safety of peficitinib (ASP015K) in patients with rheumatoid arthritis and an inadequate response to methotrexate: results of a phase III randomised, double-blind, placebo-controlled trial (RAJ4) in Japan. *Annals of the Rheumatic Diseases, 78*(10), 1305–1319.

Tanaka, Y., & Izutsu, H. (2020). Peficitinib for the treatment of rheumatoid arthritis: an overview from clinical trials. *Expert Opinion on Pharmacotherapy, 21*(9), 1015–1025.

Taylor, P. C., Keystone, E. C., Van Der Heijde, D., Weinblatt, M. E., del Carmen Morales, L., Reyes Gonzaga, J., … Beattie, S. (2017). Baricitinib versus placebo or adalimumab in rheumatoid arthritis. *New England Journal of Medicine, 376*(7), 652–662.

Tripathi, S. K., Chen, Z., Larjo, A., Kanduri, K., Nousiainen, K., Äijo, T., … Laajala, E. (2017). Genome-wide analysis of STAT3-mediated transcription during early human Th17 cell differentiation. *Cell Reports, 19*(9), 1888–1901.

Tsai, C.-H., Liu, S.-C., Wang, Y.-H., Su, C.-M., Huang, C.-C., Hsu, C.-J., & Tang, C.-H. (2017). Osteopontin inhibition of miR-129-3p enhances IL-17 expression and monocyte migration in rheumatoid arthritis. *Biochimica et Biophysica Acta (BBA)-General Subjects, 1861*(2), 15–22.

Tugwell, P., Pincus, T., Yocum, D., Stein, M., Gluck, O., Kraag, G., … Wells, G. (1995). Combination therapy with cyclosporine and methotrexate in severe rheumatoid arthritis. *New England Journal of Medicine, 333*(3), 137–142.

Van Der Linden, M. P., Le Cessie, S., Raza, K., Van Der Woude, D., Knevel, R., Huizinga, T. W., & van der Helm-van Mil, A. H. (2010). Long-term impact of delay in assessment of patients with early arthritis. *Arthritis & Rheumatism, 62*(12), 3537–3546.

van Everdingen, A. A., Jacobs, J. W., Siewertsz Van Reesema, D. R., & Bijlsma, J. W. (2002). Low-dose prednisone therapy for patients with early active rheumatoid arthritis: clinical efficacy, disease-modifying properties, and side effects: a randomized, double-blind, placebo-controlled clinical trial. *Annals of Internal Medicine, 136*(1), 1–12. https://doi.org/10.7326/0003-4819-136-1-200201010-00006

Van Vollenhoven, R. F., Fleischmann, R., Cohen, S., Lee, E. B., García Meijide, J. A., Wagner, S., … Koncz, T. (2012). Tofacitinib or adalimumab versus placebo in rheumatoid arthritis. *New England Journal of Medicine, 367*(6), 508–519.

Wang, H.-Z., Wang, H.-H., Huang, S.-S., Zhao, H., Cao, Y.-G., Wang, G.-Z., … Liu, Y.-H. (2014). Inhibitory effect of baicalin on collagen-induced arthritis in rats through the nuclear factor–κB pathway. *Journal of Pharmacology and Experimental Therapeutics, 350*(2), 435–443.

Wei, K., Korsunsky, I., Marshall, J. L., Gao, A., Watts, G. F. M., Major, T., … Brenner, M. B. (2020). Notch signalling drives synovial fibroblast identity and arthritis pathology. *Nature, 582*(7811), 259–264. https://doi.org/10.1038/s41586-020-2222-z

Weinblatt, M. E., Fleischmann, R., Huizinga, T. W., Emery, P., Pope, J., Massarotti, E. M., ... Duncan, B. (2012). Efficacy and safety of certolizumab pegol in a broad population of patients with active rheumatoid arthritis: results from the REALISTIC phase IIIb study. *Rheumatology, 51*(12), 2204–2214.

Whitaker, R. H., & Cook, J. G. (2021). Stress relief techniques: p38 MAPK determines the balance of cell cycle and apoptosis pathways. *Biomolecules, 11*(10), 1444.

Williams, H. J. (1992). Comparison of auranofin, methotrexate, and the combination of both in the treatment of rheumatoid arthritis. *Arthritis and Rheumatism, 35*, 259–269.

Willkens, R., & Stablein, D. (1996). Combination treatment of rheumatoid arthritis using azathioprine and methotrexate: a 48 week controlled clinical trial. *The Journal of Rheumatology. Supplement, 44*, 64–68.

Willkens, R. F., Urowitz, M. B., Stablein, D. M., Mckendry Jr, R. J., Berger, R. G., Box, J. H., ... Marks, C. R. (1992). Comparison of azathioprine, methotrexate, and the combination of both in the treatment of rheumatoid arthritis. *Arthritis & Rheumatism: Official Journal of the American College of Rheumatology, 35*(8), 849–856.

Xuan, W., Feng, X., Qian, C., Peng, L., Shi, Y., Xu, L., ... Tan, W. (2017). Osteoclast differentiation gene expression profiling reveals chemokine CCL4 mediates RANKL-induced osteoclast migration and invasion via PI3K pathway. *Cell Biochemistry and Function, 35*(3), 171–177.

Xuzhu, G., Komai-Koma, M., Leung, B. P., Howe, H. S., McSharry, C., McInnes, I. B., & Xu, D. (2012). Resveratrol modulates murine collagen-induced arthritis by inhibiting Th17 and B-cell function. *Annals of the Rheumatic Diseases, 71*(1), 129–135.

Yang, X., Yang, J., & Zou, H. (2013). Baicalin inhibits IL-17-mediated joint inflammation in murine adjuvant-induced arthritis. *Journal of Immunology Research, 2013*(1), 268065.

Yang, Y., Zhang, X., Xu, M., Wu, X., Zhao, F., & Zhao, C. (2018). Quercetin attenuates collagen-induced arthritis by restoration of Th17/Treg balance and activation of Heme Oxygenase 1-mediated anti-inflammatory effect. *International Immunopharmacology, 54*, 153–162.

Section B

Signaling Mechanisms Involved in the Pathogenesis of Rheumatoid Arthritis

5 Phytochemicals Targeting Nuclear Factor Kappa B and Matrix Metalloproteinase Pathways to Treat Rheumatoid Arthritis

Hamna Mehmood, Ammara Saleem, Aisha Mobashar, Muhammad Furqan Akhtar, Justyna Wozniak, and Joham Javed

WHAT WILL YOU LEARN?

This chapter will cover crucial details regarding the matrix metalloproteinase (MMP) (Malemud) and nuclear factor kappa B (NF-κB) (Huang et al., 2017) pathways linked with inflammatory diseases. The involvement of MMPs and NF-κB in the pathogenesis of rheumatoid arthritis, as well as phytochemicals that target these pathways and their preclinical research, will also be addressed.

5.1 NUCLEAR FACTOR-κB

Nuclear factor kappa β (NF-κB) controls cytokine production, DNA transcription, cell proliferation, and cell survival. It plays a role in how cells respond to various stimuli, including stress, free radicals, exposure to heavy metals, UV light, and bacterial or viral antigens. NF-κB is essential for controlling the immunological response to various infections. Improper regulation of NF-κB can result in abnormal immune development, cancer, inflammation, and autoimmune conditions.

5.2 STRUCTURE OF NF-κB

NF-κB comprises five different but structurally related members, classified into two classes.

5.2.1 Class I

- NF-κB1 (also named as p50)
- NF-κB2 (also named as p52)

5.2.2 Class II

- RelA (also named as p65)
- RelB
- c-Rel

DOI: 10.1201/9781003535201-7

The Rel homology domain, a conserved area in the N-terminal domain that is roughly 300 amino acids long, is shared by all members of the NF-κB family. This domain is crucial for nuclear translocation, DNA binding, and dimerization of NF-κB proteins. NF-κB1 (p50/p105) and NF-κB2 (p52/p100) are produced from their precursors (p105 and p100) through limited proteolytic processing. The p105 and p100 proteins contain ankyrin repeat domains in their C-terminal portions, which function as inhibitory regions sequestering these proteins in the cytoplasm. The processing of these precursors involves the ubiquitin-proteasome pathway, resulting in the generation of the active forms p50 and p52, respectively. The transactivation domain found in the C-terminal region of the RelA (p65), RelB, and c-Rel proteins is crucial for the transcriptional activation of target genes. Unlike p50 and p52, these proteins do not arise from larger precursors but are synthesized as full-length proteins.

5.3 ACTIVATION OF NF-κB

When a cell is at rest, its cytoplasm contains the inactivated form of NF-κB, which is bound to an inhibitory protein known as inhibitor of nuclear factor-κB kinase (IKK). IκB proteins keep NF-κB inactive within the cytoplasm by hiding its nuclear localization signal. The IκB proteins conceal a nuclear localization signal of NF-κB, keeping it in an inactive state inside the cytoplasm. The IKK (IκB kinase) complex is made up of three subunits:

- Catalytic IKKα
- Catalytic IKKβ
- Regulatory subunit IKKγ (referred to as NEMO or NF-κB essential modulator).

IKK complex is activated in response to stress signals, infections, and cytokines. The IKKβ subunit (in the conventional pathway) or IKKα subunit (in the noncanonical or alternative pathway) is phosphorylated during this activation. The activated IKK complex then phosphorylates IκB proteins, targeting them for ubiquitination and subsequent degradation by the proteasome. As a result of this activated proteolytic mechanism, NF-κB translocates into the nucleus where it regulates the expression of various proinflammatory cytokines and TNF-α as described in Figure 5.1.

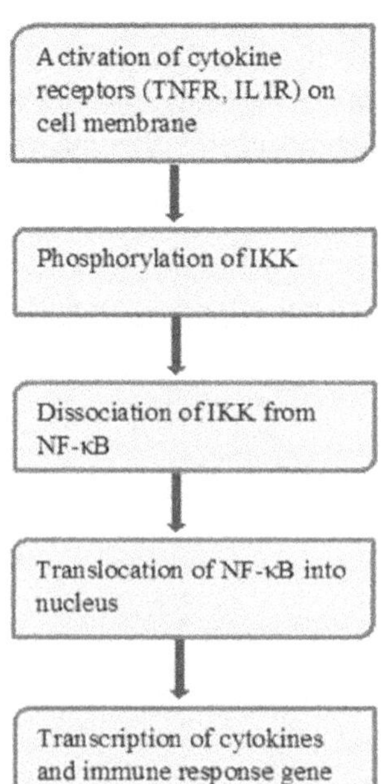

FIGURE 5.1 NF-κB activation by phosphorylation of nuclear factor-κB kinase inhibitor (IKK).

TABLE 5.1
Distinct Features of Canonical Pathway and Noncanonical Pathway

Feature	Canonical Pathway	Noncanonical Pathway
IκB degradation	Yes	No IκB-dependent degradation
Function diversity	Diverse function	Specific functions
Activation speed	Rapid and transient	Slow and persistent
Triggering stimuli	Response to TNFα receptors, B-cell receptor (BCR), pattern-recognition receptor (PRR), and IL-1	Response to LTβ, CD4 cells, B-cell activating factor (BAFF)
Complexes activated	RelA/p50 or cRel/p50	RelB/p52

5.4 NF-κB SIGNALING PATHWAY

The NF-κB signaling pathway is vital for immunological functioning, homeostasis, chronic inflammation, autoimmune disorders, tumor development, and aging. Research into the involvement of NF-κB in immune-mediated diseases and carcinogenesis suggests that this pathway could be a viable therapeutic target for a range of disorders.

NF-κB activation occurs via two distinct pathways as described in Table 5.1. These pathways differ in signaling mechanisms and biological functions.

i. Canonical/classical pathway
ii. Noncanonical/alternate pathway

5.4.1 Canonical/Classical Pathway

Proinflammatory cytokines like TNFα receptors, B-cell receptors (BCRs), IL-1 receptors, pattern-recognition receptors (PRRs), and toll-like receptors (TLRs) initiate the canonical pathway. This pathway typically results in the activation of NF-κB complexes comprising RelA/p65 or cRel/p50. Tumor growth factor-β-activated kinase 1 and other upstream kinases phosphorylate IκBα at particular NH2-terminal serine residues, initiating the canonical pathway. After phosphorylation, IκBα is ubiquitinated and directed to the 26S proteasome for degradation. The result is the release of the NF-κB subunits, which then relocate to nucleus to support gene transcription of target genes involved in inflammatory response (Kaeley, Ahmad, Pathania, & Kakkar, 2019).

NF-κB can activate numerous genes in response to infection. TLRs are responsible for detecting microbial substances such as bacterial cell walls and nucleic acids. TLRs initiate the signaling pathway leading to the production of proinflammatory cytokines, TNF-α, and macrophage inflammatory protein-1α (MIP-1α) as described in Figure 5.2.

5.4.2 Noncanonical/Alternate Pathway

RelB/p52 complexes are specifically activated by the noncanonical (alternative) route. Unlike the canonical pathway, it is not activated by TNFα (Figure 5.3). Rather, triggers like B-cell activating factor (BAFF), receptor activator of nuclear factor kappa-B ligand (RANKL), CD40 ligand (CD40L), and lymphotoxin β (LTβ) cause it to occur (Yu, Lin, Zhang, Zhang, & Hu, 2020).

Numerous genes involved in the cascade response to infection can be activated by NF-κB. TLRs play a crucial role in this process by detecting microbial substances such as bacterial cell walls and nucleic acids. Upon detection of pathogens, TLRs initiate signaling pathways for proinflammatory cytokines such as TNF-α and chemokines like MIP-1α (Ishihara et al., 2022).

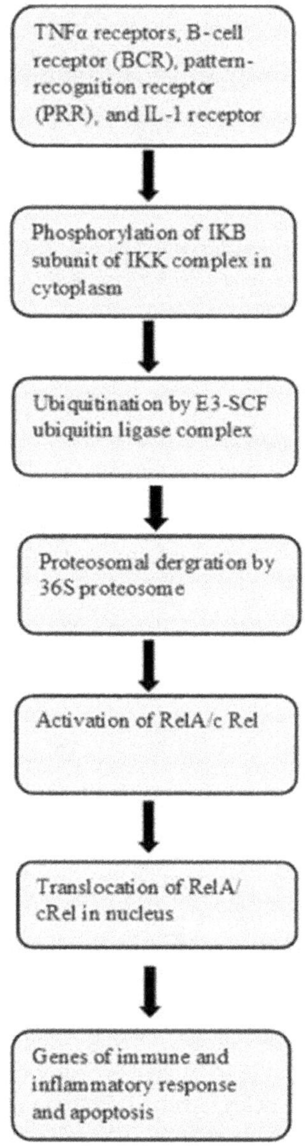

FIGURE 5.2 NF-κB activation through the canonical pathway.

5.5 METALLOPROTEINASES

The zinc-containing endopeptidases known as matrix metalloproteinases (MMPs) are responsible for remodeling an extracellular matrix and are activated by calcium. They are essential for numerous biochemical and physiological processes regulated by hormones, growth factors, and cytokines. MMPs have the ability to break down almost every extracellular matrix component, which is crucial for processes such as angiogenesis, tissue remodeling, and wound healing (Mukherjee & Das, 2024). Dysregulation of MMP expression can lead to abnormal breakdown of the extracellular matrix, contributing to chronic degenerative disorders and vascular diseases. Furthermore, the advancement of cancer and neurodegeneration is also linked to atypical MMP activation (Bassiouni, Ali, & Schulz, 2021).

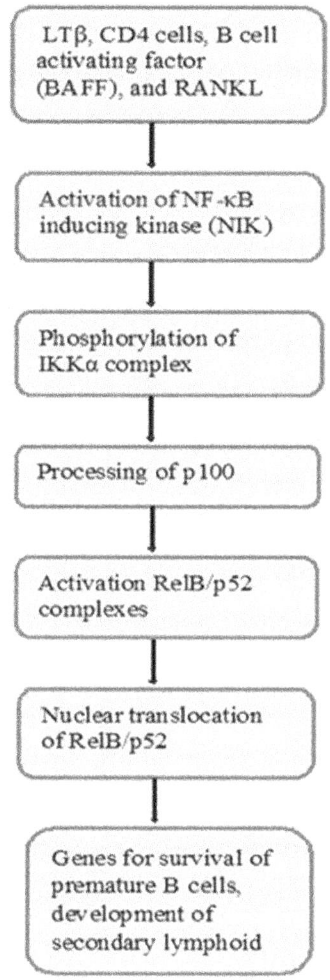

FIGURE 5.3 Mechanism of the noncanonical pathway of NF-κB activation.

5.6 CLASSIFICATION OF MMPs

MMPs are either classified based on their cellular distribution or on their structure:

5.6.1 Based on Cellular Distribution

5.6.1.1 Membrane-Type Matrix Metalloproteinases (MT-MMPs)

They can break down collagen and activate other proteases and cell surface components. They include MMP14, 15, 16, 17, 24, and 25.

5.6.1.2 Collagenases

Collagenases specifically degrade triple-helical fibrillar collagen, an essential component of bone and ligaments (MMP1, 8, 13, and18).

5.6.1.3 Gelatinases

Gelatinases such as MMP2 and 9 degrade gelatin and basal lamina molecules, influencing processes such as neurogenesis and angiogenesis.

TABLE 5.2
Classification of Matrix Metalloproteinases Based on Structure

MMPs	Type	Example
Minimal domain MMPs	Secreted	MMP-7,MMP-26
Simple hemopexin domain containing MMPs	Secreted	MMP-1,3,8,10,12,13,18,19,22,20, and 27
Gelatin binding MMPs	Secreted	MMP-2,MMP-9
Furin-activated MMPs	Secreted	MMP-11,MMP-28
Vitronectin-activated MMPs	Secreted	MMP-21
Transmembrane MMPs	Membrane-type MMPs	MMP-14,15,16, and 24
Glycosylphosphatidyl inositol MMPs	Membrane-type MMPs	MMP-17,MMP-25
Type II transmembrane MMPs	Membrane-type MMPs	MMP-23

5.6.1.4 Stromelysins

Laminin, fibronectin, gelatin, and proteoglycans are broken down by stromelysins (MMP 3, 10, and 11).

5.6.1.5 Matrilysins

Matrilysins include MMP-7 and 26 secreted by epithelial cells and macrophages. MMP-26 breaks down both extracellular matrix and nonextracellular matrix constituents, including α-1 protease inhibitor and selectins (Cabral-Pacheco et al., 2020).

5.6.2 Based on Structure

Based on their structural characteristics, MMPs can be classified into eight classes as described in Table 5.2. Out of eight, three classes are membrane-bound, while the remaining five are found in the extracellular matrix (Almutairi, Kalloush, Manoon, & Bardaweel, 2023).

5.7 BASIC STRUCTURE OF MMPs

Structural features that are common among all MMPs are given in Figure 5.4, and these include

1. A variable-length single N-terminal signal peptide that directs the MMP toward secretion.
2. A prodomain of approximately 80 amino acids maintains MMPs in their inactive form as zymogens.
3. A zinc-ion-containing catalytic domain of almost 160 amino acids with three α-helixes, three calcium ions, and five β-sheets.
4. The hinge region which is a variable-length linker that connects the catalytic domain to the hemopexin-like domain.
5. Hemopexin-like domain with four large surface area β-propeller structures that facilitate substrate protein interaction (Laronha & Caldeira, 2020).

5.7.1 Physiological Regulation of MMPs and TIMPs

MMPs are tightly regulated under physiological conditions at several distinct levels:

- Transcription
- Zymogen precursor activation
- Interaction with extracellular matrix components
- Inhibition by TIMPs

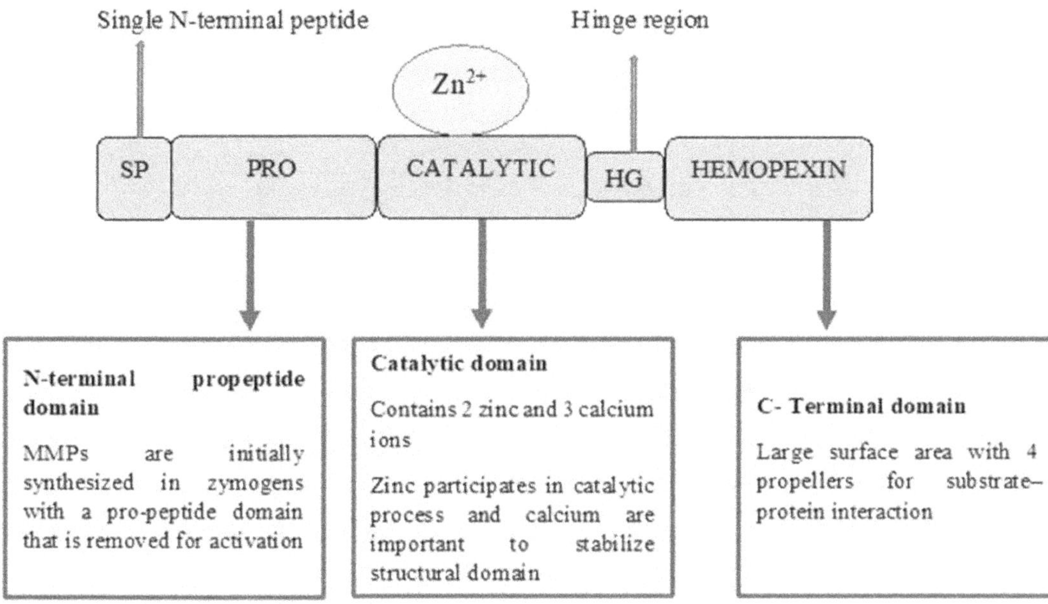

FIGURE 5.4 Basic structure of matrix metalloproteinases.

These inhibitors are responsible for tissue remodeling and restoration of extracellular matrix components. They are regulated by tissue TIMPs, which are composed of N-terminal and C-terminal domains positioned next to each other.

All four TIMPs (1–4) are highly selective and effective modulators of proteolytic activity of MMPs. The degree of extracellular matrix breakdown in the periphery of cells is regulated by a delicate balance between MMP and TIMP levels, which in turn affects cellular functions like migration, proliferation, and survival. TIMPs are smaller proteins with a size range of 22–28 kDa, varying with their degree of glycosylation. They are expressed by various cells in human tissues and body fluids (Lambert, Dassé, Haye, & Petitfrère, 2004).

5.8 ROLE OF MMPs IN INFLAMMATION

The production and functionality of MMPs are regulated by many signaling pathways. The mitogen-activated protein kinases (MAPKs) pathway is one of these pathways. Three primary forms of MAPKs are p38, c-Jun N-terminal kinases (JNKs), and extracellular signal-regulated kinases (ERKs). Stimuli such as cytokines and growth hormones activate these signaling pathways. It involves the cascade of kinases that sequentially phosphorylate serine and threonine residues in the target protein. Phosphorylation results in transcription factor regulation, which is frequently connected to various biological processes (Ohshiba, Miyaura, Inada, & Ito, 2003). For example, it has been reported that in ovarian cancer cells, downregulating the ERK pathway lowers the transcriptional gelatinases (Shan et al., 2023). MMP-1 expression can be elevated by external stimuli such as ultraviolet radiation and reactive oxygen species through activation of these pathways. Additionally, research has demonstrated that p38 regulates the expression of many MMPs in cancer cell lines, such as those from hepatocellular carcinoma (SK-Hepl and SNU-387). Activation of MMPs can also enhance metastasis, as seen with JNK activation, which increases MMP-9 levels and cell invasion in prostate cancer via androgen receptor activation (Lee, 2021).

MMPs control activation and inactivation of IL-1β, a cytokine that promotes inflammation. For instance, MMP-7 and MMP-12 cleave proTNF-α, enabling macrophages to release TNF-α in active

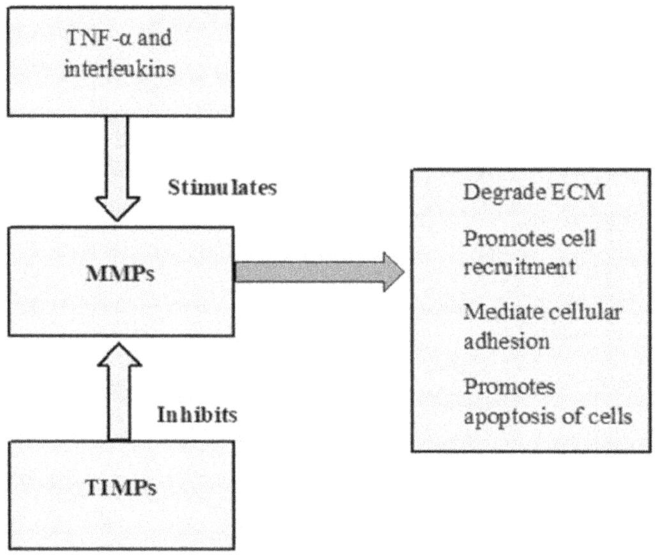

FIGURE 5.5 Physiological regulation of MMPs and TIMPs.

form and promote inflammation. While MMPs are secreted by cells, they can also re-enter cells after secretion and accumulate intracellularly. Proteolytic cleavage, post-translational modification, and oxidative stress all contribute to the activation of intracellular MMPs as described in Figure 5.5. Once activated, they interact with intracellular molecules to regulate signaling processes related to cellular metabolism and pathogenic mechanisms. Furthermore, secreted MMPs contribute significantly to inflammation-related disorders by modulating cytokines and chemokines, thus playing roles in both acute and chronic inflammation (Lee & Kim, 2022).

5.9 NF-κB IN RHEUMATOID ARTHRITIS

A frequent characteristic of the human synovium in rheumatoid arthritis is the presence of activated NF-κB. It targets proinflammatory cytokines that are constitutively elevated in RA. The cells from the synovial epithelium of RA patients contain activated forms of p50 and p65, but the cells from normal synovium do not contain these transcription factors (Serasanambati & Chilakapati, 2016).

The survival, development, and production of antibodies by B cells and plasma cells depend on the alternative NF-κB pathway as described in Figure 5.6. These cells maintain and promote long-term inflammation in the RA synovium. Components of the classical NF-κB pathway, particularly c-Rel and RelA, play significant roles in T-cell receptor signaling. Dysregulated NF-κB signaling can lead to unintended T-cell activation, resulting in immunological and inflammatory reactions. New pharmacological agents that inhibit the NF-κB pathway long-term are being explored for the management of rheumatoid arthritis (Barrow, 2021).

The formation of osteoclasts and bone remodeling are influenced by receptor activators of NF-κB (Mueller et al., 2021)/receptor activator of NF-kB ligand (RANKL)/Osteoprotegerin signaling. The homotrimeric protein RANKL is generated by activated T cells and osteoblasts. RANKL activates MAPK to induce the transcription factor activator protein-1 (AP-1) and IKK to activate NF-κB. The TNF family contains the homotrimeric transmembrane receptor RANK, which is expressed by mature osteoclasts and dendritic cells. RANK lacks the intrinsic protein kinase activity necessary to trigger the signaling cascade. Instead, RANK binds with tumor necrosis factor receptor 6, which is required for bone health. By binding to its receptor on the surface of osteoclasts, RANKL promotes the maturation and differentiation of osteoclasts. In RA patients, RANKL is overexpressed

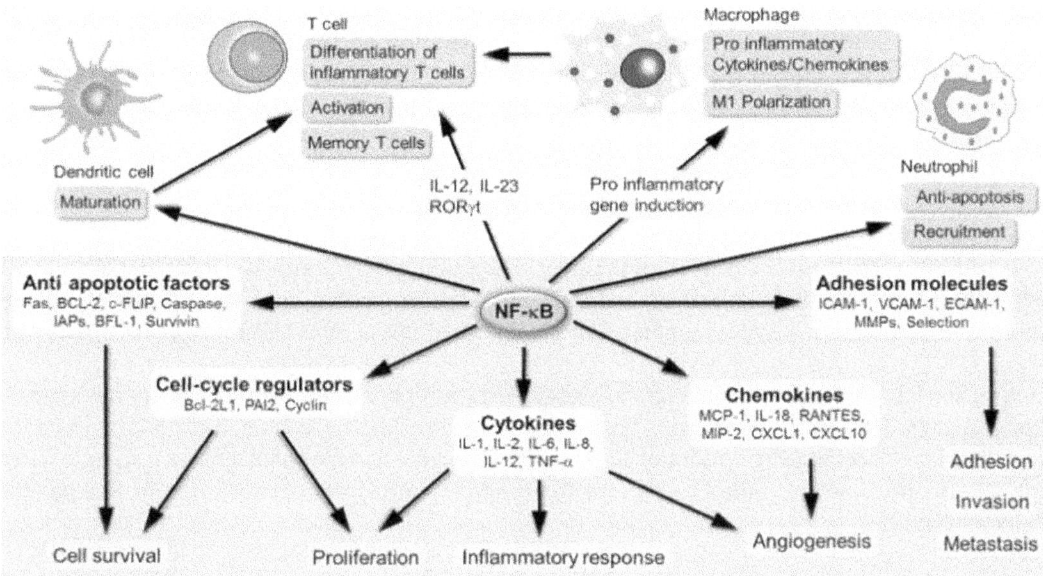

FIGURE 5.6 Activation of NF-KB transcription factor.

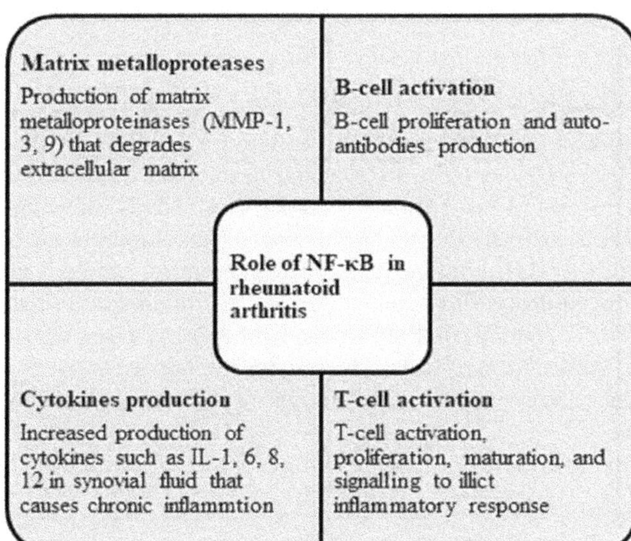

FIGURE 5.7 Role of NF-κB in rheumatoid arthritis.

in the synovium and contributes to bone erosions that cause long-lasting bone destruction (Hussein & Aboukhamis, 2022). The role of NF-KB is described in Figure 5.7.

5.10 MMPs IN RHEUMATOID ARTHRITIS

RA is characterized by abnormal activation of cells within the synovial fluid, leading to progressive joint damage with loss of bone and cartilage. MMPs play an essential part in the pathophysiology of RA. In the afflicted joints, synovial cells secrete various proteases, including MMPs, which break down the collagen and proteoglycans that constitute the extracellular matrix. The elevated levels of MMPs including 1, 13, 3, and 9 are associated with the deterioration of cartilage in RA joints. Collagen is cleaved by MMP 1 and 13, while aggrecan-containing proteoglycan is the target of MMP-3 and MMP-9. Degeneration of proteoglycans at the surface and disintegration of collagen fibrils in the deep zones lead to destruction of articular cartilage (Kumar et al., 2023). The cellular invasion of cartilage is caused by MMP 14, which is abundantly expressed at the interface between pannus and cartilage. In addition to being involved in extracellular matrix remodeling, MMP 14 also has a significant function in the angiogenic response in RA (Pulik, Łęgosz, & Motyl, 2023).

MMP-MT1 (MMP-14) is involved in the matrix degradation by fibroblasts and osteoclast-mediated bone resorption in RAs. It activates other MMPs, which further contribute to the degradation of bone and cartilage. RA synovium also exhibits elevated levels of MMP-13 expression (Li et al., 2022).

Clinical trials have been conducted on several MMP inhibitors, showing successful outcomes in animal models. However, these results have not always translated into human trials. For instance, the collagenase-specific MMP inhibitor Ro32–3555 (Trocade) underwent extensive testing to determine its effectiveness, but failed to prevent the progression of joint deterioration in RA patients (Malemud, 2019).

5.11 PHYTOCHEMICALS THAT TARGET NF-κB AND MMPs

There are different phytochemicals that target NF-κB and MMPs to halt inflammation as shown in Figure 5.8.

5.11.1 Resveratrol

Resveratrol is a polyphenol that is found in natural plants, particularly *Polygonum cuspidatum*. It participates in the treatment of numerous chronic inflammatory diseases, diabetes mellitus, coronary heart disease, and obesity due to its diverse pharmacological activities. In a preclinical study, rats were treated with resveratrol suspension (20 mg/kg, orally) daily, showing its therapeutic activity in thioacetamide-induced liver fibrosis and cyclophosphamide-induced liver toxicity by suppression of TNF-α and NF-kB (Ebrahim et al., 2022). Resveratrol can both suppress inflammation and activate Sirt1, a crucial deacetylase that suppresses the inflammatory response and maintains peripheral T-cell tolerance. Among Sirt1 substrates is p65/RelA, a key regulator of inflammatory cytokine signaling and leukocyte activation. The resveratrol-induced activation of Sirt1 inhibits the acetylation of RelA. Thus, the expression of inflammatory proteins produced by NF-κB is decreased (Malaguarnera, 2019).

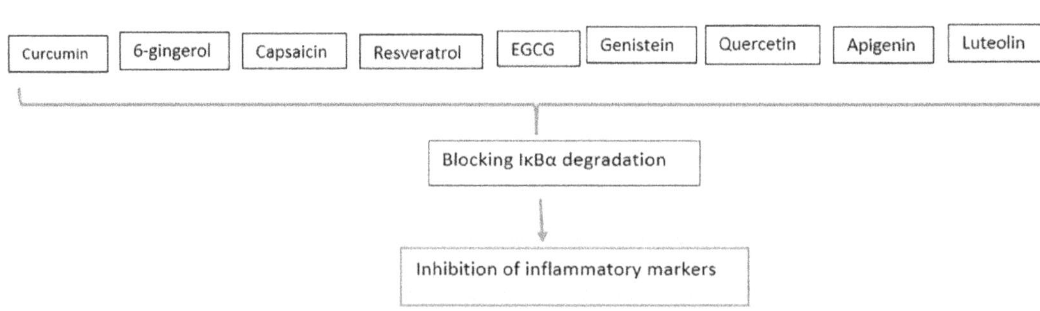

FIGURE 5.8 Different phytochemicals that target NF-KB and MMPs to exhibit antiarthritic action.

5.11.2 Myricetin

Myricetin is a flavonoid glycoside found in fruits, vegetables, berries, nuts, and herbs like *Comptonia peregrina* and *Morella cerifera*. Anti-inflammatory, hepatoprotective, antidiabetic, and antioxidant properties of myricetin in the treatment of disease are supported by research. It reduces inflammation in lipopolysaccharide (LPS)-induced inflammation models by blocking the IKK/NF-κB pathway, which in turn downregulates the production of TNF-α, IL-6, and IL-1β. It has been demonstrated to prevent cytokine-induced cholangiocarcinoma cell invasion and migration (Song et al., 2021). Furthermore, myricetin administration successfully inhibited the upregulation of genes linked to inflammation and metastasis, such as MMP-9, COX-2, and intercellular adhesion molecule-1 (Rahmani et al., 2023).

5.11.3 Silymarin

Silymarin is a flavonoid extracted from seeds and fruit of milk thistle plant, *Silybum marianum*. Its pharmacological properties include anti-inflammatory, antiapoptotic, antimetastatic, and antiangiogenesis actions (Dorjay, Arif, & Adil, 2018). Silymarin demonstrates its anti-inflammatory activity by downregulating proinflammatory mediators such as COX, lipoxygenase, NOS, and various cytokines. When 25 mg/kg of sylimarin was administered to rats in the hepatotoxicity model caused by D-galactosamine, the expressions of NF-κB mRNA and protein were significantly reduced (Surai, Surai, & Earle-Payne, 2024).

5.11.4 Oridonin

The diterpenoid oridonin is derived from the herbal plant *Rabdosia rubescens*. It has been used to treat cancer and inflammation in traditional East Asian medicine (Li et al., 2023). It demonstrates anti-inflammatory potential by the expression of TLR4 and p38-mitogen-activated protein kinase (p38-MAPK), inhibiting the NF-κB signaling pathway, and lowering serum cytokine secretion, which includes TNF-α and IL-6 (Liu, Xu, Zhou, & Shen, 2021).

5.11.5 Baicalein

Baicalein is an active constituent extracted from the roots of *Scutellariae radix*. It exhibits antiviral, antioxidant, and anti-inflammatory pharmacological actions. Baicalein functions as an antitumor agent by scavenging free radicals, reducing oxidative stress, and lowering the expression of MMP-9 and MMP-2, both gelatinases, thereby attenuating MAPK activities that control the cell cycle. The expression of gelatinases has been associated with a high potential for metastasis in several human carcinomas (Liu et al., 2016).

5.11.6 Ginsenosides

Ginsenosides, derived from ginseng, possess antioxidant, anti-inflammatory, and antimetastatic properties. They can suppress the expression of several MMPs, including 1, 2, 3, 7, 9, and 13. Additionally, ginsenosides showed anti-inflammatory qualities by inhibiting iNOS, NF-κB, and MAPK pathways in RAW264.7 cells (Yi, 2019).

5.11.7 Astragalin

Astragalin is a natural flavonoid obtained from *Astragalus memeranaceus*, *Cuscuta australis*, *Cuscuta chinensis*, and *Centella asiatica* (Tian, Wei, Hu, & Zhao, 2022). It has several pharmacological characteristics, such as coagulation-promoting, anti-inflammatory, antioxidant, and antitumor effects. By preventing the phosphorylation of IκBα, IKKα/β, and p65 proteins, it prevents NF-κB from translocation. By activating the classical pathway, TLR4 initiates intracellular signal transduction and releases more inflammatory cytokines, such as nitric oxide. NF-κB, IKKβ, and IκB mRNA expression was dramatically reduced by astragalin in mice with acute colitis induced with dextran sodium sulfate. It also decreased colon tissue's expression of TLR4 mRNA. Furthermore, astragalin has a significant effect on SIRT1. It deacetylates the p65 subunits, thereby inhibiting NF-κB. This activation of SIRT1 contributes to its anti-inflammatory effects (Ruan et al., 2024).

5.11.8 Quercetin

Quercetin, a naturally occurring dietary flavonoid, exhibits anti-inflammatory, anticancer, and antioxidant qualities. It controls the p38 MAPK signaling pathway that triggers breast cancer cells to undergo apoptosis. It also inhibits MMP-9 activation through downregulation of the Akt/MMP-9 and ERK/AP-1 pathways (Hsieh et al., 2022).

5.11.9 Piperine

Black pepper (*Piper nigrum*) and long pepper (*Piper longum*) are the sources of piperine, which is an alkaloid. It has antiallergic, hepatoprotective, and neuroprotective properties. It significantly reduces the levels of TLR4, NF-κB p65, and IL-1β, which are involved in the NF-kB/TLR4 signaling cascade. Following an ischemic stroke in rats, piperine increased neurogenesis and angiogenesis by modifying the NF-kB/TLR4 inflammatory signaling cascade (Yang, Sheng, Lu, & Zhao, 2021). NF-κB/AP-1 and PKCα/ERK1/2 pathways are two of the main signaling pathways that control tumor cell invasion. In an *in vitro* tumor model produced by phorbol 12-myristate 13-acetate, piperine inhibited the PKCα/ERK1/2 and NF-κB/AP-1 pathway, leading to the downregulation of MMP-9 production. Furthermore, it downregulates the expression of MyD88 and miR-127 in an LPS-induced in vitro model of osteoarthritis (Tripathi, Ray, & Mishra, 2022).

5.11.10 Berberine

Berberine is an isoquinoline alkaloid obtained from *Coptis chinensis*. It significantly reduces the synthesis of TNF-α and MMP-9, which are typically elevated with TNF-α stimulation. This reduction in MMP-9 is particularly relevant because MMP-9 is associated with tissue remodeling and invasion in various pathological conditions. Since AP-1 is a transcription factor that controls MMP-9 expression, blocking AP-1's ability to bind DNA effectively reduces MMP-9 levels and subsequent cell invasion (Li et al., 2018).

5.11.11 Evodiamine

Evodiamine is the main bioactive quinolone alkaloid extracted from *Evodia rutaecarpa*. It halts cell cycle by inhibiting invasion, metastasis, and angiogenesis in many cancer cell lines. Its derivatives, especially those with a benzoyl group at the N13-position, have shown a stronger anticancer and broad-spectrum action against MDA-MB-435 cells. These derivatives block cell invasion and migration by downregulating MMP-9. Additionally, evodiamine suppresses the production of ERK and p-p38 MAPK (Khan et al., 2021).

5.11.12 Matrine

Matrine is an alkaloid derived from the Chinese herb *Sophora flavescens*. It exhibits antiviral and anticancer properties against lung cancer and breast cancer (Lin, He, Wu, Xu, & Du, 2022). Additionally, it has been shown to block the NF-κB pathway, which prevents castration-resistant prostate cancer cells from proliferating *in vitro*. Furthermore, matrine significantly lowers NF-κB expression and MMP-9 and MMP-2 mRNA levels in MDA-MB-231 tumor cells (Zhang et al., 2020).

5.11.13 Triptolide

Triptolide is a diterpenoid triepoxide obtained from *Tripterygium wilfordi*. Because of its strong anti-inflammatory and immunosuppressive properties, it has been used extensively in the treatment of inflammatory conditions. It has been demonstrated to inhibit MMP-2 activity in the gelatin zymography test. It also reduced p-JNK, NF-κB, MMP-9, and MMP-2 levels and was shown to reduce the expression of MMP-2 and NF-κB in B16F10 cells *in vitro*. This inhibition of MMP-2 and MMP-9 expression by blocking NF-κB led to reduced cell motility and invasion (Jao et al., 2016).

Sulforaphane (SFN), a well-studied plant-derived isothiocyanate organosulfur compound, exhibits versatile actions in modulating both the redox system and immune homeostasis. SFN achieves this through intricate interactions with key signaling pathways such as Nrf2 and NF-κB. In a groundbreaking study, SFN was found to exert a novel mechanism of action by directly inhibiting NF-κB. Findings highlight SFN's unique ability to intervene at the molecular level, influencing critical pathways involved in inflammation and oxidative stress. SFN's role as a natural modulator of immune responses underscores its therapeutic potential in combating inflammatory diseases and supporting overall health (Checker, Gambhir, Thoh, Sharma, & Sandur, 2015).

Allyl-isothiocyanate (AITC), a common aliphatic isothiocyanate found in many Brassica species. Recent research highlights AITC's multifaceted role in modulating key signaling pathways. AITC promotes Nrf2's nuclear shift of Nrf2, a pivotal transcription factor known for its role in orchestrating antioxidant defenses. This results in the subsequent suppression of NF-κB, a master regulator of inflammation. This cascade of events leads to the upregulation of heme oxygenase-1, an Nrf2 target gene crucial for mitigating inflammatory responses. Additionally, findings by Wagner et al. propose that AITC suppresses inflammation through its interaction with miR-155, a microRNA implicated in regulating signals of NF-κB and Nrf2. By targeting miR-155, AITC further enhances its ability to modulate inflammatory processes at a molecular level. These discoveries underscore AITC's potential as a natural compound with therapeutic implications for inflammatory diseases, leveraging its interactions with multiple pathways to combat oxidative stress and inflammation effectively (Wagner, Boesch-Saadatmandi, Dose, Schultheiss, & Rimbach, 2012).

REFERENCES

Almutairi, S., Kalloush, H. M. D., Manoon, N. A., & Bardaweel, S. K. (2023). Matrix metalloproteinases inhibitors in cancer treatment: an updated review (2013–2023). *Molecules, 28*(14), 5567.

Barrow, M. (2021). An overview of the NF-kB mechanism of pathophysiology in rheumatoid arthritis, investigation of the NF-kB ligand RANKL and related nutritional interventions. *Autoimmunity Reviews, 20*(2), 102741.

Bassiouni, W., Ali, M. A., & Schulz, R. (2021). Multifunctional intracellular matrix metalloproteinases: implications in disease. *The FEBS Journal, 288*(24), 7162–7182.

Cabral-Pacheco, G. A., Garza-Veloz, I., Castruita-De la Rosa, C., Ramirez-Acuña, J. M., Perez-Romero, B. A., Guerrero-Rodriguez, J. F., … Martinez-Fierro, M. L. (2020). The roles of matrix metalloproteinases and their inhibitors in human diseases. *International Journal of Molecular Sciences, 21*(24), 9739.

Checker, R., Gambhir, L., Thoh, M., Sharma, D., & Sandur, S. K. (2015). Sulforaphane, a naturally occurring isothiocyanate, exhibits anti-inflammatory effects by targeting GSK3β/Nrf-2 and NF-κB pathways in T cells. *Journal of Functional Foods, 19*, 426–438.

Dorjay, K., Arif, T., & Adil, M. (2018). Silymarin: an interesting modality in dermatological therapeutics. *Indian Journal of Dermatology, Venereology and Leprology, 84*, 238.

Ebrahim, H. A., Kamar, S. S., Haidara, M. A., Latif, N. S. A., Ellatif, M. A., ShamsEldeen, A. M., … Dawood, A. F. (2022). Association of resveratrol with the suppression of TNF-α/NF-kB/iNOS/HIF-1α axis-mediated fibrosis and systemic hypertension in thioacetamide-induced liver injury. *Naunyn-Schmiedeberg's Archives of Pharmacology, 395*(9), 1087–1095.

Hsieh, H.-L., Yu, M.-C., Cheng, L.-C., Chu, M.-Y., Huang, T.-H., Yeh, T.-S., & Tsai, M.-M. (2022). Quercetin exerts anti-inflammatory effects via inhibiting tumor necrosis factor-α-induced matrix metalloproteinase-9 expression in normal human gastric epithelial cells. *World Journal of Gastroenterology, 28*(11), 1139.

Huang, H., Du, T., Xu, G., Lai, Y., Fan, X., Chen, X., … Liu, L. (2017). Matrine suppresses invasion of castration-resistant prostate cancer cells by downregulating MMP-2/9 via NF-κB signaling pathway. *International Journal of Oncology, 50*(2), 640–648.

Hussein, R., & Aboukhamis, I. (2022). The association of serum RANKL levels with disease activity and hematological parameters in Syrian patients with rheumatoid arthritis. *Biochemistry and Biophysics Reports, 32*, 101373.

Ishihara, Y., Kado, S. Y., Bein, K. J., He, Y., Pouraryan, A. A., Urban, A., … Vogel, C. F. (2022). Aryl hydrocarbon receptor signaling synergizes with TLR/NF-κB-signaling for induction of IL-22 through canonical and non-canonical AhR pathways. *Frontiers in Toxicology, 3*, 787360.

Jao, H. Y., Yu, F. S., Yu, C. S., Chang, S. J., Liu, K. C., Liao, C. L., … Chung, J. G. (2016). Suppression of the migration and invasion is mediated by triptolide in B16F10 mouse melanoma cells through the NF-kappa B-dependent pathway. *Environmental Toxicology, 31*(12), 1974–1984.

Kaeley, N., Ahmad, S., Pathania, M., & Kakkar, R. (2019). Prevalence and patterns of peripheral neuropathy in patients of rheumatoid arthritis. *Journal of Family Medicine and Primary Care, 8*(1), 22–26.

Khan, Y. H., Uttra, A. M., Qasim, S., Mallhi, T. H., Alotaibi, N. H., Rasheed, M., … Khan, S.-U.-D. (2021). Potential role of phytochemicals against matrix metalloproteinase induced breast cancer; an explanatory review. *Frontiers in Chemistry, 8*, 592152.

Kumar, S., Kumar, H., Mittal, A., Singh, P. P., Yadav, V., Kumar, D., … Mishra, V. (2023). Correlation between synovial fluid levels of matrix metalloproteinase's (MMP-1, MMP-3, and MMP-9) and TNF-α with the severity of osteoarthritis knee in rural Indian population. *Indian Journal of Orthopaedics, 57*(10), 1659–1666.

Lambert, E., Dassé, E., Haye, B., & Petitfrère, E. (2004). TIMPs as multifacial proteins. *Critical Reviews in Oncology/Hematology, 49*(3), 187–198.

Laronha, H., & Caldeira, J. (2020). Structure and function of human matrix metalloproteinases. *Cells, 9*(5), 1076.

Lee, H. S., & Kim, W. J. (2022). The role of matrix metalloproteinase in inflammation with a focus on infectious diseases. *International Journal of Molecular Sciences, 23*(18), 10546.

Lee, S. Y. (2021). Anti-metastatic and anti-inflammatory effects of matrix metalloproteinase inhibition by ginsenosides. *Biomedicines, 9*(2), 198.

Li, G. Q., Gao, S. X., Wang, F. H., Kang, L., Tang, Z. Y., & Ma, X. D. (2023). Anticancer mechanisms on pyroptosis induced by Oridonin: new potential targeted therapeutic strategies. *Biomedicine & Pharmacotherapy, 165*, 115019.

Li, J., Liu, F., Jiang, S., Liu, J., Chen, X., Zhang, S., & Zhao, H. (2018). Berberine hydrochloride inhibits cell proliferation and promotes apoptosis of non-small cell lung cancer via the suppression of the MMP2 and Bcl-2/Bax signaling pathways. *Oncology Letters, 15*(5), 7409–7414.

Li, R.-L., Duan, H.-X., Liang, Q., Huang, Y.-L., Wang, L.-Y., Zhang, Q., … Peng, W. (2022). Targeting matrix metalloproteases: a promising strategy for herbal medicines to treat rheumatoid arthritis. *Frontiers in Immunology, 13*, 1046810.

Lin, Y., He, F., Wu, L., Xu, Y., & Du, Q. (2022). Matrine exerts pharmacological effects through multiple signaling pathways: a comprehensive review. *Drug Design, Development and Therapy, 16*, 533–569.

Liu, H., Dong, Y., Gao, Y., Du, Z., Wang, Y., Cheng, P., … Huang, H. (2016). The fascinating effects of baicalein on cancer: a review. *International Journal of Molecular Sciences, 17*(10), 1681.

Liu, X., Xu, J., Zhou, J., & Shen, Q. (2021). Oridonin and its derivatives for cancer treatment and overcoming therapeutic resistance. *Genes & Diseases, 8*(4), 448–462.

Malaguarnera, L. (2019). Influence of resveratrol on the immune response. *Nutrients, 11*(5), 946.

Malemud, C. J. (2019). Inhibition of MMPs and ADAM/ADAMTS. *Biochemical Pharmacology, 165*, 33–40.

Mueller, A.-L., Payandeh, Z., Mohammadkhani, N., Mubarak, S. M., Zakeri, A., Alagheband Bahrami, A., … Shakibaei, M. (2021). Recent advances in understanding the pathogenesis of rheumatoid arthritis: new treatment strategies. *Cells, 10*(11), 3017.

Mukherjee, A., & Das, B. (2024). The role of inflammatory mediators and matrix metalloproteinases (MMPs) in the progression of osteoarthritis. *Biomaterials and Biosystems, 13, 100090.*

Ohshiba, T., Miyaura, C., Inada, M., & Ito, A. (2003). Role of RANKL-induced osteoclast formation and MMP-dependent matrix degradation in bone destruction by breast cancer metastasis. *British Journal of Cancer, 88*(8), 1318–1326.

Pulik, Ł., Łęgosz, P., & Motyl, G. (2023). Matrix metalloproteinases in rheumatoid arthritis and osteoarthritis: a state of the art review. *Reumatologia, 61*(3), 191.

Rahmani, A. H., Almatroudi, A., Allemailem, K. S., Alwanian, W. M., Alharbi, B. F., Alrumaihi, F., … Almatroodi, S. A. (2023). Myricetin: a significant emphasis on its anticancer potential via the modulation of inflammation and signal transduction pathways. *International Journal of Molecular Sciences, 24*(11), 9665.

Ruan, J., Shi, Z., Cao, X., Dang, Z., Zhang, Q., Zhang, W., … Wang, T. (2024). Research progress on anti-inflammatory effects and related mechanisms of astragalin. *International Journal of Molecular Sciences, 25*(8), 4476.

Serasanambati, M., & Chilakapati, S. R. (2016). Function of nuclear factor kappa B (NF-kB) in human diseases-a review. *South Indian Journal of Biological Sciences, 2*(4), 368–387.

Shan, L., Wang, F., Zhai, D., Meng, X., Liu, J., & Lv, X. (2023). Matrix metalloproteinases induce extra-cellular matrix degradation through various pathways to alleviate hepatic fibrosis. *Biomedicine & Pharmacotherapy, 161*, 114472.

Song, X., Tan, L., Wang, M., Ren, C., Guo, C., Yang, B., … Pei, J. (2021). Myricetin: a review of the most recent research. *Biomedicine & Pharmacotherapy, 134*, 111017.

Surai, P. F., Surai, A., & Earle-Payne, K. (2024). Silymarin and inflammation: food for thoughts. *Antioxidants, 13*(1), 98.

Tian, S., Wei, Y., Hu, H., & Zhao, H. (2022). Mixed computational-experimental study to reveal the anti-metastasis and anti-angiogenesis effects of astragalin in human breast cancer. *Computers in Biology and Medicine, 150*, 106131.

Tripathi, A. K., Ray, A. K., & Mishra, S. K. (2022). Molecular and pharmacological aspects of piperine as a potential molecule for disease prevention and management: evidence from clinical trials. *Beni-Suef University Journal of Basic and Applied Sciences, 11*(1), 16.

Wagner, A. E., Boesch-Saadatmandi, C., Dose, J., Schultheiss, G., & Rimbach, G. (2012). Anti-inflammatory potential of allyl-isothiocyanate–role of Nrf2, NF-κB and microRNA-155. *Journal of Cellular and Molecular Medicine, 16*(4), 836–843.

Yang, J., Sheng, Y., Lu, Y., & Zhao, Y. (2021). Piperine modulates NLRP3/TLR4/NF-KB signalling processes of angiogenesis and neurogenesis in rats following ischemic stroke. *Acta Poloniae Pharmaceutica, 78*(2), 1–13.

Yi, Y.-S. (2019). Ameliorative effects of ginseng and ginsenosides on rheumatic diseases. *Journal of Ginseng Research, 43*(3), 335–341.

Yu, H., Lin, L., Zhang, Z., Zhang, H., & Hu, H. (2020). Targeting NF-κB pathway for the therapy of diseases: mechanism and clinical study. *Signal Transduction and Targeted Therapy, 5*(1), 209.

Zhang, H., Chen, L., Sun, X., Yang, Q., Wan, L., & Guo, C. (2020). Matrine: a promising natural product with various pharmacological activities. *Frontiers in Pharmacology, 11*, 588.

6 Phytochemicals Targeting Arachidonic Acid Pathway to Treat Rheumatoid Arthritis

Aisha Shehzad, Aqsa Mushtaq,
Hafiza Sidra Yaseen, and Ammara Saleem

WHAT WILL YOU LEARN?

This chapter provides important information to the reader about arachidonic acid pathway and its role in inflammatory diseases and rheumatoid arthritis. The chapter covers the role of phytochemicals and preclinical studies that have investigated these interactions, highlighting their potential in developing novel therapeutic strategies in managing RA. By focusing on these biochemical interactions, this chapter suggests innovative approaches that could significantly improve the quality of life for individuals suffering from this chronic and debilitating disease.

LIST OF ABBREVIATIONS

AA	Arachidonic acid
COX	Cyclooxygenase
CYP 450	Cytochrome P450
EETs	Eicosatetraenoic acids
HETEs	Hydroxy eicosatetraenoic acids
HLA	Human leukocyte antigen
IL-1β	Interleukin-1β
IL-6	Interleukin-6
iNOS	Inducible nitric oxide synthase
LOX	Lipoxygenase
LTA4	Leukotriene A4
LTB4	Leukotriene B4
MMPs	Matrix metalloproteinases
NSAIDs	Nonsteroidal anti-inflammatory drugs
PGG2	Prostaglandin G2
PLA2	Phospholipase A2
RA	Rheumatoid arthritis
RANKL	Receptor activator of nuclear factor kappa-B ligand
SIRT1	Sirtuin 1
STAT3	Signal transducer and activator of transcription 3
TNF-α	Tumor necrosis factor-alpha

6.1 INTRODUCTION

Rheumatoid arthritis (RA) is a chronic autoimmune disease marked by synovial inflammation and the formation of auto-antibodies, like anticitrullinated protein antibodies and rheumatoid factors persistently (Wu, Yang, & Lai, 2020). These antibodies are involved in its pathogenesis,

DOI: 10.1201/9781003535201-8

contributing to the destruction and functional impairment of joints over time. RA is marked by synovitis results in the thickening of the synovial membrane, invasion of cartilage and bone by the inflamed synovium, and subsequent joint destruction. This autoimmune response results in pain, swelling, stiffness, and decreased mobility in affected joints (Conigliaro et al., 2016).

The pathogenesis of RA links with multiple factors like environment, genetic, and immunological response. Genetically, human leukocyte antigen (HLA) class II genes, predominantly HLA-DRB1, play a major role in RA. Various determinants including infections, smoking, and changes in hormones aggravate it. Immunologically, disease is characterized by stimulation of numerous immune cells, including B cells, T cells, macrophages, and fibroblast-like synoviocytes, which perpetuate the inflammatory process (Deane et al., 2017).

Prostaglandins (PGs), produced via the cyclooxygenase (COX) enzymes, are involved in the inflammatory milieu of disease. They contribute to vasodilation, increased vascular permeability, and leukocyte recruitment to the inflamed synovium. Leukotrienes, generated through the lipoxygenase (LOX) pathway, are involved in leukocyte chemotaxis and activation, further exacerbating the inflammatory response. Thromboxanes (TX), another product of the arachidonic acid (AA) pathway, contribute to platelet aggregation and vascular changes, which can affect joint microcirculation (Wang et al., 2021).

Pharmacological inhibition of the AA pathway has been a successful strategy in RA treatment because this mechanism is responsible for controlling inflammation in RA. Corticosteroids have been widely used to reduce inflammation and alleviate symptoms. Additionally, biologics targeting specific cytokines or molecules involved in the AA pathway, such as leukotriene modifiers, have shown promising results in RA treatment (Everts, Währborg, & Hedner, 2000). Nonsteroidal anti-inflammatory drugs (NSAIDs) that are inhibitors of COX enzymes and selective COX-2 inhibitors relieve inflammation and pain. Moreover, research into phytochemicals and other natural compounds that modulate the AA pathway is ongoing, offering potential new treatments that could provide better efficacy and safety profiles (Everts et al., 2000).

6.2 ARACHIDONIC ACID PATHWAY IN INFLAMMATORY DISEASES

The AA pathway contributes to pathophysiology of numerous ailments like RA, cardiovascular diseases, and asthma. Leukotrienes are produced in asthma by causing bronchoconstriction, mucus secretion, and airway inflammation. Elevated levels of leukotrienes like LTC4, LTD4, and LTE4 are associated with asthma exacerbations. In RA, the AA pathway is upregulated, leading to increased concentration of proinflammatory eicosanoids such as prostaglandin E2 (PGE2) and leukotriene B4 (LTB4). These mediators contribute to chronic inflammation, pain, and joint destruction. Proinflammatory eicosanoids, such as TXA2, contribute to atherosclerosis and thrombosis by promoting platelet aggregation and vascular inflammation. Increased levels of proinflammatory PGs and leukotrienes are found in the intestinal mucosa of inflammatory bowel disease patients, contributing to chronic gut inflammation (Everts et al., 2000).

6.3 ARACHIDONIC ACID PATHWAY IN RA PATHOGENESIS

The AA pathway is a crucial biochemical cascade involved in inflammation and immune responses. It begins with the production of AA from phospholipids, a process mediated via the enzyme phospholipase A2 (PLA2). Once released, AA serves as a substrate for further metabolism through the primary enzymatic pathways: LOX, COX, and cytochrome P450 (CYP450) enzymes as shown in Table 6.1. Each of these pathways generates various distinct bioactive lipids, which play critical roles in modulating inflammatory processes and immune system functions (Sharma, Rani, Dhalwal, Shinde, & Mahadik, 2009).

AA accelerates the release of matrix metalloproteinase 9 (MMP9), which damages an extracellular matrix, leading to the damage of joints. Because mediators of AA, LTB4, 5-hydroxyeicosatetraenoic

TABLE 6.1
Mediators Involved in Arachidonic Acid Pathway

Component	Role in Arachidonic Acid Pathway	Impact on Rheumatoid Arthritis
Arachidonic Acid	Released from cell membrane phospholipids by phospholipase A2 (PLA2)	Serves as a precursor for proinflammatory eicosanoids
Cyclooxygenase (COX) Enzymes	COX-1 and COX-2 convert arachidonic acid into prostaglandin H2 (PGH2)	COX-2 increases the production of proinflammatory prostaglandins during inflammation
Prostaglandins (e.g., PGE2)	Synthesized from PGH2 via specific synthases	PGE2 promotes inflammation, pain, and fever, contributing to joint damage
Thromboxanes (e.g., TXA2)	Produced from PGH2 by thromboxane synthase	TXA2 causes platelet aggregation and vasoconstriction, exacerbating inflammation
Lipoxygenase (LOX) Enzymes	Convert arachidonic acid into leukotrienes	Leukotrienes (e.g., LTB4) attract immune cells, increasing tissue damage and pain
Leukotrienes (e.g., LTB4)	Produced from arachidonic acid by 5-LOX enzymes	LTB4 is a potent chemotactic agent for neutrophils, promoting inflammatory responses
Cytokines	Influence the activity of enzymes in the arachidonic acid pathway	Proinflammatory cytokines like TNF-α and IL-1β upregulate COX-2 and PLA2, amplifying the inflammatory response
Phospholipase A2 (PLA2)	Releases arachidonic acid from membrane phospholipids	PLA2 activation initiates the inflammatory cascade in rheumatoid arthritis

acid (5-HETE), PGE2, 8-HETE, 12-HETE, 15-HETE, and TXA2 were increased in arthritis. These eicosanoids produced by AA regulate inflammation and inflammatory cell infiltration. LTB4 plays a role in the chemotaxis of polymorphonuclear neutrophil and upsurge acute inflammation. PGE2 contributes to synovial hyperplasia, immune system inflammation, leukocyte infiltration, and release of cytokines like interleukin (IL)-23-dependent IL-17. 15-HETE upsurges the level of MMP9 and MMP2, which lead to angiogenesis (Table 6.2). The levels of 15-lipoxygenase and 5-lipoxygenase (5-LOX) were found to be greater in RA in contrast to osteoarthritis (Goldring, 2003).

6.3.1 Cyclooxygenase Pathway

The COX pathway is a central component of AA cascade, involving two key isoenzymes: COX-1 and COX-2 (Saleem, Saleem, Akhtar, Shahzad, & Jahan, 2020). These enzymes transform AA into prostaglandin G2 (PGG2) and consequently prostaglandin H2 (PGH2). PGH2 plays a role in the making of various PGs and thromboxanes, which are the lipids associated with numerous physiological and pathological processes in the body as described in Figure 6.1.

COX-1 is constitutively found in most tissues and is primarily linked with functioning of normal physiological processes. It protects the mucosa of stomach, supporting kidney function and regulating the aggregation of platelets. For example, TXA2 derived from PGH2 through COX-1 activity is essential for platelet aggregation and vasoconstriction, thereby contributing to hemostasis. On the other hand, COX-2 is an inducible enzyme that is upregulated in response to stimuli such as cytokines, bacterial endotoxins, and growth factors (Saleem et al., 2019). While COX-2 is minimally expressed under normal conditions, its expression is significantly increased during

TABLE 6.2
Phytochemicals Targeting Arachidonic Acid Pathway in Rheumatoid Arthritis

Phytochemical	Class	Botanical Source	Mechanism of Action	References
Curcumin	Polyphenol	*Curcuma longa*	1. Downregulation of iNOS and 5-LOX expression 2. Inhibits COX-2 and reduces the production of PGE2 3. Suppresses the activation of nuclear factor-kappa B (NF-κB), MAPKs, and STAT3, reducing the expression of pro-inflammatory cytokines such as TNF-and IL-1β	Aggarwal and Harikumar (2009)
Resveratrol	Polyphenol	Grapes, berries, and peanuts	1. Inhibits both COX-2 and 5-LOX and reduces the production of prostaglandins and leukotrienes 2. Activates sirtuin 1 (SIRT1), 3. Inhibits NF-κB signaling, reducing the expression of proinflammat ory cytokines	Salehi et al. (2018)
Quercetin	Flavonoid	Fruits, vegetables, and grains	1. Inhibits both COX-2 and 5-LOX pathways 2. Inhibits enzymes such as PLA2. 3. Scavenges free radicals and reduces oxidative stress, which can exacerbate inflammation 4. Decreases the levels of PGE2 and LTB4	Zhao, Guo, Schrodi, and He (2021)
Omega-3 fatty acids	Fatty acid	Fish oil	1. Inhibits COX and LOX, leading to decreased synthesis of PGE2 and LTB4	Aggarwal and Harikumar (2009)
Epigallocatechin Gallate (EGCG)	Catechin	Green tea	1. Blocks major enzymes like PLA2, COX-2, and LOX, 2. Downregulates the expression of proinflammatory cytokines. e.g., TNF-α and IL-1β. 3. Has antioxidant activities	Aggarwal and Harikumar (2009)

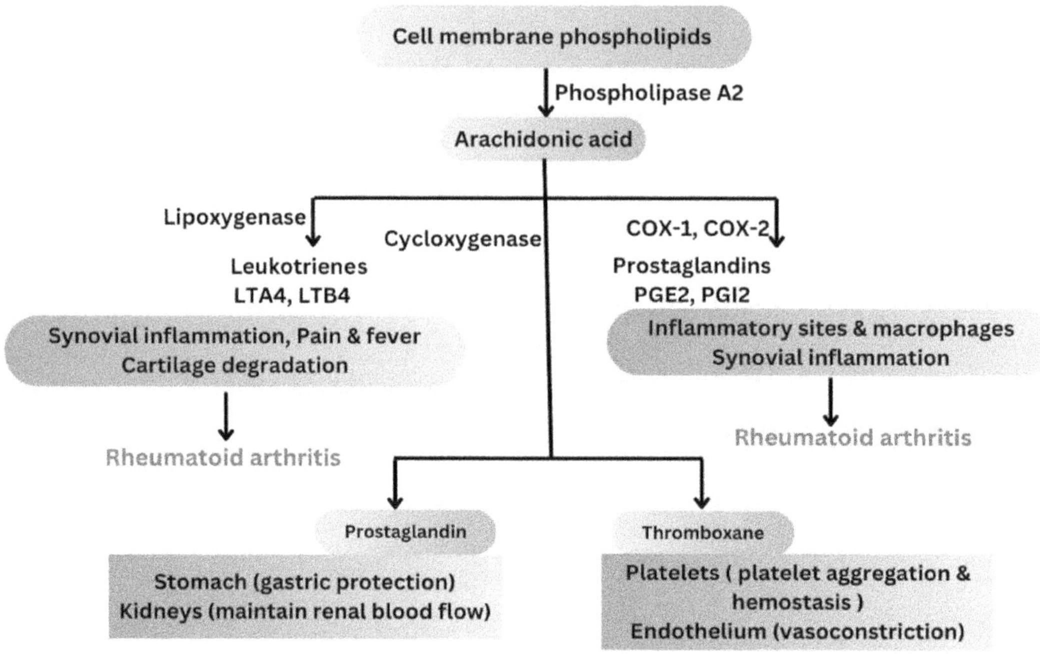

FIGURE 6.1 Role of arachidonic pathway in pathophysiology of rheumatoid arthritis.

inflammation, injury, and certain disease states. This upregulation leads to the production of large amounts of PGs that mediate inflammatory responses (Korbecki, Baranowska-Bosiacka, Gutowska, & Chlubek, 2014).

One of the key products of the COX-2 pathway is PGE2, which is involved in RA by promoting inflammation, pain, and joint damage. It contributes to the inflammatory process by inducing vasodilation, increasing vascular permeability, and enhancing the recruitment and activation of immune cells. Additionally, PGE2 sensitizes nociceptors, leading to increased pain perception. In the joints of RA patients, elevated levels of PGE2 are associated with synovial inflammation, cartilage degradation, and bone erosion (Korbecki et al., 2014).

The distinct roles of COX-1 and COX-2 have significant implications for therapeutic interventions. NSAIDs block both COX-1 and COX-2, which are useful in pain and inflammation. However, the nonselective stopping of COX-1 results in side effects like diminished renal function and gastrointestinal bleeding and due to the disruption of its protective physiological functions. For minimizing these side effects, selective COX-2 inhibitors (Coxibs) were used (Akhtar et al., 2021). The drugs specifically target COX-2, leading to alleviation of inflammation and pain while minimizing gastrointestinal side effects. Nonetheless, the use of coxibs has been linked to a greater risk of cardiovascular events, highlighting the complexity of modulating the COX pathway (Bingham, 2002).

6.3.2 Lipoxygenase Pathway

The LOX pathway is another critical arm of AA cascade, primarily involving the enzyme 5-LOX. This pathway plays a significant role in the synthesis of leukotrienes that play a role in inflammation and immune responses. In this mechanism, first AA converts into LTA4 by 5-LOX. LTA4 serves as a key intermediate and can be further metabolized into various other leukotrienes with distinct biological functions (Nakamura & Shimizu, 2023).

One of the primary metabolites of LTA4 is LTB4, a highly potent chemoattractant. LTB4 plays a crucial role in recruiting neutrophils and other immune cells to sites of inflammation, such as the inflamed synovium in RA. By attracting these immune cells, LTB4 exacerbates the inflammatory response, contributing to the chronic inflammation and joint damage characteristic of RA. The accumulation of neutrophils in the synovium leads to the production of reactive oxygen species and proteolytic enzymes, further damaging the joint tissues (Horn et al., 2015).

In addition to LTB4, LTA4 can also be converted into a group of compounds known as cysteinyl leukotrienes, which include LTD4, LTC4, and LTE4. The leukotrienes are formed through the addition of a cysteine-containing peptide to LTA4. Cysteinyl leukotrienes are potent mediators of greater vascular permeability, smooth muscle contraction, and edema. In the context of RA, they contribute to the swelling and stiffness of affected joints by promoting fluid accumulation and tissue edema (Christie & Henderson, 2023). Moreover, cysteinyl leukotrienes play significant roles in other inflammatory conditions, such as asthma, where they induce bronchoconstriction and mucus production (Capra et al., 2007).

The LOX pathway's involvement in inflammation makes it a critical objective in treating RA and various inflammatory ailments. Inhibitors of 5-LOX, as well as leukotriene receptor antagonists, have been developed to modulate this pathway and reduce inflammation. For example, zileuton, a 5-LOX inhibitor, can decrease the production of leukotrienes and has been used in the treatment of asthma. Similarly, leukotriene receptor antagonists like montelukast block the actions of cysteinyl leukotrienes and are also used in managing asthma and allergic rhinitis (Aggarwal & Harikumar, 2009; Saleem, Mubeen, Akhtar, & Zeb, 2022).

Phytochemicals that inhibit the LOX pathway offer promising potential for developing new treatments with fewer side effects. These compounds may provide novel approaches to managing chronic inflammatory diseases by targeting specific steps within the LOX pathway, thereby reducing leukotriene production and mitigating inflammation. Overall, the LOX pathway's role in mediating inflammatory responses underscores its importance in the pathophysiology of RA and other related conditions. By understanding the specific mechanisms through which leukotrienes contribute to inflammation and tissue damage, researchers can develop targeted therapies aimed at modulating this pathway. Such therapies hold the promise of improving the management and outcomes of patients with RA and other inflammatory diseases, ultimately enhancing their quality of life (Bingham, 2002; Zheng, Li, Hong, & Yang, 2019).

6.3.3 Cytochrome p450 Pathway

The cytochrome P450 pathway represents a critical and multifaceted component of the AA pathway. This pathway involves metabolism of AA by various CYP enzymes, leading to the production of bioactive lipids such as epoxyeicosatrienoic acids (EETs) and HETEs. These metabolites play intricate roles in modulating inflammation and immune responses, demonstrating both proinflammatory and anti-inflammatory effects. EETs are primarily recognized for their anti-inflammatory properties. They are synthesized by CYP epoxygenases from AA and have been shown to reduce inflammation. EETs exert their effects by promoting vasodilation, enhancing blood flow, and reducing leukocyte adhesion and infiltration into tissues. Additionally, EETs can stop the release of nuclear factor kappa B (NF-κB), a major transcription factor plays a role in inflammation. By regulating these processes, EETs help protect tissues from inflammatory damage and promote healing. In RA, anti-inflammatory actions of EETs may help mitigate synovial inflammation and joint destruction, offering potential therapeutic benefits (Bingham, 2002).

HETEs, on the other hand, have more diverse and context-dependent effects. HETEs are produced through the hydroxylation of AA by different CYP enzymes. Depending on the specific enzyme involved and the site of hydroxylation, HETEs can have varying impacts on inflammation. For instance, 12-HETE and 15-HETE can exhibit proinflammatory effects by promoting leukocyte chemotaxis and enhancing the inflammatory response. Conversely, other HETEs such as 20-HETE possess both pro- and anti-inflammatory properties, related to tissue context and local inflammatory milieu (Rodríguez-Carrio et al., 2021).

The dual nature of HETEs underscores the complexity of the CYP pathway in regulating inflammation. In RA, the proinflammatory HETEs can exacerbate synovial inflammation and play a role in the chronicity of ailment. However, the anti-inflammatory HETEs may counterbalance these effects, highlighting the potential for selective targeting of specific HETE pathways to achieve therapeutic benefits (Aggarwal & Harikumar, 2009).

The CYP pathway's involvement in producing both EETs and HETEs illustrates its nuanced role in inflammation and immune regulation. Therapeutic strategies aimed at modulating this pathway must consider the balance among proinflammatory and anti-inflammatory effects. By enhancing the production of EETs or mimicking their actions through synthetic analogs could provide a novel approach to reduce inflammation and tissue damage in RA. Conversely, selectively inhibiting the pro-inflammatory HETEs may offer another avenue for therapeutic intervention (Hoxha & Zappacosta, 2020).

Additionally, natural compounds and phytochemicals that influence the CYP pathway are being explored for their potential to offer safer and more effective treatments for inflammatory diseases. The cytochrome P450 pathway is a vital component of the AA cascade with significant implications for inflammation and immune regulation. Its metabolites, EETs, and HETEs, play complex roles in modulating inflammatory responses, making this pathway an important focus for treating RA and other inflammatory maladies (Christmas, 2015).

6.4 ROLE OF PROINFLAMMATORY MEDIATORS IN RA

The bioactive lipids produced by the AA pathway, including PGs and leukotrienes, are crucial mediators of inflammation in RA. These compounds are deeply involved in various pathological processes that characterize the diseases as discussed below.

6.4.1 Synovial Inflammation

PGE2 and LTB4 play pivotal roles in promoting synovial inflammation. They enhance the infiltration of immune cells, like macrophages, neutrophils, and lymphocytes, in synovial membrane. This influx of immune cells leads to synovial hyperplasia, where the synovial lining thickens abnormally. Over time, this results in the formation of pannus, an aggressive tissue that invades and destroys adjacent cartilage and bone. PGE2, in particular, contributes to vasodilation and increased vascular permeability, aiding the movement of immune cells to the place of inflammation. LTB4 acts as a potent chemoattractant, further recruiting and activating neutrophils and other immune cells within the synovium (Afnan, Saleem, & Akhtar, 2023; Gandhi et al., 2022).

6.4.2 Cartilage Degradation

The proinflammatory mediators produced in RA significantly contribute to cartilage degradation. PGE2 and LTB4 stimulate synovial fibroblasts and chondrocytes to produce MMPs and other catabolic enzymes. MMPs, such as MMP-1 (collagenase-1) and MMP-13 (collagenase-3), degrade extracellular matrix components of cartilage, involving collagen and proteoglycans. The enzymatic breakdown of cartilage matrix results in a lack of cartilage integrity and function, which is a hallmark of RA progression (Mehmood, Saleem, Akhtar, & Mobashar, 2025).

6.4.3 Bone Erosion

Proinflammatory mediators such as PGE2 and LTB4 also play crucial roles in bone erosion, a significant cause of disability in RA patients. These mediators promote the differentiation and activation of osteoclasts, which are involved in bone resorption. PGE2 enhances osteoclastogenesis by enhancing the level of receptor activator of nuclear factor kappa-B ligand (RANKL) on osteoblasts and stromal cells (Chen et al., 2023). RANKL interacts with its receptor, RANK, on osteoclast

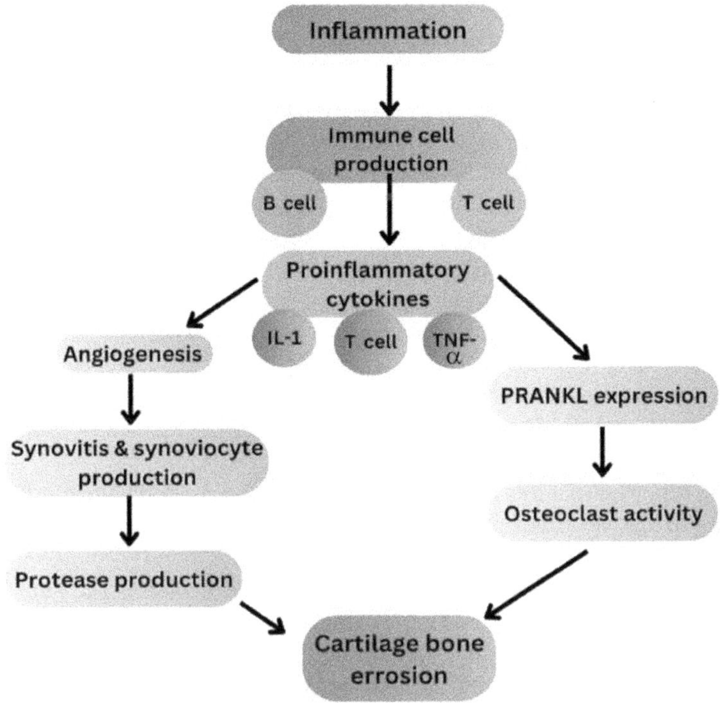

FIGURE 6.2 Role of proinflammatory mediators in cartilage and bone erosion.

precursors, stimulating their maturation and activity. LTB4 similarly contributes to osteoclast differentiation and bone resorption as shown in Figure 6.2. The increased osteoclast activity leads to the erosion of bone at the sites of inflammation, contributing to joint destruction and deformities in RA (Akaogi, Nozaki, Satoh, & Yamada, 2006).

6.4.4 Pain Sensation

Pain is a predominant symptom in RA, significantly affecting patients' quality of life. PGE2 plays a central role in mediating pain and hyperalgesia (increased sensitivity to pain) in RA. It sensitizes nociceptors, the sensory neurons responsible for transmitting pain signals to the central nervous system. By lowering the threshold for nociceptor activation, PGE2 enhances pain perception in response to inflammatory stimuli. This process involves the modulation of ion channels and receptors on nociceptive neurons, leading to increased neuronal excitability. The persistent production of PGE2 in inflamed joints results in chronic pain and discomfort for RA patients (Javed, Saleem, & Akhtar, 2023).

6.5 PHYTOCHEMICALS TARGETING ARACHIDONIC ACID PATHWAY AND PRECLINICAL STUDIES

Phytochemicals, naturally occurring compounds found in plants, modulate the AA pathway and decrease inflammation in RA. These compounds often have fewer side effects in comparison with synthetic drugs and can target multiple inflammatory pathways simultaneously as shown in Table 6.2 (Canistro et al., 2021). Numerous groups such as alkaloids, polyphenols, coumarins, terpenoids, and flavonoids present in plants demonstrate an anti-inflammatory effect by stopping the AA pathway.

A study revealed that mangiferin found in the Ongael island alleviates the levels of tumor necrosis factor-alpha (TNF-α), IL-6, RANKL, and IL-1β in arthritic mice via inhibiting NF-κB and extracellular signal regulated kinase (ERK)1/2 (Tsubaki et al., 2015). Genitein is present in soy and prevents inflammation via diminishing IFN-γ while elevating IL-4 in the collagen-induced arthritic (CIA) animal model (Wang et al., 2008). P-coumaric acid present in wheat, grapes, apples, spinach, and tomatoes lessens TNF-α within synovial fluids in arthritic rats (Pragasam, Venkatesan, & Rasool, 2013). Triptolide decreases IL-18 as well as receptors of IL-18 in synovial fibroblasts caused via phorbol 12-myristate 13-acetate (PMA) (Lu et al., 2008). Furthermore, triptolide displays an immunosuppressant activity via reducing CD_4^+ T cells and enhancing CD_8^+ T cells within Peyer's patch of arthritic rats stimulated by collagen (Zhou et al., 2006). β-cryptoxanthin, which is a carotenoid, alleviates polyarthritis in arthritic animals by targeting COX pathways (Pattison et al., 2005).

Javaid et al. reported an antinociceptive, anti-inflammatory, and anti-arthritic potential of diosgenin in arthritic rats by downregulating COX-2, IL-6, NF-κβ, and TNF-α. Moreover, diosgenin reduced oxidative stress in arthritic animals in contrast to methotrexate-treated rats (Javed et al., 2023).

Afnan et al. reported that chrysin-treated pain and inflammation in arthritic rats by the downregulation of COX-2, IL-6, NF-κβ, and TNF-α in treated arthritic rats as compared to methotrexate and disease control (Afnan et al., 2023).

Mehmood et al. reported that diosmin-loaded nanoparticles treated adjuvant-induced arthritis in rats by modulating oxidative stress, pro-inflammatory cytokines, NF-κβ, and COX-2 in contrast to disease control (Mehmood et al., 2025).

A previous *in vitro* study revealed that the methanol extract of *Colchicum luteum* considerably stops LOX pathways and is responsible for the anti-inflammatory property (Akbar & Akbar, 2020). Hydroethanolic extract of *Colchicum luteum* decreases inflammatory mediators in the arthritic model induced by collagen with no toxic effects. The compound-15 (4-(4,7-dimethoxy-1,3-benzodioxol-5-yl)-2-oxo pyrrolidine-3-carboxylic acid) identified in this study inhibits TNF-α and COX-2. Therefore, that extract can be used in place of NSAIDs in RA management. On the other hand, more clinical findings are necessary for proving its efficacy in human beings (Abbasi et al., 2024).

Odontites vulgaris, frequently employed in Mongolian medicine, has chemical constituents such as phenyl ethanol glycosides, flavonoids, and iridoids, which reduce IL-1β, NO, IL-6, and TNF-α (Gong et al., 2020). Therefore, it is effective in the management of RA via inhibiting PI3K-Akt (phosphatidylinositol-3 kinases-protein kinase B) (Ji et al., 2021).

6.5.1 Curcumin

Curcumin, the active component of turmeric (*Curcuma longa*), is beneficial due to anti-inflammatory and antioxidant activities. Curcumin inhibits COX-2 and reduces the production of PGE2. It also suppresses the activation of NF-κB, a major transcription factor that contributes to inflammation (Aggarwal & Harikumar, 2009). Curcumin exerts its effects through multiple mechanisms such as curcumin reduces the release of AA from membrane phospholipids by inhibiting PLA2 activity. It decreases the formation of proinflammatory eicosanoids via inhibiting the expression of COX-2 and 5-LOX. It also inhibits NF-κB, MAPKs, and STAT3, decreasing the expression of proinflammatory cytokines and mediators (Pan, Lai, & Ho, 2010). It has a potent antioxidant property, which aids in mitigating oxidative stress and inflammation (Aggarwal & Harikumar, 2009).

In preclinical models of RA, curcumin has shown significant anti-inflammatory effects. In a CIA model in rats, curcumin administration significantly reduced paw swelling, synovial inflammation, and cartilage erosion. These effects were attributed to decreased levels of PGE2 and COX-2 as compared to diseased rats (Wang et al., 2019).

6.5.2 Resveratrol

Resveratrol, a polyphenol, is present in grapes, berries, and peanuts. It has anti-inflammatory and immunomodulatory effects, inhibiting both COX and LOX enzymes and reducing the synthesis of PGs and leukotrienes. Resveratrol lessens the formation of proinflammatory eicosanoids by inhibiting these enzymes like COX-2 and 5-LOX (Meng et al., 2021). It activates sirtuin 1 (SIRT1), which has anti-inflammatory effects and modulates metabolic pathways. It inhibits NF-κB signaling, decreasing the levels of pro-inflammatory cytokines. In a rat model of RA, resveratrol administration significantly reduced joint swelling, inflammation, and cartilage damage. These effects were associated with decreased levels of PGE2, LTB4, and reduced expression of COX-2 and 5-LOX (Yang et al., 2018). Another study found that resveratrol treatment in mice with CIA reduced the formation of proinflammatory cytokines like IL-6 and IL-17 and inhibited osteoclast differentiation and bone resorption (Salehi et al., 2018).

6.5.3 Quercetin

Quercetin, a flavonoid, is present in fruits, vegetables, and grains. It possesses potent anti-inflammatory and antioxidant effects and blocks both COX and LOX pathways, reducing the production of Leukotrienes (LTs) and PGs (Goyal & Agrawal, 2022). Quercetin reduces the release of AA and formation of proinflammatory eicosanoids by inhibiting these enzymes including PLA2, COX-2, and 5-LOX. It scavenges free radicals and diminishes oxidative stress, which can exacerbate inflammation. In a rat model of RA, quercetin administration significantly reduced paw swelling, synovial inflammation, and cartilage degradation (Haleagrahara et al., 2017). These effects were attributed to decreased levels of PGE2 and LTB4 and reduced expression of COX-2 and 5-LOX. Another study in a mouse model of RA found that quercetin treatment reduced the formation of proinflammatory cytokines and chemokines and inhibited the relocation of immune cells into the synovium (Zhao et al., 2021).

6.5.4 Omega-3 Fatty Acids

It is present in fish oil and exhibits anti-inflammatory effects by restricting AA for the same enzymes, resulting in the production of less inflammatory eicosanoids. It inhibits COX and LOX, leading to decreased synthesis of PGE2 and LTB4 and stimulating the formation of anti-inflammatory mediators. Several studies have confirmed the efficacy of omega-3 fatty acids in alleviating inflammation and disease severity in RA models. Supplementation with omega-3 fatty acids decreases the need for NSAIDs in patients with RA (Aggarwal & Harikumar, 2009).

6.5.5 Epigallocatechin Gallate (EGCG)

It is a catechin present in green tea, demonstrating strong anti-inflammatory activity. It blocks major enzymes like PLA2, COX-2, and LOX enzymes, in this manner alleviating the formation of proinflammatory eicosanoids. It downregulates the level of proinflammatory cytokines such as TNF-α and IL-1β. It has antioxidant activities that mitigate oxidative stress, a contributor to inflammation and joint damage in RA. A research demonstrated that EGCG mitigates joint inflammation and slow disease progression in animal models of RA. Another research reported that EGCG-treated mice exhibited reduced joint swelling and histopathological scores in adjuvant-induced arthritis as compared to disease control (Aggarwal & Harikumar, 2009).

6.5.6 Gingerol

Gingerol, a bioactive compound in ginger, may help prevent RA via inhibiting COX-2 and 5-LOX enzymes, thus reducing the production of proinflammatory cytokines, PGs, and leukotrienes. It prevents the stimulation of NF-κB via reducing the activation of inflammatory genes (van Breemen, Tao, & Li, 2011). It also has antioxidant activity and inhibits proinflammatory cytokines like TNF-α, IL-1β, and IL-6. Also, bioactive compounds like 8-gingerol, 10-shogaol, 10-gingerol, and gingerdione present in it cannot block COX-1, but nevertheless stop COX-2 that plays a role in RA (Al-Nahain, Jahan, & Rahmatullah, 2014; Dugasani et al., 2010). A recent *in vitro* investigation exhibited that synovial cells separated from synovial fluid treated with ginger significantly suppress the levels of IL-1 and IL-6 similar to betamethasone (Ribel-Madsen et al., 2012).

6.6 CONCLUSION

The AA pathway is fundamentally critical in the pathogenesis of RA, largely through the generation of proinflammatory mediators like PGs and leukotrienes. These mediators are key players in several pathological processes such as synovial inflammation, cartilage degradation, bone erosion, and pain, all of which contribute to the chronic and debilitating nature of RA. Phytochemicals including curcumin, resveratrol, diosgenin, chrysin and quercetin, gingerols, omega-3 fatty acids, EGCG, and boswellic acid play a key role in regulating the AA pathway by preventing enzymes like COX-2 and 5-LOX. These naturally occurring compounds offer a multifaceted approach in managing RA by targeting multiple enzymes and signaling pathways involved in AA metabolism. Their multitargeted actions can offer a comprehensive approach to reduce inflammation and prevent joint damage. However, the exact mechanisms of action of compounds in the context of RA require further elucidation.

6.7 FUTURE RESEARCH DIRECTIONS

Continued research is essential to fully realize the therapeutic potential of phytochemicals targeting the AA pathway in clinical settings. Key areas for future investigation are as follows:

Mechanistic studies: Detailed investigations are required to explore precise molecular mechanisms through which these phytochemicals modulate the AA pathway and related inflammatory processes in RA.

Clinical trials: Well-designed clinical trials are crucial to assess efficacy, safety, and optimal dosages of compounds in RA patients. Such trials will help determine their potential as complementary or alternative treatments to conventional RA therapies.

Bioavailability and formulation: Improving the bioavailability and stability of these phytochemicals is vital to enhance their therapeutic effectiveness. Research into novel delivery systems, such as nanoparticles or liposomal formulations, could significantly impact their clinical utility.

Synergistic effects: Investigating the potential synergistic effects of combining phytochemicals with existing RA treatments could lead to more effective and comprehensive therapeutic strategies.

6.8 TAKE HOME MESSAGE

- This suggests that phytochemicals could serve as valuable complementary therapies, providing a natural means to manage RA and potentially reduce the need for conventional medications that often have significant side effects.
- Continued research and clinical trials are essential to find out the optimal dosages, safety, and long-term efficacy of these compounds.
- Integrating phytochemicals into RA treatment regimens holds promise for enhancing therapeutic outcomes and offering patients a more holistic approach to disease management.

REFERENCES

Abbasi, H., Sharif, M., John, P., Bhatti, A., Hayat, M. Q., & Mansoor, Q. (2024). Phytochemical, cytoprotective profiling, and anti-inflammatory potential of Colchicum luteum in rheumatoid arthritis: an experimental and simulation study. *Nutrients, 16*(23), 4020.

Afnan, A., Saleem, A., & Akhtar, M. F. (2023). Chrysin, a 5, 7-dihydroxyflavone restrains inflammatory arthritis in rats via subsiding oxidative stress biomarkers and inflammatory cytokines. *Inflammopharmacology, 31*(4), 1863–1878.

Aggarwal, B. B., & Harikumar, K. B. (2009). Potential therapeutic effects of curcumin, the anti-inflammatory agent, against neurodegenerative, cardiovascular, pulmonary, metabolic, autoimmune and neoplastic diseases. *The International Journal of Biochemistry & Cell Biology, 41*(1), 40–59.

Akaogi, J., Nozaki, T., Satoh, M., & Yamada, H. (2006). Role of PGE2 and EP receptors in the pathogenesis of rheumatoid arthritis and as a novel therapeutic strategy. *Endocrine, Metabolic & Immune Disorders-Drug Targets (Formerly Current Drug Targets-Immune, Endocrine & Metabolic Disorders), 6*(4), 383–394.

Akbar, S., & Akbar, S. (2020). Colchicum luteum Baker (Colchicaceae). In Shahid Akbar (Ed.), *Handbook of 200 Medicinal Plants: A Comprehensive Review of Their Traditional Medical Uses and Scientific Justifications* (pp. 691–694). Publisher Springer Nature.

Akhtar, M. F., Khan, K., Saleem, A., Baig, M. M. F. A., Rasul, A., & Abdel-Daim, M. M. (2021). Chemical characterization and anti-arthritic appraisal of Monotheca buxifolia methanolic extract in Complete Freund's Adjuvant-induced arthritis in Wistar rats. *Inflammopharmacology, 29*, 393–408.

Al-Nahain, A., Jahan, R., & Rahmatullah, M. (2014). Zingiber officinale: a potential plant against rheumatoid arthritis. *Arthritis,* 2014(1), 159089.

Bingham, C. O. (2002). The pathogenesis of rheumatoid arthritis: pivotal cytokines involved in bone degradation and inflammation. *The Journal of Rheumatology Supplement, 65*, 3–9.

Canistro, D., Chiavaroli, A., Cicia, D., Cimino, F., Curro, D., Dell'Agli, M., ... Martinelli, G. (2021). The pharmacological basis of the curcumin nutraceutical uses: an update. *Pharmadvances, 3*(2), 421–466.

Capra, V., Thompson, M. D., Sala, A., Cole, D. E., Folco, G., & Rovati, G. E. (2007). Cysteinyl-leukotrienes and their receptors in asthma and other inflammatory diseases: critical update and emerging trends. *Medicinal Research Reviews, 27*(4), 469–527.

Chen, X., Chen, J., Miao, C., Yin, G., Zhang, Z., Sun, R., & Ni, S. (2023). Acetyl zingerone ameliorates osteoarthritis by inhibiting chondrocyte programmed cell death. *Molecular Medicine Reports, 28*(5), 1–15.

Christie, P. E., & Henderson, W. R. (2023). Lipid inflammatory mediators: leukotrienes, prostaglandins, platelet-activating factor. In B. Zweiman, L. B. Schwartz (Eds.), Inflammatory Mechanisms in Allergic Diseases (pp. 233–254). Taylor and Francis.

Christmas, P. (2015). Role of cytochrome P450s in inflammation. *Advances in Pharmacology, 74*, 163–192.

Conigliaro, P., Chimenti, M., Triggianese, P., Sunzini, F., Novelli, L., Perricone, C., & Perricone, R. (2016). Autoantibodies in inflammatory arthritis. *Autoimmunity Reviews, 15*(7), 673–683.

Deane, K. D., Demoruelle, M. K., Kelmenson, L. B., Kuhn, K. A., Norris, J. M., & Holers, V. M. (2017). Genetic and environmental risk factors for rheumatoid arthritis. *Best Practice & Research Clinical Rheumatology, 31*(1), 3–18.

Dugasani, S., Pichika, M. R., Nadarajah, V. D., Balijepalli, M. K., Tandra, S., & Korlakunta, J. N. (2010). Comparative antioxidant and anti-inflammatory effects of [6]-gingerol,[8]-gingerol,[10]-gingerol and [6]-shogaol. *Journal of Ethnopharmacology, 127*(2), 515–520.

Everts, B., Währborg, P., & Hedner, T. (2000). COX-2-Specific inhibitors–the emergence of a new class of analgesic and anti-inflammatory drugs. *Clinical Rheumatology, 19*, 331–343.

Gandhi, Y., Kumar, R., Grewal, J., Rawat, H., Mishra, S. K., Kumar, V., ... Sharma, P. (2022). Advances in anti-inflammatory medicinal plants and phytochemicals in the management of arthritis: a comprehensive review. *Food Chemistry Advances, 1*, 100085.

Goldring, S. (2003). Pathogenesis of bone and cartilage destruction in rheumatoid arthritis. *Rheumatology, 42*(*suppl_2*), ii11–ii16.

Gong, X., Wang, J., Zhang, M., Wang, P., Wang, C., Shi, R., ... Li, M. (2020). Bioactivity, compounds isolated, chemical qualitative, and quantitative analysis of Cymbaria daurica extracts. *Frontiers in Pharmacology, 11*, 48.

Goyal, A., & Agrawal, N. (2022). Quercetin: a potential candidate for the treatment of arthritis. *Current Molecular Medicine, 22*(4), 325–335.

Haleagrahara, N., Miranda-Hernandez, S., Alim, M. A., Hayes, L., Bird, G., & Ketheesan, N. (2017). Therapeutic effect of quercetin in collagen-induced arthritis. *Biomedicine & Pharmacotherapy, 90*, 38–46.

Horn, T., Adel, S., Schumann, R., Sur, S., Kakularam, K. R., Polamarasetty, A., … Heydeck, D. (2015). Evolutionary aspects of lipoxygenases and genetic diversity of human leukotriene signaling. *Progress in Lipid Research, 57*, 13–39.

Hoxha, M., & Zappacosta, B. (2020). CYP-derived eicosanoids: implications for rheumatoid arthritis. *Prostaglandins & Other Lipid Mediators, 146*, 106405.

Javed, M., Saleem, A., & Akhtar, M. F. (2023). Diosgenin, a steroidal sapogenin, arrests arthritis through modulation of inflammatory cytokines and oxidative stress biomarkers in Wistar rats. *Inflammopharmacology, 31*(4), 1951–1966.

Ji, M., Wang, C., Yang, T., Meng, X., Wang, X., & Li, M. (2021). Integrated phytochemical analysis based on UPLC–MS/MS and network pharmacology approaches to explore the effect of Odontites vulgaris Moench on rheumatoid arthritis. *Frontiers in Pharmacology, 12*, 707687.

Korbecki, J., Baranowska-Bosiacka, I., Gutowska, I., & Chlubek, D. (2014). Cyclooxygenase pathways. *Acta Biochimica Polonica, 61*(4), 639–649.

Lu, Y., Wang, W.-J., Leng, J.-H., Cheng, L.-F., Feng, L., & Yao, H.-P. (2008). Inhibitory effect of triptolide on interleukin-18 and its receptor in rheumatoid arthritis synovial fibroblasts. *Inflammation Research, 57*, 260–265.

Mehmood, H., Saleem, A., Akhtar, M. F., & Mobashar, A. (2025). Amelioration of Freund's adjuvant-induced polyarthritis by Diosmin-loaded nanoparticles in Wistar rats; a mechanistic study. *Inflammopharmacology, 33*(6), 3099–3117.

Meng, T., Xiao, D., Muhammed, A., Deng, J., Chen, L., & He, J. (2021). Anti-inflammatory action and mechanisms of resveratrol. *Molecules, 26*(1), 229.

Nakamura, M., & Shimizu, T. (2023). Therapeutic target of leukotriene B4 receptors, BLT1 and BLT2: insights from basic research. *Biochimie, 215*, 60–68.

Pan, M.-H., Lai, C.-S., & Ho, C.-T. (2010). Anti-inflammatory activity of natural dietary flavonoids. *Food & Function, 1*(1), 15–31.

Pattison, D. J., Symmons, D. P., Lunt, M., Welch, A., Bingham, S. A., Day, N. E., & Silman, A. J. (2005). Dietary β-cryptoxanthin and inflammatory polyarthritis: results from a population-based prospective study. *The American Journal of Clinical Nutrition, 82*(2), 451–455.

Pragasam, S. J., Venkatesan, V., & Rasool, M. (2013). Immunomodulatory and anti-inflammatory effect of p-coumaric acid, a common dietary polyphenol on experimental inflammation in rats. *Inflammation, 36*, 169–176.

Ribel-Madsen, S., Bartels, E. M., Stockmarr, A., Borgwardt, A., Cornett, C., Danneskiold-Samsøe, B., & Bliddal, H. (2012). A synoviocyte model for osteoarthritis and rheumatoid arthritis: response to ibuprofen, betamethasone, and ginger extract—a cross-sectional in vitro study. *Arthritis,* 2012(1), 505842.

Rodríguez-Carrio, J., Coras, R., Alperi-López, M., López, P., Ulloa, C., Ballina-García, F. J., … Suárez, A. (2021). Profiling of serum oxylipins during the earliest stages of rheumatoid arthritis. *Arthritis & Rheumatology, 73*(3), 401–413.

Saleem, A., Mubeen, A., Akhtar, M. F., & Zeb, A. (2022). Polystichum braunii ameliorates airway inflammation by attenuation of inflammatory and oxidative stress biomarkers, and pulmonary edema by elevation of aquaporins in ovalbumin-induced allergic asthmatic mice. *Inflammopharmacology, 30*(2), 639–653.

Saleem, A., Saleem, M., Akhtar, M. F., Shahzad, M., & Jahan, S. (2020). Moringa rivae leaf extracts attenuate Complete Freund's adjuvant-induced arthritis in Wistar rats via modulation of inflammatory and oxidative stress biomarkers. *Inflammopharmacology, 28*, 139–151.

Saleem, A., Saleem, M., Akhtar, M. F., Sharif, A., Javaid, Z., & Sohail, K. (2019). In vitro and in vivo anti-arthritic evaluation of Polystichum braunii to validate its folkloric claim. *Pakistan Journal of Pharmaceutical Sciences, 32,* 1–7.

Salehi, B., Mishra, A. P., Nigam, M., Sener, B., Kilic, M., Sharifi-Rad, M., … Sharifi-Rad, J. (2018). Resveratrol: a double-edged sword in health benefits. *Biomedicines, 6*(3), 91.

Sharma, K., Rani, R., Dhalwal, K., Shinde, V., & Mahadik, K. (2009). Natural compounds as anti-arthritic agents-a review. *Pharmacognosy Reviews, 3*(5), 22.

Tsubaki, M., Takeda, T., Kino, T., Itoh, T., Imano, M., Tanabe, G., … Nishida, S. (2015). Mangiferin suppresses CIA by suppressing the expression of TNF-α, IL-6, IL-1β, and RANKL through inhibiting the activation of NF-κB and ERK1/2. *American Journal of Translational Research, 7*(8), 1371.

van Breemen, R. B., Tao, Y., & Li, W. (2011). Cyclooxygenase-2 inhibitors in ginger (Zingiber officinale). *Fitoterapia, 82*(1), 38–43.

Wang, B., Wu, L., Chen, J., Dong, L., Chen, C., Wen, Z., … Wang, D. W. (2021). Metabolism pathways of arachidonic acids: mechanisms and potential therapeutic targets. *Signal Transduction and Targeted Therapy, 6*(1), 94.

Wang, J., Zhang, Q., Jin, S., He, D., Zhao, S., & Liu, S. (2008). Genistein modulate immune responses in collagen-induced rheumatoid arthritis model. *Maturitas, 59*(4), 405–412.

Wang, Q., Ye, C., Sun, S., Li, R., Shi, X., Wang, S., … Shi, Q. (2019). Curcumin attenuates collagen-induced rat arthritis via anti-inflammatory and apoptotic effects. *International Immunopharmacology, 72*, 292–300.

Wu, C.-Y., Yang, H.-Y., & Lai, J.-H. (2020). Anti-citrullinated protein antibodies in patients with rheumatoid arthritis: biological effects and mechanisms of immunopathogenesis. *International Journal of Molecular Sciences, 21*(11), 4015.

Yang, G., Chang, C.-C., Yang, Y., Yuan, L., Xu, L., Ho, C.-T., & Li, S. (2018). Resveratrol alleviates rheumatoid arthritis via reducing ROS and inflammation, inhibiting MAPK signaling pathways, and suppressing angiogenesis. *Journal of Agricultural and Food Chemistry, 66*(49), 12953–12960.

Zhao, J., Guo, S., Schrodi, S. J., & He, D. (2021). Molecular and cellular heterogeneity in rheumatoid arthritis: mechanisms and clinical implications. *Frontiers in Immunology, 12*, 790122.

Zheng, L.-X., Li, K.-X., Hong, F.-F., & Yang, S.-L. (2019). Pain and bone damage in rheumatoid arthritis: role of leukotriene B4. *Clinical and Experimental Rheumatology, 37*(5), 872–878.

Zhou, J., Xiao, C., Zhao, L., Jia, H., Zhao, N., Lu, C., … Lu, A.-P. (2006). The effect of triptolide on CD4+ and CD8+ cells in Peyer's patch of SD rats with collagen induced arthritis. *International Immunopharmacology, 6*(2), 198–203.

7 Phytochemicals Targeting JAK/STAT Pathway to Treat Rheumatoid Arthritis

Sara Zahid, Sampath Chinnam, Sana Kashan, and Ammara Saleem

WHAT WILL YOU LEARN IN THIS CHAPTER?

The readers will learn about the mechanism of the JAK/STAT pathway and its role in inflammatory afflictions and in the pathogenesis of rheumatoid arthritis. This chapter will provide knowledge about phytochemicals tormenting the JAK/STAT pathway and relevant clinical studies. To improve the therapeutic efficacy phytocompounds, we summarize many drug delivery strategies (DDS) that address the physico-chemical constraints of intrinsic JAK/STAT inhibitors.

ABBREVIATIONS

ADAM	Altered disintegrin protein and metalloproteinase
COX-2	Cyclooxygenase -2
CUR	Curcumin
DCs	Dendritic cells
EGF	Epidermal growth factor
FGF-2	Fibroblast growth factor-2
HO-1	Heme oxygenase-1
ICA	Icariin
IFN	Interferon
IL	Interleukin
JAK/STAT	Janus kinase–signal transducer and activator of transcription
MAPK	Mitogen-activated protein kinase
MDC	Monodansylcadaverine
MMP	Matrix metalloproteinase
PI3-K	Phosphatidylinositol-4–5-bisphosphate -3-kinase
PLGA	Polylactic glycolic acid
QCT	Quercetin
RA	Rheumatoid arthritis
ROS	Reactive oxygen species
SAPK	Stress-activated protein kinase
SMI1	Small molecule inhibitor-1
TNF	Tumor necrosis factor
TyK2	Tyrosine kinase 2
VEGF	Vascular endothelial growth factor

DOI: 10.1201/9781003535201-9

7.1 INTRODUCTION

An inflammatory systemic, polyarticular, metastatic, and chronic progressive musculoskeletal disease disrupting the synovial joints is known as rheumatoid arthritis (RA) (Saleem, Saleem, Akhtar, Shahzad, & Jahan, 2019). Significant tissue damage related to RA can also happen in the kidneys, skin, eyes, lungs, heart, and blood vessels, among other organs. The clinical and molecular hallmarks of RA include aberrant humoral, cellular, and innate immunity. As a result, aberrant proliferation kinetics lead to an aberrant survival of the key cellular components of the RA disease pathology, including synovial tissue fibroblasts, triggered white blood cells, i.e., mast cells, neutrophils, macrophages, T lymphocytes, B lymphocytes, and accessory-antigen-presenting cells. In RA, the synovium of the synovial joints becomes hyperplastic. This alteration results from increased quantities of different chemokines and adhesion proteins stimulating the migration, circulation, and adherence of both sparked immune and nonimmune cells (Saleem, Saleem, & Akhtar, 2020; Zhang et al., 2023). Apart from that, it has been verified that RA requires substantially higher expression of circulatory proinflammatory cytokines, including tumor necrosis factor-alpha (TNF-α), tumor growth factor beta (TGF-β), interleukin (IL) (1ß, IL-6, IL-7, IL-8, IL-12/IL-23, IL-15, IL-17, IL-18, IL-32), and interferon-γ (IFN-γ), which are produced by different specialized cells, as well as growth factors, including fibroblast growth factor-2 (FGF-2), epidermal growth factor (EGF), and vascular endothelial growth factor, which is mainly produced by synovial-like fibroblasts and inflammatory macrophages (Aletaha & Smolen, 2018; Radu & Bungau, 2021). These proinflammatory cytokines, along with growth factors such as FGF-2, vascular endothelial growth factor, and EGF, lead to synovial joint annihilation. Therefore, the overall changes that result from these different factors in RA synovial joints are crucial to understanding the RA process. These changes include the following: a reduction in the production of extracellular matrix derived from cartilage; an increased repetition of apoptotic chondrocytes; resistance to apoptosis in synovial tissue; an upregulated genetic expression of matrix metalloproteinase (MMP); and altered disintegrin protein and metalloproteinase (ADAM) gene expression (Gao, Zhang, & Liu, 2024; Kurowska-Stolarska & Alivernini, 2022; Maeda et al., 2022; Zhang et al., 2023).

Numerous signal transduction pathways have been linked to the advancement of RA. For instance, IL-1ß (homolog of interleukin protein) is known to primarily activate the cellular crosstalk of stress and mitogen-activated protein kinase (SAPK/MAPK) pathway and IFN-γ and IL-6 to exclusively instigate the Janus kinase–signal transducer and activator of transcription (JAK/STAT) pathway (Ciobanu et al., 2020). There is also convincing evidence that IL-6 and IFN-γ activate MAPK signaling. It is found that TNF-α can also activate JAK/STAT, even though it was primarily reported to stimulate the SAPK/MAPK pathway (Liao, Zheng, Lu, & Shen, 2024). Specifically, recombinant human TNF-α instigated human chondrocytes to phosphorylate the STAT3 protein (p-STAT3), without altering the content of STAT3. The distinct role of JAK/STAT pathway activation in the progression of RA was further confirmed when tofacitinib, a JAK3-selective small molecule inhibitor (SMI), was approved by the U.S. Food and Drug Administration (FDA) for use as a medicinal treatment for RA. Tofacitinib's effective addition to the arsenal of RA treatments has aided in the advancement of specific JAK/STAT pathway proteins JAK1 & 2, TYK2-, and pan-JAK selective modulators for the treatment of RA (Dreo et al., 2024; Jain et al., 2025; Liao et al., 2024).

This chapter focuses on the JAK/STAT cellular signaling role in RA. The authors specifically talked about the molecular mechanisms underlying the JAK-selective SMIs' actions. The authors also highlighted a few significant gaps in our understanding of how these JAK-selective SMIs truly control the synovial joint alterations consistent with the course of RA. In addition, developed phytochemical formulations for RA are the subject of the review, with special attention paid to clinical trials, regulatory issues, current difficulties, and potential future developments.

7.2 RISK ELEMENTS AND SIGNS

Although RA primarily affects the joints, it can also show damaging effects on the eyes, skin, alveolar cells, liver, heart, and bones, among other organ systems. Inflammation, edema, fever, discomfort, and stiffness in the afflicted joints are typical symptoms of RA. These processes usually affect the tiny joints in the fore and hind limbs (hands and feet), but they can also damage larger joints such as knees and shoulders (Gao et al., 2024).

7.3 PATHOGENESIS OF RHEUMATOID ARTHRITIS

In the beginning stages of RA, there is constant cellular activation, which causes autoimmunity in the joints and various other organs. The disease's medical emblems and symptoms usually appear after joint damage and synovial inflammation. Fibroblast-like synoviocytes or FLS are essential to pathogenic processes. There are three stages identified in the development and progression of RA: the chronic provocative stage, which is interceded by inflammation-associated cytokines including interleukins IL-1β, IL-6, VEGF, EGFR, TGF-β, and TNF-α; the nonspecific inflammatory stage, which is enhanced by the activation of inflammatory T-helper cells in the synovium; and the tissue destruction phase (Kurowska-Stolarska & Alivernini, 2022; Maeda et al., 2022; Zhang et al., 2023).

7.4 AUTOIMMUNE RESPONSE AND INFLAMMATION

There is evidence connecting the generation of autoantibodies to severe symptoms such as joint damage and higher death rates in RA. The production of immunological complex bodies through autoantibodies, contrary to specific citrullinated peptides (ACPAs) containing citrulline-based antigens, is most likely the source of this inflammation. Complement activation results from these complexes binding to rheumatoid factors. One significant development in recent years has been the ability to identify autoimmune reactions to citrullinated self-proteins (Vander Cruyssen et al., 2005).

The combination of both the immune systems, i.e., innate and adaptive immune mechanisms, induces cartilage damage and restoration in and around the surrounding bone marrow and synovium. This cascade fuels circulating leukocyte migration into the inflammatory joint and causes persistent inflammation in RA. Angiogenesis is required for this process to occur because the hypertrophic joint needs oxygen and nutrients. Angiogenesis is triggered by proangiogenic factors. In the synovium intima, FLS develop into a very different intrusive phenotype that encourages extracellular matrix invasion and worsens joint damage (Elshabrawy et al., 2015).

7.5 SIGNALING CASCADES CROSSTALKING THE PATHOGENESIS OF RHEUMATOID ARTHRITIS

RA progresses through several signal transduction channels. The aberrant signals (such as the JAK/STAT pathway) are frequently used as targets for medication research.

7.5.1 The JAK/STAT Signaling Cascade

The JAK–STAT signaling cascade is one of the most robust for cytokine crosstalk. It explains that TNF-α swiftly induces the expression of target regulatory genes, such as those of the IFN, calmodulin, gp130, and common γ chain families, as well as receptor tyrosine kinases associated with G protein-coupled receptors (Chaimowitz, Smith, & Forbes Satter, 2024). This signaling pathway is thought to be particularly significant in controlling inflammation and immunological function and to have a major contribution in cellular differentiation, active proliferation, apoptosis, and immune function (Hu et al., 2023). It has been observed by several recent investigations that RA results in aberrant stimulation of the JAK–STAT signaling cascade. There are four members of the JAK family: TyK2 (tyrosine kinase 2), JAK1, JAK2, and JAK3. These four members are highly conserved in their whole evolutionary process and have varying molecular weights (Kiu & Nicholson, 2012). While JAK3 is only liable to be expressed in endothelial cells, blood, and vascular smooth muscle, TyK2, JAK1, and JAK2 are broadly distributed in multiple tissues and systems. Research has indicated that JAK contributes significantly to RA. Both transcriptional activation and signal transduction are mediated by a class of cytoplasmic proteins called STAT. Various STAT genes including STAT 1–4, STAT5A & 5B, and STAT6 are discovered to be members of the STAT family. Five domains of STAT such as the N-terminal conserved domain, the helix domain, the DNA-binding domain, the ligation domain, the SH2 domain, and other six highly conserved functional domains make up STAT (Liongue & Ward, 2013; Lv et al., 2024; Murray, 2007).

7.5.2 Preclinical and Clinical Studies

The JAK/STAT pathway appears to play a role in aberrant hyperplasia related to RA FLS as well as synovial inflammation and bone degradation. Among these elements, RA's pathological foundation is synovitis (Hu et al., 2023). In RA synovium, various inflammatory responses have been noted, including the activation of genes related to adhesion molecules, transcription factors, and cytokines, which are tightly associated with bone destruction in particular signaling pathways. In the pathophysiology of RA, vitiligo, and psoriasis, JAK1 may be implicated in signal transduction associated with several cytokines, such as IL-6 and IFN-γ, which combine and make receptor complex molecules that ultimately lead to the activation of JAK1 kinase. T-cell differentiation, proliferation, and growth are significantly influenced by JAK3, which takes part in the signaling pathways linked with interleukin family, i.e., IL-21, IL-15, IL-9, IL-7, IL-4, and IL-1 (Malemud, 2018). Selective inhibition of tyrosine kinase 2 (Tyk2) embarks a significant role in the restoration of cartilage in RA. It is likely that IFNs, IL-6, IL-10, IL-12, and IL-23 activate the powerful kinase Tyk2. JAK2 causes STAT3 and STAT5 to become downstream activated. It transmits signals through various receptors, including IFN-γ R2, IL-6R, and IL-12Rβ, which are involved in inflammatory and autoimmune reactions. Rats with adjuvant-induced arthritis or with collagen-induced arthritis and other animal models have also manifested similar expression patterns (Ciobanu et al., 2020; Crispino & Ciccia, 2021) (Figure 7.1).

Using CEP-33779, a highly selective JAK2 inhibitor, Kristine et al. intervened in collagen antibody and collagen-intoxicated arthritis in mice. Following treatment, serum levels of cytokines

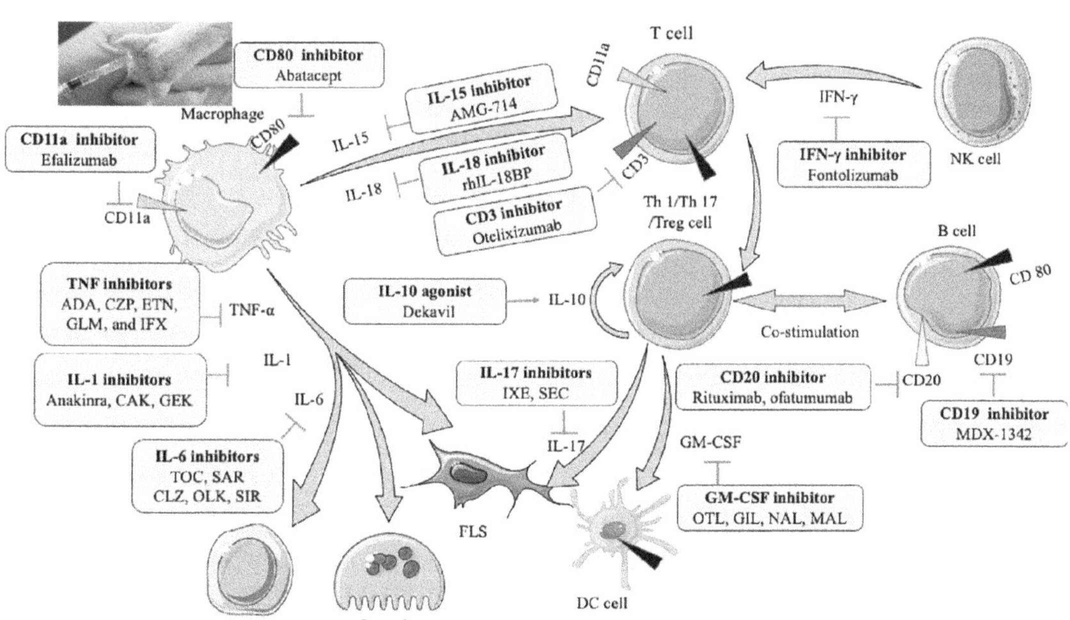

FIGURE 7.1 Therapeutic targets of rheumatoid arthritis.

(TNF-γ, IL-12, and TNF-γ) as well as IL-2, IL-12, and p-STAT 3 significantly decreased in the synovial fluid of albino mice. CEP-33779 dramatically reduced a number of histological parameters such as matrix defacement and erosion, subchondral osteolysis (bone cells breakdown), osteogenesis (synthesis of bone cells), synovial hyperplasia, vasculitis, and synovial inflammation, which are meant to reveal improvement in arthritis. These findings suggested that JAK2 participated in the inflammatory signaling and pathophysiology of RA and the fact that blocking JAK2 could be responsible for treating RA by preventing the production of cytokines and the activation of both T and B cells (Stump et al., 2011; Yamaguchi et al., 2024).

STAT1 engages in bodily functions via IFN-γ-mediated signaling cascades and is primarily stimulated by cytokines such as IFN-γ, IL-10, and IL-6. STAT1 has been observed to have dual pathogenic and preventive functions in the development of RA synovitis. Nevertheless, STAT1 expression typically increases in seditious arthritic damage, suggesting that STAT1's pro-apoptotic and anti-inflammatory actions are inadequate to offset its pro-inflammatory activities. Only IFN-κ/β and IFN-λ are responsible for the activation of STAT2. Through the formation of a complicated heterodimer transcription complex of interferon-stimulated gene factor by STAT1 and interferon regulatory factor 9 (IRF-9), it has been evaluated that STAT2 is significantly involved in RA-associated inflammation. CD4+ T-cell development into Th1 and Th17 cells results in the STAT4, contributing to RA inflammation by controlling the IL-12 to IL-23 ratio (Kasperkovitz et al., 2004). According to multiple meta-analyses, a particular single-nucleotide polymorphism at rs7574865 of the STAT4 gene might be connected with the responsiveness of inflammation in RA. Nonetheless, further scrutiny is required to validate this finding. Further elements include the transcription factor STAT5, which is involved in the development of Treg cells together with Foxp3. Numerous investigations have revealed that STAT5's impact can be the reverse of STAT3's impact (Kasperkovitz et al., 2004; Tong, Zhang, Tong, & Liu, 2013).

Similarly, in B lymphocytes, IFN-α activates IL-13, whereas IL-4 activates STAT6. Lack of IL-4 and STAT6 can markedly worsen arthritis-mediated pain in mouse models of proteoglycan-induced arthritis. The main downregulated transcription factor of the gp130 receptor, STAT3, is upregulated by cytokines like IL-6, IL-10, COX-2, and IFN-α/β. It might worsen the clinical symptoms of RA, encourage chronic arthritis, and control the aberrant progression and endurance traits of RA synovial cells (Ebrahimiyan, Mostafaei, Aslani, Jamshidi, & Mahmoudi, 2019).

A researcher noted that STAT3 has DNA-attachment properties in mononuclear cells of synovial fluid from inflammatory arthritis patients (Walker & Smith, 2005). Furthermore, Lee et al. discovered that the transcription factor, STAT3, may stimulate the synthesis of antibodies, boost T-cell activity, and reduce FLS apoptosis, proposing that STAT3 is engaged in numerous pathways related to RA pathogenesis. p-STAT3 was discovered to be strongly expressed in the cartilage and synovium of mice given collagen. Furthermore, following STAT3 inhibitor therapy, there is an aberrant decline in the levels of the inflammatory cytokines IL-6 and IL-17 in serum. STAT3 was discovered to be highly upregulated and expressed in both RA CD4+ T and synovial cells (Ding et al., 2023). STAT3 activation gave synovial cells tumor-like features, accelerated cell division, extended survival, and increased infiltration of adjacent joint tissues. Notably, RA helper cell differentiation is significantly influenced by STAT3 (Alka & Mishra, 2025; Di Benedetto et al., 2021).

7.6 ANTIARTHRITIC EFFECTS OF PHYTOACTIVE COMPOUNDS TARGETING JAK/STAT SIGNALING CASCADE

Phytochemicals have been emphasized recently as viable/alternative options for RA treatment. Plant-derived natural chemicals are thought to have both therapeutic and preventive benefits when taken as dietary supplements for RA. Additionally, it has been proposed that phytochemicals can strengthen the osteocyte barriers by regulating cytokines and reducing oxidative stress, among other modes of action. The ability of phytochemicals to reduce the generation of cytokines in RA has received the majority of attention thus far (Alka & Mishra, 2025). As was covered in the previous

TABLE 7.1
Phytoactive Compounds Affecting JAK/STAT Signaling Pathway in the Recovery of Rheumatoid Arthritis

Phytochemicals		Targets of JAK/ STAT Pathway
Polyphenols	Chrycin	JAK1, STAT3
	Curcumin	JAK1,2 STAT3
	Garcinol	STAT3, 5A
	Silibinin	JAK2 STAT3
	EGCG	JAK3 STAT1,3
	Chalcons	STAT3
	Cardamonin	STAT3
	Apigenin	JAK1, 2 STAT3
	Resveratol	STAT1 STAT3
	Quercetin	STAT3
	Bergamottin	JAK1, 2 STAT3
	Bavachin	STAT3
Steroids	Diosgenin	JAK1, 2 STAT3
Terpenoids	Cucurbitacins	JAK, JAK2 STAT3, STAT5

section, blocking the JAK–STAT pathway stops some proinflammatory cytokine signaling pathways, suggesting a potential novel treatment for inflammatory RA. Here, we summarize the current research studies suggesting that phytochemicals may be able to treat/manage arthritis in an animal model by influencing the JAK–STAT system (Di Benedetto et al., 2021; Sivasakthi, Sanmuga Priya, & Senthamil Selvan, 2021; Yin et al., 2021). Table 7.1 shows the therapeutic targets in the JAK/STAT pathway that can be inhibited to cure osteocytes and cartilage destruction.

7.6.1 Curcumin

Curcumin (CUR) obtained from *Curcuma longa L.* [family: Zingiberaceae] is a polyphenol molecule that can reduce inflammation, boost immunity, and act as an antioxidant. Additionally, it can be used to prevent and treat disorders associated with aging, including RA, chronic respiratory conditions, and hypertension. Thus, curcumin can be assumed as a promising agent for RA treatment (Kou et al., 2023). Furthermore, in vitro research done on mouse primary chondrocytes has shown that CUR can treat RA by modulating pathways linked to apoptosis, which enhances the condition of chondrocytes in the RA model (Henrotin et al., 2010). The CUR-treated exosomes of mesenchymal stem cells were found to preserve viability of primary chondrocytes, downregulate the DNA methylation of specific miRNA genes like miR-124 and miR-143 promoters at specific regions, resulting in the inhibition of regulated expression of their respective target genes as well

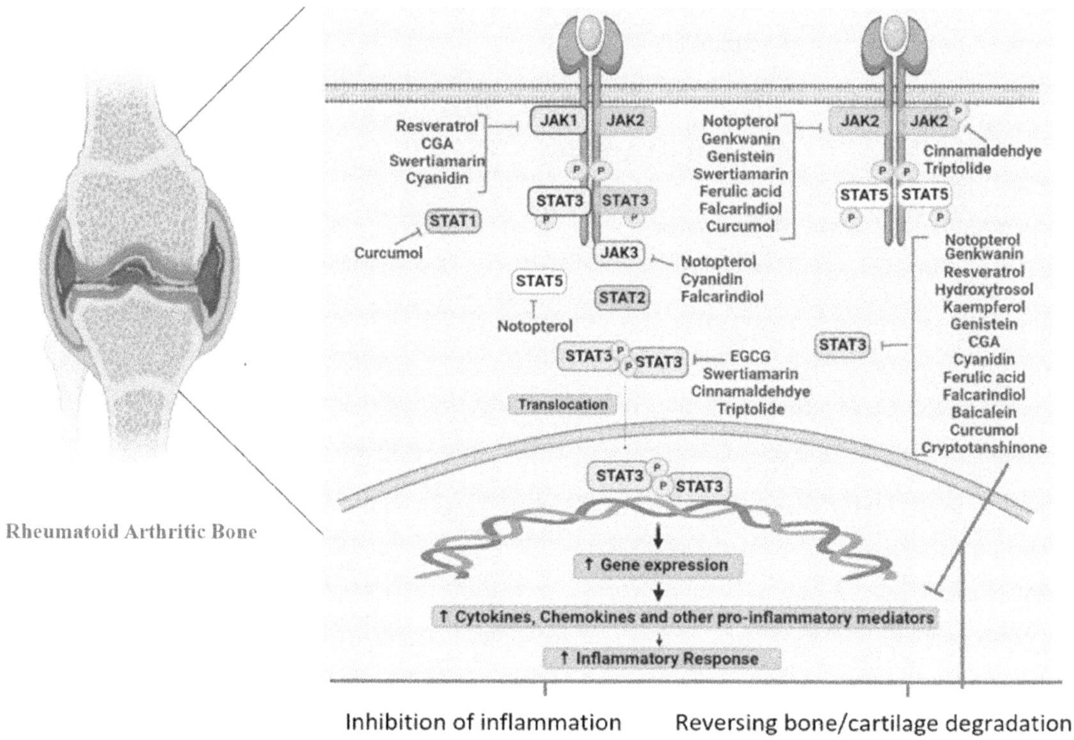

FIGURE 7.2 Diagram showing how the activation of the Janus kinase (JAK)/signal transducer and activator of transcription (STAT) signaling pathway causes the overexpression of genes related to rheumatoid arthritis and survival.

as rho-associated coiled-coil-containing protein kinase 1 (ROCK1) and NF-kB that block cellular apoptosis, and speedup the pathological process of RA observed through vibrant techniques of luciferase assay, flow cytometry, immunohistochemistry, and most importantly through TUNEL staining (Ashrafizadeh et al., 2020). In an earlier study, CUR was provided to human chondrocytes and found that through CCK8 expression analysis, immunofluorescence, and immunoprecipitation, certain caspase 3 and MMP9 genetic expression levels in chondrocytes are directly suppressed by CUR by obstructing the attachment of DNA with p65-NF-kB, controlling apoptosis, and ultimately alleviating osteoarthritis (Henrotin et al., 2010) (Figure 7.2). While numerous preclinical investigations have demonstrated the clinical and biochemical activity of CUR, randomized, double-blinded, placebo-controlled experimental trials are required to validate in vitro results. Experimental clinical research currently remains comparatively scarce at the moment.

Numerous preclinical studies demonstrate the safety and tolerability of curcumin, even at very high dosages. Curcumin has not been licensed as a medication despite having strong pharmacological effects and safety; its bioavailability is thought to be a possible barrier to the full translation of in vitro advantages into clinical implications. Curcumin exhibits decreased levels of systemic bioavailability, biological distribution, and oral absorption. As a result, some research focuses on developing solutions for these issues. Among these techniques are the utilization of phospholipid complexes, liposomes, nanoparticles, and micelles to increase curcumin's biological distribution, absorption, and bioavailability (Ashrafizadeh et al., 2020).

7.6.2 Resveratrol (RES)

Numerous plants contain the polyphenol molecule resveratrol (Vitaceae; *Vitis vinifera* L.). Numerous biological actions of RES, such as antiviral, anti-inflammatory, antiaging, and antioxidant properties, have been demonstrated. It is extensively utilized in the management and prevention of RA, liver disorders, and cardiovascular diseases. Numerous research studies on the use of RES to treat RA involve several mechanisms including the JAK/STAT pathway (Ma, Wang, Dong, Li, & Cai, 2015). A previous study reported antiarthritic activity of RES, in which LPS-induced inflammation in RAW 264.7 murine macrophages treated with RES and inflammatory response was reduced by RES due to downregulation of inflammatory cytokines and moderation of JAK/STAT and NF-KB signaling (Ma et al., 2015).

7.6.3 Icariin

The primary active component of Epimedium, Icariin [Berberidaceae; *Epimedium brevicornu Maxim*] (ICA), is well known for its antioxidant as well as anti-inflammatory qualities. Early research on ICA mostly concentrated on improving antiaging and reproductive properties. With the advancement of contemporary pharmacological and pharmaceutical research, it has been discovered that ICA is good for several disorders, including those of the neurological, cardiovascular, and musculoskeletal systems (Guo et al., 2021). A researcher used varying doses of ICA on chondrocytes and discovered that the compound can control cell death via several mechanistic pathways, hence playing a preventative and therapeutic function in RA. It was found that ICA reduced chondrocyte apoptosis, selectively elevated CYTOR expression in chondrocytes, and promoted chondrocyte activity produced by IL-1β (Figure 7.2). The protective effect of ICA on chondrocytes is reversed by CYTOR knockdown. Thus, by controlling CYTOR, ICA treatment can prevent chondrocyte death (Chi, Gao, Shu, & Lu, 2014).

Another study demonstrated that ICA 10 μM prevented apoptosis by controlling the JAK/STAT cascade, thus safeguarding the cartilage in the joints. ICA reversed the overexpression of p-p65 and the dilapidation of STAT 5, reduced reactive oxygen species (ROS), as well as nitric oxide, therefore blocking the formation of cell apoptotic proteins and catabolism. It also downregulated the expression of JAK 2 and JAK 5 caused by TNF-α in chondrocytes. Thus, by blocking the JAK/STAT signaling system, ICA can prevent chondrocytes from undergoing apoptosis and relieve the degenerative process of RA (Guo et al., 2021).

7.6.4 Quercetin

Fruits and vegetables are a great source of quercetin (QCT), a kind of flavone known for its anti-inflammatory, antiviral, and antioxidant characteristics. Because of its great efficacy and minimal side effects, QCT has garnered a lot of interest and is essential in the treatment of disorders associated with aging. QCT possesses anti-inflammatory and antiapoptotic properties, among other biological activities, that are effective in the fight against RA. Nonetheless, QCT's antiapoptotic mechanism is intricate and entwined with multiple signaling pathways (Goyal & Agrawal, 2022). By using network pharmacology, Wang et al. reported that QCT suppresses the p38 MAPK JAK/STAT signaling pathway-mediated chondrocyte death triggered by IL-1β in RA (Wang, Xie, et al., 2021). Feng et al. reported by in vitro and in vivo research that QCT treats RA by overexpressing the IRT1/AMPK signaling cascade, increasing the degenerative process of RA by blocking endoplasmic reticulum oxidative stress and death of associated chondrocytes in arthritic rats and RA articular chondrocytes (Wang, Xie, et al., 2021) (Figure 7.2). A researcher reported the protective impact of QCT against RA using in vitro and in vivo methods. Tissue staining and whole blood biochemical analysis demonstrated that there is an increase in the number of M2 macrophages in the synovium and in the expression levels of TGF-β1 and TGF-β2 in synovial fluid (Mamani-Matsuda et al., 2006). Moreover, using flow cytometry, qRT-PCR, and Western blotting, it was demonstrated that QCT inhibits the JAK/STAT pathway in apoptotic rat chondrocytes, reduces intracellular ROS, and restores mitochondrial membrane potential (MMP). In rat models of osteoarthritis, intra-articular injection of QCT reduced chondrocyte death and cartilage deterioration. Consequently, the two studies mentioned above demonstrate that QCT has an antiapoptotic effect on RA chondrocytes through both in vitro and in vivo investigations, making them important in future research for treating RA (Tang et al., 2023; Zhang et al., 2024).

7.6.5 Loganin

Iridoid glycoside, Loganin (Log) [Cornaceae; *Cornus officinalis* Siebold and Zucc] was isolated from *Corni fructus*. Because of its wide range of pharmacological actions, which include anti-RA properties, the Log has drawn a lot of interest in therapeutics. According to studies, Log exhibited anti-inflammatory, antioxidant, antiapoptosis, and antiarthritic effects by modulating the phosphatidylinositol-4,5-bisphosphate 3-kinase (PI3K)/protein kinase B (Akt) and signaling pathways for transcription 3 (STAT3)/NF-κB, JAK/STAT3 (Yang, Gu, Zhao, & Zhang, 2019).

7.6.6 Andrographolide

The primary active ingredient in *Andrographis paniculata* is andrographolide (AG) [Acanthaceae; *Andrographis paniculata* (Burm.f.) Wall. ex Nees]. It is a naturally occurring antibiotic. It is an effective treatment for inflammatory disorders and can eliminate heat, eliminate toxins, relieve pain and inflammation, and more. The biological activities displayed by AG are diverse, and this has long been a popular area of study. Anti-inflammatory, antioxidative stress, antiproliferative, and antiapoptotic properties are possible roles for AG. According to numerous studies, AG can reduce RA by controlling apoptosis (Ala'a, Canatan, & Ezeamuzie, 2009).

A previous study finding demonstrated a decrease in apoptosis and an improvement in cartilage degradation in treated animals by AG. The expression levels of JAK protein in the mice's proximal tibia were then assessed, and it was discovered that oral AG could counteract the RA-induced downregulation of JAK. Similarly, in vitro tests were conducted on primary chondrocytes obtained from mice. According to their findings, AG can resist arthritis by suppressing JAK protein, which in turn prevents apoptosis (Gupta, Mishra, Kumar, Singh, & Ganju, 2020).

7.6.7 Geniposide

A naturally occurring substance derived from Gardenia, geniposide [Rubiaceae; *Gardenia jasminoides J.Ellis*] (GE) has pharmacological actions such as antiproliferative, antiapoptotic, anti-inflammatory, and anti-angiogenic abilities. Several biological actions of GE have been shown in the treatment of liver, brain, cardiovascular, tumor, and other disorders. Furthermore, GE has anti-inflammatory and antiapoptotic properties that make it useful in the treatment of RA (Chen et al., 2015). In vivo experimental analysis revealed that GE can progressively mitigate the degenerative process of RA in albino Wistar rats. GE suppresses JAK and STAT protein expression in vitro. In addition, it stimulates rat chondrocytes activated with IL-1β to express COL2A1 (Huang et al., 2025). Furthermore, their research discovered that GE raises the expression of the antiapoptotic protein Bcl2 while suppressing the production of proapoptotic nuclear transcription factors encoding proteins like Bax, Cyt c, and caspase 3. These alterations might be caused by GE blocking the JAK/STAT pathway's activation, which would otherwise stimulate the PI3K/Akt/NF-κB cellular signaling cascade. Consequently, GE may be useful as a therapy in the treatment of RA. Nevertheless, more research is needed to determine whether GE is as effective as conventional medications and whether it would have the same impact on patient RA chondrocytes (Wang, Wu, et al., 2021).

7.6.8 Morroniside

Morroniside (MR) [Cornaceae; *Cornus officinalis* Siebold and Zucc] is an iridoid glycoside that is extracted from the botanical medicine *Cornus officinalis* with a range of biological actions, including anti-inflammatory, antioxidant, and antiapoptotic abilities (Yu et al., 2021). By influencing the anti-apoptotic activity, MR has the potential to treat RA. Conducive to assessing the therapeutic effect of MR on cartilage degradation, a researcher used previously treated RA rats with certain low and high doses of MR. It was discovered that MR greatly inhibited the growth of osteophytes and the hardness of subchondral bone, as well as attenuated the progression of RA in treated mice. Additionally, MR reduced caspase 1 expression, stopped the JAK/STAT signaling pathway from conducting, and lessened apoptosis. Therefore, by blocking JAK/STAT signaling, MR stops the deterioration of the cartilage matrix and lowers chondrocyte death. Using concentrations of 4 and 20 μg/kg, a researcher discovered that articular cartilage is protected by both low and high MR doses, while low doses of MR are more beneficial than higher ones (Yu et al., 2021). It is plausible that concurrent activation of additional signal pathways by high-dose MR could be detrimental to cartilage.

A researcher found that MR has a dose-dependent protective impact on chondrocytes and lowers apoptosis of chondrocytes at 10, 50, and 10,000 μg/kg. The disparity between the two species (rats and mice) and the various approaches utilized to imitate OA could be the reason for these two divergent experimental results (Park et al., 2021).

7.6.9 Engeletin

A naturally occurring product of *Smilax rhizomilax*, engeletin [Smilacaceae; *Smilax glabra Roxb*] (ENG) exhibits antiproliferative and anti-inflammatory properties and suppresses oxidative stress by reducing cellular lipid peroxidation (Saleem et al., 2019). It is found in many fruits and vegetables. It has a few adverse effects and can be used to treat a wide range of illnesses. The applicability of ENG is not as broad as that of other blooming chemicals, and more research is necessary. Only one study has looked at the results of utilizing ENG to treat RA to date. Wang and colleagues extracted articular chondrocytes from albino rat knee joints, created an RA rat model via anterior cruciate ligament transection, and used ENG to treat samples. By upregulating the nuclear Bcl2 expression and decreasing nuclear Bax expression, they came to know that ENG is able to retain certain specified MMP, reduce ROS generation, diminish chondrocyte apoptosis, and contribute to cartilage protection using WB, qPCR, Annexin V-FITC/PI, and flow cytometry. This cartilage defense mechanism functions by blocking the NF-κB and JAK/STAT pathways and triggering the erythroid 2-related factor 2 (Nrf2) pathway (Wang, Jiang, et al., 2021; Zhong, Huang, Chen, & Lei, 2023).

7.7 NANOPARTICLES AS NOVEL THERAPEUTIC DRUG DELIVERY SYSTEMS IN PHYTOCHEMICAL THERAPY FOR THE TREATMENT OF RHEUMATOID ARTHRITIS

Drug delivery systems (DDSs) with nanoparticles are novel and exciting. They are intended to administer a certain dosage of a medication to a targeted area of the body. They provide for the safe gradual release of drugs, hence increasing their bioavailability. For RA treatment, targeted drug delivery nanomaterials aim to deliver active phytocomponents locally to the lesion site with minimal harm to surrounding healthy cells and tissues. The FDA has approved polylactic-co-glycolic acid (PLGA) as a drug carrier, making it the most widely used nanoparticle due to its biocompatibility. Several preclinical studies have shown that combining MTX-loaded PLGA nanoparticle-coated phytoactive compounds with near-infrared radiation therapy produced a more sustained and effective therapeutic effect in reducing arthritis than MTX administration alone (Pham, 2011).

7.8 CONCLUSION AND FUTURE PROSPECTS

The incidence of RA is rising as the population ages, which will have a significant negative impact on people's lives as well as place a significant financial strain on society. A significant quantity of articular cartilage is destroyed and joint function deteriorates in cases when the patient's condition is severe, which has a major impact on day-to-day activities. Joint replacement surgery is required if conservative treatment fails. On the other hand, there is a significant surgical trauma, a short service life for the artificial joint that needs to be replaced, and a high operation expense. One of the key elements in the pathophysiology of RA is apoptosis. In the degenerative phase of RA, the prevention of apoptosis protects articular cartilage, and it has a wide range of potential therapeutic applications as an RA therapeutic target. It is anticipated that the targeted administration of medications to prevent chondrocyte death would open up new possibilities for the clinical management of RA. The unique anti-RA properties of phytochemicals are mostly derived from their inhibition of apoptosis-related processes and components. Phytochemicals have been shown to modulate multiple players in synovial inflammation, regulate immune cell function, reduce the proliferation of synoviocytes and balance the apoptotic process, reduce angiogenesis, and stop/prevent the degeneration of bone and cartilage, all of which are prominent characteristics of RA. The present data from in vitro and in vivo models suggest the potential of phytochemicals as novel disease-modifying antirheumatic medicines, but further research is needed to determine their efficacy in humans. This study emphasizes the potential application of phytochemicals as a therapeutic target for RA in the future.

A suitable dose of phytoactive compounds can influence apoptotic transcription factors and proteins, including the caspase family, Bax, and Bcl2, by modulating apoptotic signal cascades like JAK/STAT, MAPK, ERK1/2, AKT/mTOR, NF-κB, PI3K/AKT, and AMPK, noncoding RNA, or some cytokines to inhibit apoptosis in RA (Alka & Mishra, 2025). Many researchers are looking at different phytochemicals for the treatment of RA because of their low side effects, high cost-effectiveness, and wide applicability. Consequently, phytochemicals are now an essential resource for creating cutting-edge RA treatments. Utilizing nanotechnology-based drug delivery methods, some phytochemicals with limited bioavailability, poor water solubility, and high toxicity can be greatly improved. As a result, we should investigate more efficient ways to distribute different natural as well as synthetic phytochemicals for their focused and particular administration and bioavailability.

REFERENCES

Ala'a, A., Canatan, H., & Ezeamuzie, C. I. (2009). In vitro and in vivo anti-inflammatory effects of andrographolide. *International Immunopharmacology, 9*(3), 313–318.

Aletaha, D., & Smolen, J. S. (2018). Diagnosis and management of rheumatoid arthritis: a review. *JAMA, 320*(13), 1360–1372.

Alka, & Mishra, A. (2025). Targeting rheumatoid arthritis risk factors with phytochemicals: an anti-inflammatory perspective. *Inflammopharmacology, 33*(7), 3561–3582.

Ashrafizadeh, M., Rafiei, H., Mohammadinejad, R., Afshar, E. G., Farkhondeh, T., & Samarghandian, S. (2020). Potential therapeutic effects of curcumin mediated by JAK/STAT signaling pathway: a review. *Phytotherapy Research, 34*(8), 1745–1760.

Chaimowitz, N. S., Smith, M. R., & Forbes Satter, L. R. (2024). JAK/STAT defects and immune dysregulation, and guiding therapeutic choices. *Immunological Reviews, 322*(1), 311–328.

Chen, J.-Y., Wu, H., Li, H., Hu, S.-L., Dai, M.-M., & Chen, J. (2015). Anti-inflammatory effects and pharmacokinetics study of geniposide on rats with adjuvant arthritis. *International Immunopharmacology, 24*(1), 102–109.

Chi, L., Gao, W., Shu, X., & Lu, X. (2014). A natural flavonoid glucoside, icariin, regulates Th17 and alleviates rheumatoid arthritis in a murine model. *Mediators of Inflammation, 2014*(1), 392062.

Ciobanu, D. A., Poenariu, I. S., Crînguş, L.-I., Vreju, F. A., Turcu-Stiolica, A., Tica, A. A., ... Dinescu, S. C. (2020). JAK/STAT pathway in pathology of rheumatoid arthritis. *Experimental and Therapeutic Medicine, 20*(4), 3498–3503.

Crispino, N., & Ciccia, F. (2021). JAK/STAT pathway and nociceptive cytokine signalling in rheumatoid arthritis and psoriatic arthritis. *Clinical and Experimental Rheumatology, 39*(3), 668–675.

Di Benedetto, P., Ruscitti, P., Berardicurti, O., Panzera, N., Grazia, N., Di Vito Nolfi, M., ... Rucci, N. (2021). Blocking Jak/STAT signalling using tofacitinib inhibits angiogenesis in experimental arthritis. *Arthritis Research & Therapy, 23*(1), 213.

Ding, Q., Hu, W., Wang, R., Yang, Q., Zhu, M., Li, M., ... Zhu, Y. Z. (2023). Signaling pathways in rheumatoid arthritis: implications for targeted therapy. *Signal Transduction and Targeted Therapy, 8*(1), 68.

Dreo, B., Muralikrishnan, A. S., Husic, R., Lackner, A., Brügmann, T., Haudum, P., ... Stradner, M. (2024). JAK/STAT signaling in rheumatoid arthritis leukocytes is uncoupled from serum cytokines in a subset of patients. *Clinical Immunology, 264*, 110238.

Ebrahimiyan, H., Mostafaei, S., Aslani, S., Jamshidi, A., & Mahmoudi, M. (2019). Studying the association between STAT4 gene polymorphism and susceptibility to rheumatoid arthritis disease: an updated meta-analysis. *Iranian Journal of Immunology, 16*(1), 71–83.

Elshabrawy, H. A., Chen, Z., Volin, M. V., Ravella, S., Virupannavar, S., & Shahrara, S. (2015). The pathogenic role of angiogenesis in rheumatoid arthritis. *Angiogenesis, 18*(4), 433–448.

Gao, Y., Zhang, Y., & Liu, X. (2024). Rheumatoid arthritis: pathogenesis and therapeutic advances. *MedComm, 5*(3), e509.

Goyal, A., & Agrawal, N. (2022). Quercetin: a potential candidate for the treatment of arthritis. *Current Molecular Medicine, 22*(4), 325–335.

Guo, X., Ji, J., Zhang, J., Hou, X., Fu, X., Luo, Y., ... Feng, Z. (2021). Anti-inflammatory and osteoprotective effects of Chikusetsusaponin IV a on rheumatoid arthritis via the JAK/STAT signaling pathway. *Phytomedicine, 93*, 153801.

Gupta, S., Mishra, K., Kumar, B., Singh, S., & Ganju, L. (2020). Andrographolide attenuates complete Freund's adjuvant induced arthritis via suppression of inflammatory mediators and pro-inflammatory cytokines. *Journal of Ethnopharmacology, 261*, 113022.

Henrotin, Y., Clutterbuck, A., Allaway, D., Lodwig, E., Harris, P., Mathy-Hartert, M., ... Mobasheri, A. (2010). Biological actions of curcumin on articular chondrocytes. *Osteoarthritis and Cartilage, 18*(2), 141–149.

Hu, Q., Bian, Q., Rong, D., Wang, L., Song, J., Huang, H.-S., ... Wang, P.-Y. (2023). JAK/STAT pathway: extracellular signals, diseases, immunity, and therapeutic regimens. *Frontiers in Bioengineering and Biotechnology, 11*, 1110765.

Huang, M., Jiang, J., Li, Y.-J., Jiang, M.-Y., Yang, M., Deng, H., ... Chen, J.-L. (2025). Exploring the anti-inflammatory mechanism of geniposide in rheumatoid arthritis via network pharmacology and experimental validation. *Scientific Reports, 15*(1), 24812.

Jain, A., Bishnoi, M., Prajapati, S. K., Acharya, S., Kapre, S., Singhvi, G., & Palakurthi, S. (2025). Targeting rheumatoid arthritis: a molecular perspective on biologic therapies and clinical progress. *Journal of Biological Engineering, 19*, 67.

Kasperkovitz, P., Verbeet, N., Smeets, T., Van Rietschoten, J., Kraan, M., Van Der Pouw Kraan, T., ... Verweij, C. (2004). Activation of the STAT1 pathway in rheumatoid arthritis. *Annals of the Rheumatic Diseases, 63*(3), 233–239.

Kiu, H., & Nicholson, S. E. (2012). Biology and significance of the JAK/STAT signalling pathways. *Growth Factors, 30*(2), 88–106.

Kou, H., Huang, L., Jin, M., He, Q., Zhang, R., & Ma, J. (2023). Effect of curcumin on rheumatoid arthritis: a systematic review and meta-analysis. *Frontiers in Immunology, 14*, 1121655.

Kurowska-Stolarska, M., & Alivernini, S. (2022). Synovial tissue macrophages in joint homeostasis, rheumatoid arthritis and disease remission. *Nature Reviews Rheumatology, 18*(7), 384–397.

Liao, H., Zheng, J., Lu, J., & Shen, H.-l. (2024). NF-κB signaling pathway in rheumatoid arthritis: mechanisms and therapeutic potential. *Molecular Neurobiology, 62*(6), 6998–7021.

Liongue, C., & Ward, A. C. (2013). Evolution of the JAK-STAT pathway. *Jak-Stat, 2*(1), e22756.

Lv, Y., Qi, J., Babon, J. J., Cao, L., Fan, G., Lang, J., ... Wang, F. (2024). The JAK-STAT pathway: from structural biology to cytokine engineering. *Signal Transduction and Targeted Therapy, 9*(1), 221.

Ma, C., Wang, Y., Dong, L., Li, M., & Cai, W. (2015). Anti-inflammatory effect of resveratrol through the suppression of NF-κB and JAK/STAT signaling pathways. *Acta Biochimica et Biophysica Sinica, 47*(3), 207–213.

Maeda, K., Yoshida, K., Nishizawa, T., Otani, K., Yamashita, Y., Okabe, H., ... Saito, M. (2022). Inflammation and bone metabolism in rheumatoid arthritis: molecular mechanisms of joint destruction and pharmacological treatments. *International Journal of Molecular Sciences, 23*(5), 2871.

Malemud, C. J. (2018). The role of the JAK/STAT signal pathway in rheumatoid arthritis. *Therapeutic Advances in Musculoskeletal Disease, 10*(5–6), 117–127.

Mamani-Matsuda, M., Kauss, T., Al-Kharrat, A., Rambert, J., Fawaz, F., Thiolat, D., ... Mossalayi, M. D. (2006). Therapeutic and preventive properties of quercetin in experimental arthritis correlate with decreased macrophage inflammatory mediators. *Biochemical Pharmacology, 72*(10), 1304–1310.

Murray, P. J. (2007). The JAK-STAT signaling pathway: input and output integration. *The Journal of Immunology, 178*(5), 2623–2629.

Park, E., Lee, C. G., Han, S. J., Yun, S. H., Hwang, S., Jeon, H., ... Jeong, S.-Y. (2021). Antiosteoarthritic effect of morroniside in chondrocyte inflammation and destabilization of medial meniscus-induced mouse model. *International Journal of Molecular Sciences, 22*(6), 2987.

Pham, C. T. (2011). Nanotherapeutic approaches for the treatment of rheumatoid arthritis. *Wiley Interdisciplinary Reviews: Nanomedicine and Nanobiotechnology, 3*(6), 607–619.

Radu, A.-F., & Bungau, S. G. (2021). Management of rheumatoid arthritis: an overview. *Cells, 10*(11), 2857.

Saleem, A., Saleem, M., & Akhtar, M. F. (2020). Antioxidant, anti-inflammatory and antiarthritic potential of *Moringa oleifera* Lam: an ethnomedicinal plant of Moringaceae family. *South African Journal of Botany, 128*, 246–256.

Saleem, A., Saleem, M., Akhtar, M. F., Shahzad, M., & Jahan, S. (2019). *Moringa rivae* leaf extracts attenuate Complete Freund's adjuvant-induced arthritis in Wistar rats via modulation of inflammatory and oxidative stress biomarkers. *Inflammopharmacology, 28*(1), 139–151.

Sivasakthi, P., Sanmuga Priya, E., & Senthamil Selvan, P. (2021). Molecular insights into phytochemicals exhibiting anti-arthritic activity: systematic review: John Di Battista. *Inflammation Research, 70*(6), 665–685.

Stump, K. L., Lu, L. D., Dobrzanski, P., Serdikoff, C., Gingrich, D. E., Dugan, B. J., ... Dorsey, B. D. (2011). A highly selective, orally active inhibitor of Janus kinase 2, CEP-33779, ablates disease in two mouse models of rheumatoid arthritis. *Arthritis Research & Therapy, 13*(2), R68.

Tang, M., Zeng, Y., Peng, W., Xie, X., Yang, Y., Ji, B., & Li, F. (2023). Pharmacological aspects of natural quercetin in rheumatoid arthritis. *Drug Design, Development and Therapy, 16*, 2043–2053.

Tong, G., Zhang, X., Tong, W., & Liu, Y. (2013). Association between polymorphism in STAT4 gene and risk of rheumatoid arthritis: a meta-analysis. *Human Immunology, 74*(5), 586–592.

Vander Cruyssen, B., Peene, I., Cantaert, T., Hoffman, I., De Rycke, L., Veys, E., & De Keyser, F. (2005). Anti-citrullinated protein/peptide antibodies (ACPA) in rheumatoid arthritis: specificity and relation with rheumatoid factor. *Autoimmunity Reviews, 4*(7), 468–474.

Walker, J. G., & Smith, M. D. (2005). The Jak-STAT pathway in rheumatoid arthritis. *The Journal of Rheumatology, 32*(9), 1650–1653.

Wang, H., Jiang, Z., Pang, Z., Qi, G., Hua, B., Yan, Z., & Yuan, H. (2021). Engeletin protects against TNF-α-induced apoptosis and reactive oxygen species generation in chondrocytes and alleviates osteoarthritis in vivo. *Journal of Inflammation Research, 14*, 745–760.

Wang, X.-P., Xie, W.-P., Bi, Y.-F., Wang, B.-A., Song, H.-B., Wang, S.-L., & Bi, R.-X. (2021). Quercetin suppresses apoptosis of chondrocytes induced by IL-1β via inactivation of p38 MAPK signaling pathway. *Experimental and Therapeutic Medicine, 21*(5), 468.

Wang, Y., Wu, H., Deng, R., Dai, X. J., Bu, Y. H., Sun, M. H., ... Wang, R. H. (2021). Geniposide downregulates the VEGF/SphK1/S1P pathway and alleviates angiogenesis in rheumatoid arthritis in vivo and in vitro. *Phytotherapy Research, 35*(8), 4347–4362.

Yamaguchi, Y., Kadowaki, T., Sakai, E., Noguromi, M., Oyakawa, S., & Tsukuba, T. (2024). Impaired development of collagen antibody-induced arthritis in Rab44-deficient mice. *Biomedicines, 12*(11), 2504.

Yang, Y., Gu, Y., Zhao, H., & Zhang, S. (2019). Loganin attenuates osteoarthritis in rats by inhibiting IL-1β-induced catabolism and apoptosis in chondrocytes via regulation of phosphatidylinositol 3-kinases (PI3K)/Akt. *Medical Science Monitor: International Medical Journal of Experimental and Clinical Research, 25*, 4159.

Yin, Q., Wang, L., Yu, H., Chen, D., Zhu, W., & Sun, C. (2021). Pharmacological effects of polyphenol phytochemicals on the JAK-STAT signaling pathway. *Frontiers in Pharmacology, 12*, 716672.

Yu, H., Yao, S., Zhou, C., Fu, F., Luo, H., Du, W., ... Wu, C. (2021). Morroniside attenuates apoptosis and pyroptosis of chondrocytes and ameliorates osteoarthritic development by inhibiting NF-κB signaling. *Journal of Ethnopharmacology, 266*, 113447.

Zhang, F., Jonsson, A. H., Nathan, A., Millard, N., Curtis, M., Xiao, Q., ... Weisenfeld, D. (2023). Deconstruction of rheumatoid arthritis synovium defines inflammatory subtypes. *Nature, 623*(7987), 616–624.

Zhang, F., Zhang, Y., Zhou, J., Cai, Y., Li, Z., Sun, J., ... Hao, G. (2024). Metabolic effects of quercetin on inflammatory and autoimmune responses in rheumatoid arthritis are mediated through the inhibition of JAK1/STAT3/HIF-1α signaling. *Molecular Medicine, 30*(1), 170.

Zhong, X., Huang, R., Chen, X., & Lei, Y. (2023). A review on the pharmacological aspects of engeletin as natural compound. *Drug Design, Development and Therapy, 17*, 3833–3843.

8 Phytochemicals Targeting PI3K/Akt/mTOR Signaling Pathway to Treat Rheumatoid Arthritis

Ramla Shabbir, Rukhsar Imran, Shaneel Kousar, Yusuf Tatur, Muhammad Irfan, Mohammad Saleem, and Ammara Saleem

WHAT WILL YOU LEARN?

In this chapter, you will

- Explore the intricate world of cellular signaling pathways, where the phosphatidylinositol-3 kinases/protein kinase B/mammalian target of rapamycin (PI3K/Akt/mTOR) cascade emerges as a key player in the pathophysiology of rheumatoid arthritis (RA).
- Discover how this pathway is essential in regulating the aberrant cellular activities that lead to inflammation and damage to tissues in RA.
- Explore the possible consequences of phytochemicals on the dysregulated PI3K/Akt/mTOR axis.
- Through preclinical research and clinical trials, review the most recent findings about the safety and efficacy of phytochemical therapies in the treatment of RA.

LIST OF ABBREVIATIONS

AD	Alzheimer's disease
Bad	Bcl-2-associated death promoter
bFGF	Basic fibroblast growth factor
CaMMB-AMPK-mTOR	Calmodulin-dependent protein kinase kinase beta-5-adenosine monophosphate-activated protein kinase-mTOR
CD	Crohn's disease
DMARDs	Disease-modifying antirheumatic drugs
EGFR	Epidermal growth factor receptor
FOXO	Forkhead box protein O
GPCRs	G protein-coupled receptors
GSK3β	Glycogen synthase kinase-3 beta
IκBα	Nuclear factor of kappa light polypeptide gene enhancer in B-cells inhibitor alpha
IKK	Inhibitor of nuclear factor-κB (IκB) kinase
MMP	Matrix metalloproteinase
mTOR	Mammalian target of rapamycin
NLRP3	Nucleotide-binding domain, leucine-rich–containing family, pyrin domain-containing-3

DOI: 10.1201/9781003535201-10

NSAIDs	Nonsteroidal anti-inflammatory drugs
PD	Parkinson's disease
PGE2	Prostaglandin E2
PI3K	Phosphatidylinositol-3 kinases
PIP2	Phosphatidylinositol 4,5-bisphosphate
PIP3	Phosphatidylinositol 3,4,5-trisphosphate
PKB	Protein kinase B
PTEN	Phosphatase and tensin homolog
RA	Rheumatoid arthritis
RA-FLS	Rheumatoid arthritis -fibroblast-like synoviocytes
RANKL	Receptor activator of nuclear factor kappa beta
ROS	Reactive oxygen species
RSK2	Ribisomal S6 kinase 2
SLE	Systemic Lupus Erythematosus
UC	Ulcerative colitis

8.1 INTRODUCTION OF PI3K/AKT/MTOR (PAM) PATHWAY

The phosphatidylinositol-3 kinases/protein kinase B/mammalian target of rapamycin (PI3K/Akt/mTOR—PAM) pathway is a crucial signal transduction system that regulates numerous physiological and cellular functions. This contributes to the pathogenesis of various diseases including cancer (breast, colon, and prostate cancers) and RA. This pathway is inhibited by various factors such as phosphatidylinositol-3,4,5-trisphosphate 3-phosphatase (PTEN) and glycogen synthase kinase-3 beta (GSK-3β) (Mohseni, Casolaro, Bermúdez-Humarán, Keyvani, & Taghinezhad-S, 2021). This section will delve into the route's comprehensive molecular characteristics, including its physiological functions and regulatory mechanisms, as well as their significance in health and disease. We shall emphasize its relevance in rheumatoid arthritis (RA).

8.2 BACKGROUND AND SIGNIFICANCE

The mTOR and PI3K pathways serve a substantial part in governing various cellular activities, including growth, survival, and metabolism. After being thought of as separate routes at first, researchers revealed that they collaborate through the metabolic processes of an essential enzyme known as Akt. Despite their original designation as independent routes, our work focuses on the complex relationships between these pathways and their cooperative contribution to cellular signaling. PKB, additionally referred to as Akt, is a key survival kinase that, like PI3K and mTOR, regulates various cellular functions including growth, replication, cell death, and metabolism (Weichhart & Säemann, 2008). Dysregulation of these pathways may lead to the onset and progression of diseases including diabetes, sarcoma, and cardiovascular diseases (Sun et al., 2020).

This pathway regulates cell growth and survival (Afify, Oo, Hassan, Seno, & Seno, 2021). As compared to other signaling routes, its components are more complex. The three primary components of the system are mTOR, protein kinase B (Akt), and phosphoinositide 3-kinase (PI3K).

A group of phospholipases indirectly influence intracellular vesicular signaling and trafficking by phosphorylating specific inositol lipids within cells (Akinleye et al., 2013). PI3Ks mediate an array of significant events in the inflammatory reaction to injury and infection. Considering the substrate selectivity and structural characteristics, PI3Ks have been classified into three distinct groups. Class I PI3K isoforms transform phosphatidylinositol 4,5-bisphosphate (PIP2) into phosphatidylinositol 3,4,5-trisphosphate (PIP3) to regulate cell motility, growth, survival, and differentiation. PIP3 functions as a "second messenger" through its interaction with various cell surface receptors. Based on regulatory mechanisms, IA and IB are subclasses in which Class I PI3Ks are categorized. p110 is a catalytic component of class IA PI3Ks, while regulatory subunit is p85.

These subunits are heterodimers. Genes PIK3CA, PIK3CB, and PIK3CD encode Class IA PI3Ks, thus producing catalytic isoforms p110α, p110β, and p110δ. They interact with regulatory subunits resulting in heterodimers, p85 for class IA and p110γ or p87 for class IB. The genes PIK3CG, PIK3R5, and PIK3R6 encode class IB isoforms p110g, p101, and p87, respectively.

Class I PI3Ks are associated with critical cellular functions such as translation, cell proliferation, metabolism of glucose, movement of the cytoskeleton, survival, and change, often becoming activated in malignant tumors.

Class II PI3Ks regulate activities like angiogenesis, growth, relocation, primary cilium function, metabolism of glucose, and survival of cells. The genes PIK3C2A, PIK3C2B, and PIK3C2G, respectively, encode the catalytic isoforms PI3KC2α, PI3KC2β, and PI3KC2γ.

Class III PI3Ks are heterodimers consisting of regulatory and accessory subunits VPS15 (encoded by PIK3R4) and the catalytic subunit VPS34 (encoded by PIK3C3).

They play critical roles in autophagy, endosomal transport, and engulfment of particles (Peng, Wang, Zhou, Mei, & Zeng, 2022; Hawkins, Stephens, & Lipids, 2015). PTEN, the ortholog of phosphatase and tensin that was removed from chromosome 10, however, reverses this process. PTEN exerts a significant negative control on the PI3K pathway by directly opposing PI3K. Additionally, reports indicate a direct interaction between PI3K and PTEN, which leads to the enhancement of PTEN activity (Wang, 2021).

8.2.1 Protein Kinase B (Akt)

PKB, which is a serine/threonine kinase, is crucial for the functioning of the PAM pathway. AKT proteins are categorized into three subtypes: AKT1 (PKBα), AKT2 (PKBβ), and AKT3 (PKBγ). AKT1 is extensively present across several tissues and is intricately involved in promoting anti-apoptotic processes, proliferation, and survival of cells. In contrast, AKT2 is predominantly found in metabolic tissues like the liver, pancreas, and muscles, which are highly responsive to insulin. Deletions of AKT2 can lead to a type 2 diabetes phenotype characterized by impaired glucose absorption.

Additionally, AKT3, present in the brain and testes, is linked with neuroinflammation and neuronal abnormalities. These kinds of circumstances have been associated with its absence (Miricescu et al., 2021). The molecular framework consists of three key domains: the pleckstrin homology (PH) domain at the N-terminus, the regulatory domain at the C-terminus, and the central catalytic kinase domain. PIP3 interacts with the PH domain during activation, triggering structural modifications that expose the Thr309 residue in the AKT1 kinase domain, permitting phosphorylation. The mTOR complex 2 (mTORC2) then boosts AKT1 phosphorylation at Ser473 in its C-terminal region. AKT2 and AKT3 possess equivalent phosphorylation sites, with AKT2 phosphorylated at Thr309 and Ser474 and AKT3 phosphorylated at Thr305 and Ser472 (Zhong & Goodwin, 2024).

8.2.2 Mammalian Target of Rapamycin (mTOR)

It is a serine/threonine protein kinase. mTOR, a key member of the phosphatidylinositol 3-kinase-related kinase family, coordinates a network of signaling pathways that govern development and metabolism in response to an array of environmental stimuli, including glucose, growth hormones, and amino acids. It influences a wide range of biological activities, including cell cycle regulation, consumption of energy, protein synthesis, cell proliferation, survival, and senescence. In mammals, mTOR initiates the two distinct complexes: mTOR complex 1 (mTORC1) and mTOR complex 2.

These two complexes vary from one another in terms of their cellular activity, substrates, upstream signals, and rapamycin sensitivity. mTORC1, comprising mTOR, regulatory-associated protein of mTOR (Raptor), inhibitory proteins Deptor and PRAS40, and adaptor protein mLST8, controls processes that include the synthesis of lipids, mitochondrial metabolism, and autophagy.

It is activated by various signals, which include energy status, growth hormones, amino acids, cytokines, and oxygen. The primary upstream signaling pathway for mTORC1 activation involves PI3K–AKT, which ultimately inhibits the tuberous sclerosis complex, allowing mTORC1 activation. mTORC2, consisting of mTOR, Protor, the adaptor protein Rictor, mSIN1, Deptor, and mLST8, is less understood but is thought to be activated by growth factors and PI3K signaling. It is very important in cell survival and nutrient uptake by phosphorylating and activating AKT kinase, as well as other targets like glucocorticoid-induced protein kinase 1 (SGK1) and protein kinase C. The interplay between mTORC1 and mTORC2 is complex. AKT, activated by mTORC2, can in turn activate mTORC1 by inhibiting TSC1/2. The indirect modulation of AKT is achieved by TSC1/2 through the positive regulation of mTORC2. Additionally, mTORC1 negatively regulates mTORC2 activity through downstream targets like S6K. Overall, mTORC1 and mTORC2 signaling pathways are precisely regulated in response to metabolic variations and external stimuli, thereby maintaining cellular homeostasis (Vakrakou et al., 2022).

The PI3K/Akt pathway's major downstream target, mTOR, controls the translational rate and the initiation step of numerous cellular activities during protein synthesis—the ribosomal protein S6 phosphorylation by mTOR resulting in activation of translation and the promotion of ribosome complex assembly. pS6K is a positive regulator of protein synthesis. MTOR activity is frequently determined through the phosphorylation state of S6K (Giacoppo, Pollastro, Grassi, Bramanti, & Mazzon, 2017). Several medical conditions, including aging, arthritis, insulin resistance, osteoporosis, malignancies, and neurological disorders, have been revealed to be strongly associated with mTOR dysregulation. Indeed, mTORC1 primarily regulates cell growth and metabolism, orchestrating processes such as protein synthesis, lipid synthesis, and autophagy in response to various environmental cues. Instead, mTORC2 is more intensely associated with cell survival, proliferation, and cytoskeletal organization, playing a role in preserving cellular integrity and functionality. This distinction underscores the diverse roles played by mTORC1 and mTORC2 in coordinating cellular responses to internal and external stimuli, ultimately contributing to overall cellular homeostasis and function (Panwar et al., 2023).

8.3 REGULATION OF PATHWAY

The activation of signaling can be initiated by various substances such as insulin, glucose, and numerous growth factors such as insulin-like growth factor receptor, platelet-derived growth factor receptor, and epidermal growth factor receptor, along with cytokines. G protein-coupled receptors (GPCRs) or receptor tyrosine kinases (RTKs) are usually phosphorylated by these substances, which results in the activation of PI3K and the synthesis of phospholipids. These signals then trigger the activation of two downstream effectors of PI3K, mTORC1 and AKT (Sun et al., 2020). Autophosphorylation, dimerization, conformational shifts, and lipid kinase interaction ligand-bound RTKs all experience PI3K, which results in PI3K activation and recruitment (Figure 8.1) (Mahajan & Mahajan, 2012).

8.4 PHYSIOLOGICAL FUNCTIONS

The pathway influences the proliferation of cells, growth, survival, migration, invasion, apoptosis, DNA repair, metabolism of glucose, and motility, among other pleiotropic impacts on cellular physiology (Miricescu et al., 2020). Extracellular growth ligands typically engage with their respective receptors, such as RTKs, B- and T-cell receptors, integrins, cytokine receptors, and GPCRs, to initiate PI3K activation. PI3K binds to the receptor upon ligand–receptor interaction, facilitated directly (through its regulatory subunit) or indirectly (via adaptor molecules like insulin receptor substrate proteins), leading to activation of PI3K. Furthermore, the catalytic subunit (p110) can also directly or indirectly contribute to PI3K activation through its interactions with small GTPases such as Ras and Ras-related protein Rab-5A. Once activated, Akt exerts its effects depending on

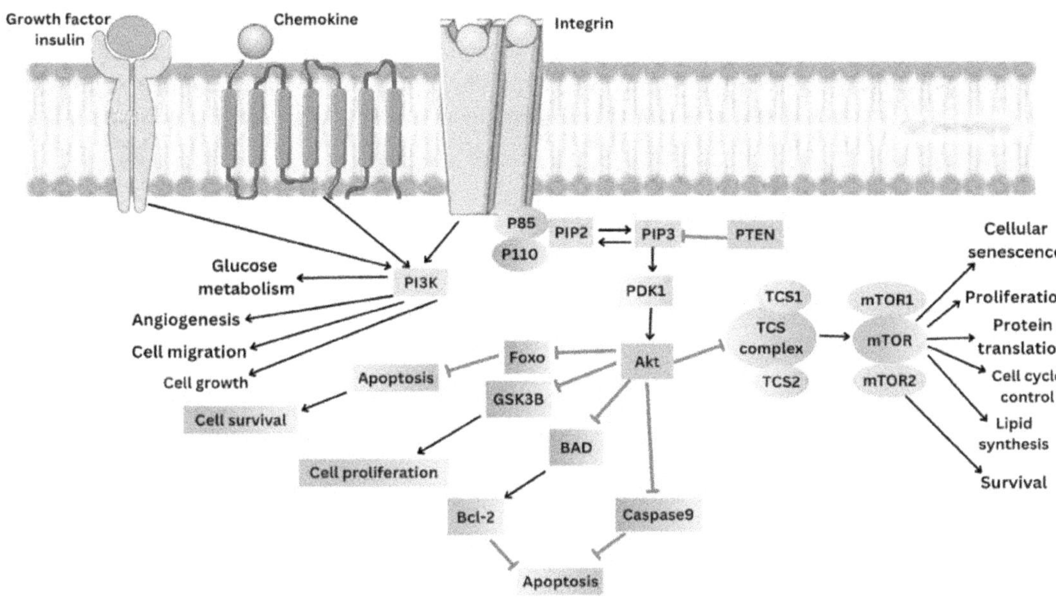

FIGURE 8.1 The PI3K/Akt/mTOR pathway is presented. G protein-coupled receptors (GPCRs), receptor tyrosine kinase (RTK), and GTP-binding proteins attaching to adaptor proteins are some of the components that make up the PI3K pathway. The catalytic subunit (p110) and regulatory subunit (p85) make up the structure of PI3K. After PI3K phosphorylates PIP2, PIP3 is generated. Subsequently, PIP3 activates Akt, a key downstream effector, and 3-phosphoinositide-dependent kinase 1 (PDK1). Akt phosphorylation promotes cell survival and proliferation. PTEN restricts PIP3 from activating Akt by dephosphorylating it. Akt phosphorylation activates mTOR, one of the primary downstream effectors. mTOR directly initiates processes such as cell-cycle-regulating proteins, lipid synthesis, and protein translation. Through its inhibition of FOXO, GSK3B, Bcl-2-related BAD, and Caspase 9, AKT prevents apoptosis and increases cell survival. Lines that end in flat lines indicate inhibitory events, whereas arrows indicate stimulatory events.

the target substrate and subsequent events involved in phosphorylation. These effects span a wide range of cellular functions, including apoptosis, glucose metabolism, proliferation, migration, and gene transcription. For instance, Akt inhibits forkhead box protein O (FOXO), thereby suppressing its pro-apoptotic and catabolic effects. It also inhibits GSK3β, thereby preventing apoptosis and promoting glycogen synthesis. Additionally, Akt directly inhibits the activities of Bcl-2-associated death promoter (Bad) and caspase-9, which play a crucial role in apoptosis, thus prolonging cell viability. Overall, activated Akt orchestrates a coordinated regulation of its substrates to inhibit apoptosis and promote the survival of cells, proliferation, and metabolism (Camaya, Donnelly, & O'Brien, 2022). Further phosphorylation of downstream substrates by activated mTOR implicates several biological processes. Direct control over protein synthesis is attributed to mTORC1. It also plays a significant part in energy homeostasis, ribosome biogenesis, protein synthesis, nucleotide metabolism, and cell-cycle progression. Furthermore, mTORC1 negatively controls catabolic processes like autophagy, which in turn controls the anabolism and catabolism in response to stimuli from the environment. mTORC2 predominantly controls cell migration, cytoskeletal remodeling, proliferation, and survival (Popova & Jücker, 2021).

8.5 IMPLICATIONS IN HEALTH AND DISEASES

PI3/Akt/mTOR dysregulation is a contributory factor for numerous diseases, notably cancer, neuron degeneration, aging, diabetes mellitus, metabolic diseases, complications of wound healing, RA, and psychiatric problems (Ramasubbu & Devi Rajeswari, 2023). Highlighting its significance as a target for therapy, aberrantly active PAM pathway signaling is implicated in numerous malignancies, such as cervical, breast, glioblastoma, gastric, pancreatic, leukemogenesis, prostate cancer, and non-small cell lung cancer (Afify et al, 2021). Comprehending the significance of this pathway in the etiology of numerous diseases yields a significant understanding of pathology and expedites the advancement of targeted treatment strategies.

8.6 PAM SIGNALING AXIS IN THE PATHOGENESIS OF INFLAMMATORY DISORDERS

Chronic inflammation, or aberrant inflammatory responses, is a prominent feature of a group of clinical diseases referred to as inflammatory diseases. It is challenging to identify the precise root and impact of inflammatory diseases because of the complex connection between chronic inflammation and their pathophysiology. For instance, obesity triggers inflammation, but long-term inflammation can contribute to diabetes associated with obesity by developing insulin resistance (Ahmed, 2011).

Here are examples of some inflammatory diseases regarding the role of the pathway in their pathogenesis.

8.6.1 Neurodegenerative Diseases

The PAM intracellular signaling cascade is vital in dictating the fate of cells in both chronic and acute conditions such as Alzheimer's disease (AD), epilepsy, and Parkinson's disease (PD). These neurodegenerative disorders often progress due to the activation of apoptosis and autophagy-related programmed cell death pathways, which are under the regulation of PI3K, Akt, and mTOR signaling (Chong, Shang, Wang, & Maiese, 2012).

8.6.1.1 Alzheimer's Disease

A dementia-causing neurological disorder, amyloid plaques outside neurons and tangled neurofibrils within neurons, is coupled with a gradual decrease in cognitive abilities (Ning, Jorfi, Patel, Kim, & Tanzi, 2022). It has been demonstrated that mTOR activity inhibition affects memory consolidation. Phosphorylated Akt substrates, including tau protein, GSK-3β, and mTOR, are found in increased levels in AD, suggesting their potential contribution to the disease progression. In AD patients, activation of p70S6K is related to the formation of hyperphosphorylated tau and neurofibrillary accumulation, whereas inhibiting mTOR has been shown to enhance memory and decrease β-amyloid levels in animal models. Maintaining mTOR activation is crucial for preventing AD, as reduced activity in peripheral lymphocytes correlates with disease progression, in addition to decreased synaptic plasticity and long-term potentiation in AD models (Chong et al., 2012).

8.6.1.2 Parkinson's Disease

A neurological movement disorder brought on by substantia nigra's dopamine-producing neuronal cells. Symptoms include bradykinesia, stiffness, and resting tremor. PD may result from decreased mTOR activity, with stress response protein REDD1 playing a role in neuronal degeneration. REDD1 increases stress response protein Akt, leading to mTOR signaling loss and neuronal degeneration. Dopaminergic neurons require mTOR activation to prevent cell death under oxidative stress. Protein translation is altered, and dopaminergic neurons are lost (Chong et al., 2012).

8.6.1.3 Epilepsy

Epilepsy is a neurological disorder, usually of a chronic state, characterized by recurrent seizures. In vitro models with hyperactive mTOR caused by mutations in TSC1 and TSC2 have a high rate of seizures. Early rapamycin reduction in mTOR signaling in animal models of TS may halt seizures, neuronal disarray, and astrogliosis, indicating that aberrant mTOR activity disrupts normal brain function and causes epilepsy. Additionally, acquired epilepsy has been associated with mTOR activation. Neuronal cell degeneration, neurogenesis, mossy fiber sprouting, and when spontaneous epilepsy starts to occur, all are reduced when mTOR activity is inhibited during kainate-induced epilepsy (Chong et al., 2012).

8.6.2 Systemic Lupus Erythematosus

An autoimmune, life-threatening disease can damage several organs. It is primarily characterized by atypical T-lymphocytes, hyperactivity of B cells, and the production of high amounts of autoantibodies. The hallmarks of the disease are the existence of autoantibodies and the deposition of immunological complexes. PI3K activity in SLE increases, but inhibitors may reduce inflammation and alleviate symptoms. PIP3 is produced by PI3Ks and activates Akt via PIP3. Akt expression is elevated in SLE peripheral blood T and B cells, with a correlation between activation degree and disease severity (Pan, Lu, Wang, Xu, & Yang, 2020).

8.6.3 Cancer

Cancer can be defined as the rapacious development, multiplication, and infiltration of malignantly altered cancer cells beyond their normal bounds into surrounding tissue. The PAM intracellular signaling pathway is frequently dysregulated in various human cancers, contributing to cancer development, proliferation, invasion, metastasis, and treatment resistance. Mutations in PIK3CA, PIK3R1, and PTEN, a tumor suppressor gene, increase PI3K/AKT pathway activity, thus promoting tumorigenesis. Akt, a crucial component, induces eNOS localization, stimulates angiogenesis, and enhances NF-kappa B transcriptional activity (Pulumati, Pulumati, Dwarakanath, Verma, & Papineni, 2023; Rascio et al., 2021).

8.6.4 Inflammatory Bowel Disease

Inflammatory bowel disease, which includes Crohn's disease (CD) and ulcerative colitis (UC), is represented by persistent inflammation in the gastrointestinal tract, that is, periodically remitted (Wang et al., 2023). The PAM pathway is significant in inflammation-related diseases like CD and UC. NF-kβ, a transcription factor, is a major modulator of inflammation, triggering the generation of pro-inflammatory cytokines and activating CD4 and CD8 T-cell functions. The key regulator of NF-kB is PI3K/Akt, which is inhibited by CCAAT/enhancer binding protein β and enhanced by AKT-induced mTOR. Modulating the PI3K/AKT pathway can be beneficial for UC (Jalil et al., 2023).

8.7 PAM SIGNALING IMBALANCE IN THE PATHOGENESIS OF RA

In RA, the PAM pathway significantly influences the proliferation of FLS and it is involved in the differentiation of osteoclasts, which are concerned with bone degradation and joint damage. After activation, the PAM pathway initiates the release of numerous inflammatory mediators such as interleukin (IL)-6, IL-1β, IL-17, and TNF-α that further contribute to bone damage. Various molecules, including different growth factors and cytokines, trigger and activate the signaling pathway by binding to tyrosine kinases, activated receptor RTKs, and GPCRs, which subsequently activate PI3K, leading to phospholipid generation, which further initiates sequential PI3K effectors, such as mTORC1 and AKt. When PI3K is activated, PIP2 converts to PIP3, activating AKT, a critical effector in PI3K signaling. AKT, a pivotal mediator in PI3K signaling, is drawn to the cell membrane's inner side through its PH domains, in addition to phosphoinositide-dependent kinase-1, which triggers Thr 308 phosphorylation. It also activates the second pathway through mTORC2, phosphorylating Ser473 of AKT's hydrophobic domain. Once activated, AKT relocates to various cellular

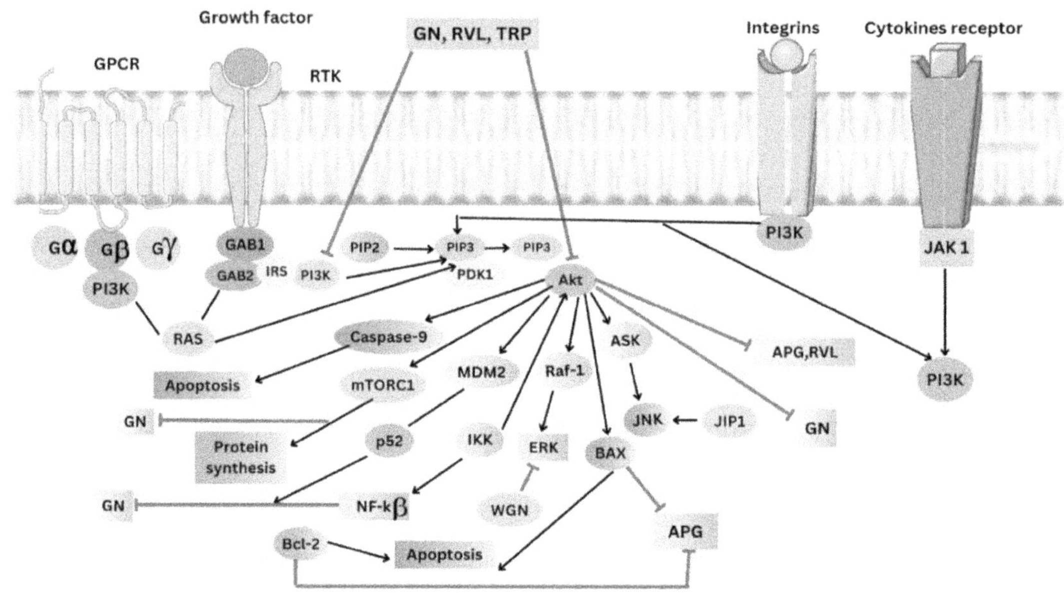

FIGURE 8.2 Potential phytochemical targets in RA and Akt/PI3K signaling. Numerous processes are regulated by the Akt/PI3 K pathway, such as growth, metabolism, proliferation, death, and the cell cycle. Normally, RTKs, cytokine receptors, and GPCRs control it. Akt, a crucial protein that regulates essential functions, is further drawn to the plasma membrane by the conversion of PIP2 to PIP3. JIP1 JNK-interacting protein; ASK apoptosis signal-regulating kinase; mTORC1, GAB1 growth factor receptor-bound protein 1; PI3K/Akt phosphoinositide 3-kinase/protein kinase B, MDM2 mouse double minute 2 homologs; mTORC1, GAB1 growth factor receptor-bound protein 2.

compartments and stimulates numerous downstream effectors, such as transcription factors, E3 ubiquitin ligases, small G protein regulators, cell cycle regulators, and protein kinases (Sun et al., 2020). Growth factor receptor-bound protein, an adapter protein within the GAB family, harbors binding sites such as PI3K and SH2 domain-containing protein tyrosine phosphatase-2. It functions as a downstream effector of membrane receptors stimulated by antigens, cytokines, hormones, and growth factors, regulating numerous signaling cascades (Figure 8.2) (Liu et al., 2017).

8.7.1 Deciphering the Therapeutic Potential of Phytochemicals

The robust vitamin and mineral profiles of plant-based foods, such as grains, vegetables, and fruits, are well known. The exploration of phytochemicals' potential health benefits stemmed from observational studies indicating that consuming plant-based foods may offer protection against various chronic age-related diseases (Block, Patterson, Subar, & Cancer, 1992). Plant products have been the source of a significant number of bioactive molecules with therapeutic potential, many of which have subsequently been turned into medications that are consumed globally for varied ailments, including inflammatory and autoimmune diseases, infectious diseases, and cancer (Venkatesha et al., 2016).

8.8 PHYTOCHEMICALS AND RA: FOCUSING ON THE PAM PATHWAY

Several symptomatic pharmaceuticals, such as corticosteroids, NSAIDs, and DMARDs, have been shown to be effective in relieving joint pain and inflammation (Smolen & Steiner, 2003). However, due to the devastating side effects of these single-target medications, plant-derived anti-inflammatory molecules, notably phenolics and alkaloids, have attracted great interest as a safe, balanced

TABLE 8.1
Sources and MOAs of the Phytochemicals Acting on PI3k/Akt/mTOR Pathway

Phytochemical	Class	Mechanism of Action	Reference
Apigenin	Flavone	PI3K/Akt pathway activation and TNF-α-induced cell apoptosis.	Shin et al. (2009)
Celastrol	Triterpenoid	Increased endoplasmic reticulum stress causing mitochondrial membrane loss, which leads to calcium-induced mTOR pathway-mediated cell apoptosis.	Wong et al. (2019)
Genistein	Phytoestrogen	NF-κB p65 phosphorylation induction through suppression of TNF-α resulting in PI3K/Akt pathway inhibition	Li et al. (2014)
Resveratrol	Polyphenol	Reduction in the concentration of MCP-1 present in synovial fluids, downregulation of protein expression of MMP3 and IL-1β thus significantly decreases phosphorylated Akt concentration in the Akt/PI3K pathway.	Tian, Chen, Gao, Li, and Xie (2013)

therapeutic strategy for mitigating joint inflammation and destruction (Wu et al., 2017). Growing hope has recently been expressed about the potential of natural products to provide relief from inflammatory diseases. With a particular emphasis on the pivotal PAM pathway and its closely linked mediators, phytochemicals offer promising avenues for addressing ailments such as lupus, gout, Lyme disease, osteoarthritis, multiple sclerosis, and RA.

8.9 PRECLINICAL STUDIES

Natural products, plants, and herbs have been a great source of therapeutic agents in recent years. Several plant-derived compounds have been identified for their antirheumatic properties. To meet the demands of the developing world and progressing diseases, the need for the identification of new compounds that have more therapeutic effects increases as compared to the previously introduced products. Apart from the above-stated compounds, a few compounds stated in Table 8.1 are still undergoing preclinical studies.

8.10 CONCLUSION

The chapter concludes by highlighting phytochemicals' vital role in modifying the PAM signaling pathway, presenting potential treatment options for RA. Several studies have shown the effectiveness of different compounds derived from plants in inhibiting key components of this system, which suppresses inflammation and the inflammation-related responses that are hallmarks of RA. Phytochemicals provide a supplementary or alternative approach to conventional RA therapies by focusing on certain molecular pathways, which may reduce side effects and improve patient outcomes. To fully comprehend the molecular interactions and maximize the therapeutic use of these organic compounds, additional studies have been conducted.

REFERENCES

Afify, S. M., Oo, A. K. K., Hassan, G., Seno, A., & Seno, M. (2021). How can we turn the PI3K/AKT/mTOR pathway down? Insights into inhibition and treatment of cancer. *Expert Review of Anticancer Therapy, 21*(6), 605–619. doi:10.1080/14737140.2021.1918001

Ahmed, A. U. J. F. i. B. (2011). An overview of inflammation: Mechanism and consequences. *Frontiers in Biology*, *6*(4), 274–281.

Akinleye, A., Avvaru, P., Furqan, M., Song, Y., Liu, D. J. J. O. h., & oncology. (2013). Phosphatidylinositol 3-kinase (PI3K) inhibitors as cancer therapeutics. *Journal of Hematology and Oncology*, *6*, 1–17.

Block, G., Patterson, B., Subar, A. J. N., & cancer. (1992). Fruit, vegetables, and cancer prevention: a review of the epidemiological evidence. *Nutrition and Cancer*, *18*(1), 1–29.

Camaya, I., Donnelly, S., & O'Brien, B. J. J. o. D. (2022). Targeting the PI3K/Akt signaling pathway in pancreatic β-cells to enhance their survival and function: an emerging therapeutic strategy for type 1 diabetes. *Journal of Diabetes*, *14*(4), 247–260.

Chong, Z. Z., Shang, Y. C., Wang, S., & Maiese, K. J. F. n. (2012). A critical kinase cascade in neurological disorders: PI3K, *Akt and mTOR*. *7*(6), 733–748.

Giacoppo, S., Pollastro, F., Grassi, G., Bramanti, P., & Mazzon, E. J. F. (2017). Target regulation of PI3K/Akt/mTOR pathway by cannabidiol in treatment of experimental multiple sclerosis. *Fitoterapia*, *116*, 77–84.

Hawkins, P., Stephens, L. J. B. E. B. A.-M., & Lipids, C. B. o. (2015). PI3K signalling in inflammation. *Biochimica et Biophysica Acta (BBA) – Molecular and Cell Biology of Lipids*, *1851*(6), 882–897.

Jalil, A. T., Hassan, N. F., Abdulameer, S. J., Farhan, Z. M., Suleiman, A. A., Al-Azzawi, A. K., … Pharmacology, C. (2023). Phosphatidylinositol 3-kinase signaling pathway and inflammatory bowel disease: current status and future prospects. *Fundamental and Clinical Pharmacology*, *37*(5), 910–917.

Li, J., Li, J., Yue, Y., Hu, Y., Cheng, W., Liu, R., … Zhang, P. (2014). Genistein suppresses tumor necrosis factor α-induced inflammation via modulating reactive oxygen species/Akt/nuclear factor κB and adenosine monophosphate-activated protein kinase signal pathways in human synoviocyte MH7A cells. *Drug Design, Development and Therapy*, *8*, 315–323.

Liu, W., Yu, W.-M., Zhang, J., Chan, R. J., Loh, M. L., Zhang, Z., … Qu, C.-K. J. L. (2017). Inhibition of the Gab2/PI3K/mTOR signaling ameliorates myeloid malignancy caused by Ptpn11 (Shp2) gain-of-function mutations. *Leukemia*,*31*(6), 1415–1422.

Mahajan, K., & Mahajan, N. P. J. J. O. C. P. (2012). PI3K-independent AKT activation in cancers: a treasure trove for novel therapeutics. *Journal of Cellular Physiology*, *227*(9), 3178–3184.

Miricescu, D., Balan, D. G., Tulin, A., Stiru, O., Vacaroiu, I. A., Mihai, D. A., … Medicine, T. (2021). PI3K/AKT/mTOR signalling pathway involvement in renal cell carcinoma pathogenesis. *Experimental and Therapeutic Medicine*, *21*(5), 1–7.

Miricescu, D., Totan, A., Stanescu-Spinu, I.-I., Badoiu, S. C., Stefani, C., & Greabu, M. J. I. J. O. M. S. (2020). PI3K/AKT/mTOR signaling pathway in breast cancer: from molecular landscape to clinical aspects. *International Journal of Molecular Sciences*, *22*(1), 173.

Mohseni, A. H., Casolaro, V., Bermúdez-Humarán, L. G., Keyvani, H., & Taghinezhad -S, S. J. G. M. (2021). Modulation of the PI3K/Akt/mTOR signaling pathway by probiotics as a fruitful target for orchestrating the immune response. *Gut Microbes*, *13*(1), 1886844.

Ning, S., Jorfi, M., Patel, S. R., Kim, D. Y., & Tanzi, R. E. J. F. i. N. (2022). Neurotechnological approaches to the diagnosis and treatment of Alzheimer's disease. *Frontier in Neuroscience*, *16*, 854992.

Pan, L., Lu, M.-P., Wang, J.-H., Xu, M., & Yang, S.-R. J. W. J. O. P. (2020). Immunological pathogenesis and treatment of systemic lupus erythematosus. *World Jornal of Pediatrics*, *16*, 19–30.

Panwar, V., Singh, A., Bhatt, M., Tonk, R. K., Azizov, S., Raza, A. S., … Therapy, T. (2023). Multifaceted role of mTOR (mammalian target of rapamycin) signaling pathway in human health and disease. *Signal Transduction and Targeted Therapy*, *8*(1), 375.

Peng, Y., Wang, Y., Zhou, C., Mei, W., & Zeng, C. J. F. I. o. (2022). PI3K/Akt/mTOR pathway and its role in cancer therapeutics: are we making headway?, *Frontier in Oncology*, *12*, 819128.

Popova, N. V., & Jücker, M. J. I. J. O. M. S. (2021). The role of mTOR signaling as a therapeutic target in cancer. *22*(4), 1743.

Pulumati, A., Pulumati, A., Dwarakanath, B. S., Verma, A., & Papineni, R. V. J. C. R. (2023). Technological advancements in cancer diagnostics: improvements and limitations. *Cancer Reports*, *6*(2), e1764.

Ramasubbu, K., & Devi Rajeswari, V. (2023). Impairment of insulin signaling pathway PI3K/Akt/mTOR and insulin resistance induced AGEs on diabetes mellitus and neurodegenerative diseases: a perspective review. *Molecular and Cellular Biochemistry, 478*(6), 1307–1324. doi:10.1007/s11010-022-04587-x

Rascio, F., Spadaccino, F., Rocchetti, M. T., Castellano, G., Stallone, G., Netti, G. S., & Ranieri, E. J. C. (2021). The pathogenic role of PI3K/AKT pathway in cancer onset and drug resistance: an updated review. *Cancers, 13*(16), 3949.

Shin, G.-C., Kim, C., Lee, J.-M., Cho, W.-S., Lee, S.-G., Jeong, M., … Lee, K. (2009). Apigenin-induced apoptosis is mediated by reactive oxygen species and activation of ERK1/2 in rheumatoid fibroblast-like synoviocytes. *Chemico-Biological Interactions, 182*(1), 29–36.

Smolen, J. S., & Steiner, G. J. N. R. D. d. (2003). Therapeutic strategies for rheumatoid arthritis. *The New England Journal of Medicine*, *2*(6), 473–488.

Sun, K., Luo, J., Guo, J., Yao, X., Jing, X., & Guo, F. (2020). The PI3K/AKT/mTOR signaling pathway in osteoarthritis: a narrative review. *Osteoarthritis Cartilage, 28*(4), 400–409. doi:10.1016/j.joca.2020.02.027

Tian, J., Chen, J.-W., Gao, J.-S., Li, L., & Xie, X. (2013). Resveratrol inhibits TNF-α-induced IL-1β, MMP-3 production in human rheumatoid arthritis fibroblast-like synoviocytes via modulation of PI3kinase/Akt pathway. *Rheumatology International, 33*, 1829–1835.

Vakrakou, A. G., Alexaki, A., Brinia, M.-E., Anagnostouli, M., Stefanis, L., & Stathopoulos, P. J. I. J. O. M. S. (2022). The mTOR signaling pathway in multiple sclerosis; from animal models to human data. *International Journal of Molecular Sciences*, *23*(15), 8077.

Venkatesha, S. H., Astry, B., Nanjundaiah, S. M., Kim, H. R., Rajaiah, R., Yang, Y., ... Moudgil, K. D. J. a. j. o. p. s. (2016). Control of autoimmune arthritis by herbal extracts and their bioactive components. *Asian Journal of Pharmaceutical Sciences*, *11*(2), 301–307.

Wang, X., Li, X., Ma, X., Zhang, L., Han, T., & Zhang, D. J. N.-S. S. A. O. P. (2023). Dihydromyricetin alleviates inflammatory bowel disease associated intestinal fibrosis by inducing autophagy through the PI3K/AKT/mTOR signaling pathway. *Naunyn-Schmiedeberg's Archives of Pharmacology,* 397, 4183–4194.

Wang, Z. J. C. (2021). Regulation of cell cycle progression by growth factor-induced cell signaling. Cells, *10*(12), 3327.

Weichhart, T., & Säemann, M. D. (2008). The PI3K/Akt/mTOR pathway in innate immune cells: emerging therapeutic applications. *Annals of the Rheumatic Disease, 67* (Suppl 3), iii70–iii74. doi:10.1136/ard.2008.098459

Wong, V. K. W., Qiu, C., Xu, S. W., Law, B. Y. K., Zeng, W., Wang, H., … Chan, T. W. (2019). Ca2+ signalling plays a role in celastrol-mediated suppression of synovial fibroblasts of rheumatoid arthritis patients and experimental arthritis in rats. *British Journal of Pharmacology, 176*(16), 2922–2944.

Wu, X., Long, L., Liu, J., Zhang, J., Wu, T., Chen, X., ... Lv, T. Z. J. M. M. R. (2017). Gambogic acid suppresses inflammation in rheumatoid arthritis rats via PI3K/Akt/mTOR signaling pathway. *Molecular Medicine Reports*, *16*(5), 7112–7118.

Zhong, H. A., & Goodwin, D. T. J. M. (2024). Selectivity studies and free energy calculations of AKT inhibitors. *Molecules*, *29*(6), 1233.

9 Phytochemicals Targeting Transforming Growth Factor (TGF)-β and Inducible Nitric Oxide Synthase to Treat Rheumatoid Arthritis

Humaira Gul, Rida Siddique, Muntaha Maqbool, Noor-ul-Eman, Atika Afzal, Faraza Javaid, Yusuf Tatur, and Ammara Saleem

WHAT DO WE LEARN IN THIS CHAPTER?

In this chapter, we will learn about a brief introduction of transforming growth factor beta (TGF-β) and inducible nitric oxide synthase (iNOS) in the therapy of rheumatoid arthritis (RA). We will delve into the role of TGF-β and iNOS in the pathogenesis of RA. We will learn the proinflammatory and anti-inflamatory role of TGF-β in RA progression by promoting cytokine expression and immune cell recruitment. We will also explore the role of iNOS and TGF-β in inflammation and tissue destruction. This chapter will discuss how targeting these two molecules can help manage RA. Furthermore, readers will find the possible therapeutic effects of various phytochemicals on such pathways. In concluding this chapter, the role of various phytochemicals to inhibit pro-inflammatory mediators like NO, IL-1β, and TNF-α; to suppress the activation of the NF-κB pathway; and to affect the levels of TGF-β and iNOS will be highlighted.

LIST OF ABBREVIATIONS

AP-1	Activating protein-1
BLFE	Berberis liquid fruit extract
CaM	Calmodulin
COX-2	Cyclooxygenase-2
ERK1/2	Extracellular signal-regulated kinases 1 and 2
FAD	Flavin adenine dinucleotide
FLS	Fibroblast-like synoviocytes
FMN	Flavin mononucleotide
IL-1	Interleukin-1
IL-1β	Interleukin-1Beta
iNOS	Inducible nitric oxide synthase
JAK-2	Janus Kinase 2
KPG	Kaempferol
LPS	Lipopolysaccharide
MAPK	Mitogen-activated protein kinase

DOI: 10.1201/9781003535201-11

NADPH	Nicotinamide adenine dinucleotide phosphate
NF-κB	Nuclear factor kappa B
NO	Nitric oxide
PCR	Polymerase chain reaction
PGE2	Prostaglandin E2
PI3K/Akt	Phosphoinositide 3-kinase/Protein kinase B
RA	Rheumatoid arthritis
STAT-3	Signal Transducer and Activator of Transcription 3
TGF-β	Transforming growth factor beta
TGF-β3	Transforming growth factor-beta 3
TNF-α	Tissue necrotic factor-alpha
Tregs	Regulatory T cells

9.1 INTRODUCTION

Known for its symmetrical erosive synovitis and involvement of the extra-articular joints, rheumatoid arthritis (RA) has an unclear origin. Despite treatment, most patients have a persistent, variable illness history that can lead to progressive joint degeneration, deformity, disability, and even early death. Approximately 9 million doctor visits and over 250,000 hospital admissions result from RA every year. It leads to disability also (Saleem, Javed, Akhtar, Mobashar, & Khan, 2025; Tahir, Saleem, & Akhtar, 2025).

With RA, bone and cartilage are eroded slowly, resulting in gradual joint degeneration. It first affects the smaller joints. Extensive involvement of the skin, eyes, kidneys, lungs, and other body organs can sometimes affect this inflammatory illness. The affected joints experience pain, swelling, stiffness, and limited range of motion. Aside from fatigue and weight loss, it can also cause cardiovascular, pulmonary, and skeletal problems in addition to joint symptoms (Radu & Bungau, 2021). The ongoing inflammation weakens the tendons and cartilage in the joints, eventually leading to bone destruction and deformity. An inflammatory disease with a high mortality rate and a high morbidity rate, RA may result in impairment and severely restrict a patient's range of motion (Saleem et al., 2025). About 1% of people worldwide have RA, with women experiencing the condition two to three times more frequently than men. In addition, age, gender, obesity, smoking, and other hereditary variables are linked as risk factors for RA (Saleem, Khalid, Akhtar, & Zeb, 2023).

Mammals contain three different isoforms of the TGF-β protein family called TGF-β1, TGF-β2, and TGF-β3. These isoforms share a significant degree of sequence similarity (84%–92%), with distinct promoter sequences, which are regulated transcriptionally. It is important to note that each TGF-β isoform is represented by a distinct gene and some structural similarities (Tzavlaki & Moustakas, 2020). The TGF-β dimer is released in an inactive form and requires activation to bind to its receptors and exert its biological functions. Thus, it requires activation to bind with its receptors. Upon activation, TGF-β binds to the TGF-β type II receptor to form a complex that engages the TGF-β type I receptor. This activation happens through the phosphorylation of serine or threonine residues on the receptor. Betaglycan is an additional third TGF-β receptor, which allows the potential affinity of TGF-β2 to the TGF-β receptor type II (Kim, Henen, & Hinck, 2019)

This protein has three main functions: inhibition of cell proliferation, promotion of extracellular matrix deposition, and complex immune-regulatory properties. It serves as a master regulator, controlling cell growth, differentiation, migration, programmed cell death (apoptosis), and extracellular matrix synthesis (Danielpour, 2005). Additionally, it modulates how cells respond to injury and regulates immune responses (Dobolyi, Vincze, Pál, & Lovas, 2012).

Nitric oxide is a versatile signaling molecule synthesized by an enzyme called nitric oxide synthase. It is involved in several physiological functions throughout the body, such as the cardiovascular, immune, and nervous systems (Groves & Wang, 2000). One considerable property of NO is its ability to diffuse readily across cell membranes, enabling it to serve as a messenger molecule.

Due to this property, it facilitates a range of biological functions, including the regulation of blood vessel activity, mitochondrial functions, T-cell signaling pathways, inflammation, oxidative stress, and apoptosis (Guerby et al., 2021).

9.2 ROLE OF TRANSFORMING GROWTH FACTOR-BETA (TGF-β) AND INDUCIBLE NITRIC OXIDE SYNTHASE (iNOS) IN RHEUMATOID ARTHRITIS

TGF-β plays a complex role in RA. Over time, RA causes joint damage as a result of chronic inflammation in the synovium. TGF-β, a cytokine with diverse functions, may play a role in RA in both promoting the repair of tissue and regulating immune responses (Carvajal Alegria et al., 2021).

One important role of TGF-β in RA is its anti-inflammatory properties. Various studies have indicated that TGF-β can facilitate tissue restoration by inducing the synthesis of extracellular matrix elements including collagen and fibronectin (Gonzalo-Gil & Galindo-Izquierdo, 2014). In addition, TGF-β is essential for the proliferation of regulatory T cells (Tregs), which are needed for immunological tolerance preservation and to avoid the occurrence of autoimmune reactions. TGF-β helps in managing overactive stimulation of the immune system by promoting the production of Tregs. The promising potential of TGF-β as a therapeutic target for regulating immune-mediated dysregulation in RA has been demonstrated, as evidenced by its regulatory function (Carvajal Alegria et al., 2021; Rico et al., 2008).

However, TGF-β does not solely play an anti-inflammatory role (Yoshimura & Muto, 2011). There are also circumstances in which it exhibits proinflammatory activity (Park et al., 2011). For instance, TGF-β has been linked to the overexpression of proinflammatory cytokines like interleukin-1 (IL-1) and tumor necrosis factor-alpha (TNF-α), both of which have been found to have a role in the development and progression of RA. Additionally, TGF-β may worsen inflammation in arthritic joints by acting as a chemoattractant for immune cell populations such as neutrophils (Carrier, Yuan, Kuchroo, & Weiner, 2007). Its multiple roles accentuate the need for a thorough understanding of its role in disease pathogenesis. Although, it stimulates tissue healing and controls immune responses. However, in some situations, it can exacerbate inflammation and cause tissue loss. TGF-β's intricate and ever-changing involvement in disease pathogenesis is made apparent by its connections with several immune cell subsets and other cell types within the RA synovium (Carvajal Alegria et al., 2021).

Various stimuli, including bacterial lipopolysaccharides (LPS) and cytokines, can induce the expression of iNOS, which does not exist constitutively in cells. Once induced, iNOS produces large amounts of NO continuously and does not depend on intracellular calcium concentrations. The increased NO production by iNOS in macrophages can have cytotoxic effects, including inhibiting DNA replication and mitochondrial electron transport enzymes, and damaging DNA (Mao, Wu, Nan, Sun, & Wang, 2023).

To develop therapeutic strategies for diseases such as cardiovascular disease, inflammation, and neurodegenerative conditions, it is critical to understand the mechanisms behind NOS-mediated NO production (Guerby et al., 2021).

The process of NO production from NOS requires the transformation of L-arginine with molecular oxygen, reduced nicotinamide adenine dinucleotide phosphate (NADPH), and L-arginine serving as substrates and co-substrates, respectively. NOS isoforms also require cofactors like flavin adenine dinucleotide (FAD), flavin mononucleotide (FMN), and 6R-l-erythro-5,6,7,8-tetrahydrobiopterin (BH_4) to fulfill their enzymatic activity. NO synthesis begins with electron transfer from NADPH to FAD and FMN by monomers of NOS, resulting in the reduction of oxygen molecules to superoxides. The reductase domain of NOS requires binding to calmodulin (CaM) to enhance electron transport, with the calmodulin-binding domains of NOS differing among isoforms. Subsequently, NO is formed through the hydroxylation of L-arginine to NV-hydroxy-L-arginine, followed by the transfer of an electron from NADPH to heme in the oxygenase domain of the opposite monomer as shown in Figure 9.1.

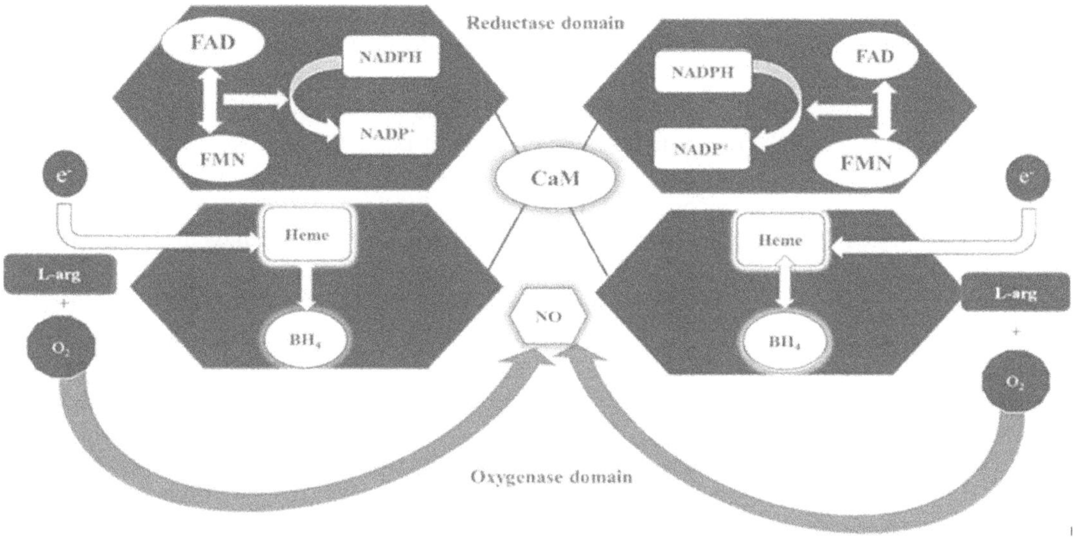

FIGURE 9.1 General mechanism of NO formation by NOSs. Calmodulin (CaM); flavin adenine dinucleotide (FAD); nicotinamide adenine dinucleotide phosphate (NADPH); L-arginine (L-arg); flavin mononucleotide (FMN); 6R-l-erythro-5,6,7,8-tetrahydrobiopterin (BH4).

9.3 ROLE OF TGF-β AND INDUCIBLE NITRIC OXIDE SYNTHASE (iNOS) IN INFLAMMATORY DISEASES

Inflammation is a complicated tissue reaction to damage. The host response is triggered by a chemical network that stimulates endothelial cells, recruits leukocytes, and stimulates phagocytic leukocytes (Javed, Saleem, & Akhtar, 2023; Puppala et al., 2024). At the start of the inflammatory process, platelets clump together and produce inflammatory mediators, such as transforming growth factors (Yoshimura, Wakabayashi, & Mori, 2010). TGF-β is involved in leukocyte adhesion and recruitment by chemotactic action, upregulation of integrin expression and adhesion, induction of type IV collagenase, auto-induction of TGF-β, and stimulation of growth factor/cytokine synthesis. Leukocyte adherence at the site of inflammation is caused by the platelet products interacting with endothelial cells and vascular leukocytes in conjunction with other inflammatory mediators. Through interactions with adhesion receptors, signaling molecules, such as growth factors and cytokines, cause localized adherence to the vascular wall. Before white blood cells extravagate into the tissues to take part in the inflammatory response, they must first adhere to the vessel wall (Puppala et al., 2024).

Leukocytes respond to chemotactic signals and gather at lesion sites, where they are inundated with inflammatory mediators. Growth factors and cytokines attach to the appropriate receptors on monocytes, lymphocytes, and neutrophils to start signal transduction that improves endocytic, biochemical, and synthetic processes (Liu et al., 2023). When activating substances are not present, monocytes undergo apoptosis, or programmed cell death, and are quickly eliminated. Prostaglandins, cytokines, reactive oxygen metabolites, and enzymes are released by activated macrophages to coordinate an inflammatory response. Nitrogen intermediates and a number of these released products may also play a role in tissue destruction. The production of growth factors and cytokines also serves as a precursor to the subsequent phase of the inflammatory response and the healing process (Bierie & Moses, 2010; Javed et al., 2023).

The iNOS plays a complicated and context-dependent role. Various research studies have shown its dual nature effect on inflammation, varying from enhancing inflammation to retarding it, depending on the specific inflammatory condition. The reduction in iNOS also reduces inflammation and tissue damage in autoimmune disorders (such as giant cell arteritis and vasculitis), allergic airway disease, and acute cardiac allograft rejection (Huang, Chen, Yang, & Hong, 2023; Kashfi, Kannikal, & Nath, 2021).

On the other hand, there are some examples where the ability of the host to express iNOS seems to benefit the host during inflammation (Huang et al., 2023; Puppala et al., 2024). For example, in sepsis and colonic mucosal injury, its expression helps stop the margination of neutrophils and reduce inflammatory neutrophil–endothelial interactions, thereby decreasing tissue damage. It promotes tissue repair and regeneration in the healing of wounds and angiogenesis (Oliveira, Assreuy, & Sordi, 2022), whereas, in some specific inflammatory conditions, the ability of the host to produce iNOS appears unimportant. Some research studies on wild-type mice showed iNOS limitations, as it may not always reduce mortality rate or tissue damage, but cause damage to liver tissues and act as a pro-inflammatory substance. However, iNOS destruction has no apparent impact on leukocyte mobilization into the peritoneal cavity (Chuah, Hertzog, & Campbell, 2022).

As a whole, iNOS plays a multifaceted role in inflammation, which changes according to the specific inflammatory context. Targeting iNOS as a therapeutic to treat various inflammatory diseases such as RA will be a valuable thing (Kashfi et al., 2021).

9.4 ROLE OF TGF-β AND iNOS IN THE PATHOGENESIS OF RHEUMATOID ARTHRITIS

Pro-inflammatory cytokines and angiogenic growth factors are secreted by rheumatoid mesenchymal fibroblasts and they affect the synovial membrane in RA (Pap, Müller-Ladner, Gay, & Gay, 2000). Rats treated with TGF-β injection experience synovial membrane damage, neutrophil recruitment, and infiltration, especially in the presence of inflammatory cells (Figure 9.2). The TGF-β functions as a proinflammatory agent as supported by several investigations. In rats with arthritis produced using a stem cell transplant model, intra-articular administration of anti-TGF-β antibodies suppresses synovial inflammation, bone resorption, and the generation of proinflammatory cytokines (Hung, Hsueh, Rethi, Lu, & Chuang, 2025). The synovial hyperplasia mechanism seems complex but different cytokines make it easy to understand. Cytokines such as basic fibroblast growth factor and survival proteins appear to be crucial in understanding (Pap et al., 2000). Synovial fibroblasts express the necessary chemokine and cytokines to play a role in the migration, activation, and survival of T cells, and B cells in turn stimulate synovial-fibroblast activation. Angiogenesis is stimulated by the synthesis of IL-18 and vascular endothelial growth factor by synovial fibroblasts. Synovial fibroblasts have a potential role in integrating the damaging and inflammatory stages of inflammatory arthritis, which are tightly controlled by cytokines (McInnes & Schett, 2007; Pap, 2004).

The iNOS plays a pivotal role in RA pathogenesis. The NO polymorphism was seen in RA (Gonzalez-Gay et al., 2004). Regulation of eNOS and nNOS is carried out by intracellular Ca^{+2} isozymes, which are controlled at the transcriptional level. Numerous cell types, such as osteoclasts, macrophages, fibrocytes, neutrophils, and osteoblasts, can produce NO in the inflamed synovium (Tran et al., 2022). It is reported that 2.5 times more NO is produced by RA patients' T cells than healthy donor T cells. When NO production is increased, then there is also an increase in cytoplasmic Ca^{+2} concentrations. Blocking of TNF causes decreases in iNOS expression in human blood mononuclear cells as shown in Figure 9.2. When iNOS is inhibited, an anti-inflammatory effect is observed. Chronic inflammatory illness may be accelerated by the release of intracellular antigens through necrosis (Nagy et al., 2010). When molecular oxygen and the cofactor NADPH are present, the iNOS enzyme produces NO. Among its various biological activities, NO stimulates matrix metalloproteinase, prevents chondrocytes from synthesizing collagen and proteoglycans, and encourages vasodilation, which permits cellular and fluid inflow into an inflammatory location. NO further reacts with reactive oxygen species (ROS) to form peroxynitrite, which promotes chondrocyte death.

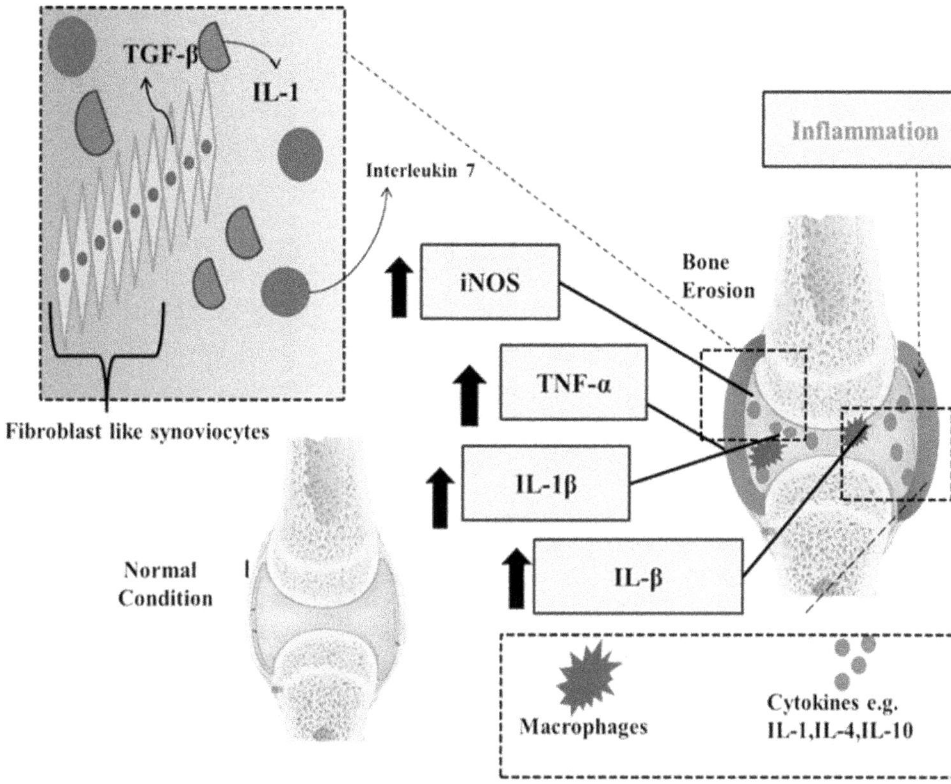

FIGURE 9.2 Role of TGF-β and iNOS in the pathogenesis of rheumatoid arthritis. Interleukin-1 (IL-1); tumor necrosis factor alpha (TNF-α); transforming growth factor (TGF)-β; inducible nitric oxide synthase (iNOS); interleukin-1 beta (IL-1β); interleukin beta (IL-β); interleukin-4 (IL-4); interleukin-10 (IL-10).

9.5 PHYTOCHEMICALS TARGETING TRANSFORMING GROWTH FACTOR AND INDUCIBLE NITRIC OXIDE SYNTHASE AND PRECLINICAL STUDIES

Many naturally occurring substances originating from plants, known as phytochemicals, have been identified in medicinal plants that possess distinct therapeutic properties. These include secondary metabolites such as terpenoids, alkaloids, polyphenols, and flavonoids, which are a good source of medications for RA, diabetes, cancer, hypertension, arrhythmia, and cardiovascular diseases (Rahman et al., 2017). Therefore, naturally occurring substances obtained from plants are extremely important to the pharmaceutical industry and may be used to treat inflammatory conditions like RA. They play a role by targeting COX-2, iNOS, and TGF-β. They show their response by regulation of mitogen-activated protein kinase (MAPK), NF-κB, and ERK1/2 pathways, by stimulation of STAT-3 and PI3K/Akt signaling, and by affecting the JAK-2 (Janus Kinase 2) gene levels, which results in reducing the degree of RA in different animal models (Yin et al., 2021) (Table 9.1).

9.5.1 Genkwanin

Genkwanin is a flavonoid separated from *Genkwa floss*. It exhibited anti-inflammatory, analgesic, and antirheumatic activities. It exhibited an antiarthritic action by downregulation of NF-κB and iNOS in LPS-induced macrophages (Sun et al., 2020). Additionally, it prevents fibroblast-like synoviocytes (FLS) from proliferating abnormally. Its anti-inflammatory properties are also linked to its control of the MAPK pathway (Bao et al., 2019).

TABLE 9.1
Phytochemicals Targeting TGF-β and iNOS for the Treatment of Rheumatoid Arthritis

Sr. #	Plant Source	Phytoconstituent	Structure	Mechanism of Action	Effect	Reference
1	*Daphne genkwa*.L	Genkwanin	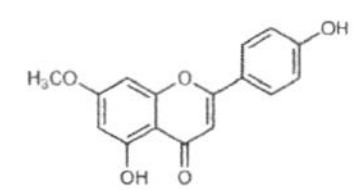	• Downregulation of COX-2 • Decreased iNOS expression in LPS-induced macrophages • Regulation of NF-κb pathway	Reduction in joint swelling and inflammation	Bao et al. (2019)
2	*Hypericum perforatum* L.	Kaempferol	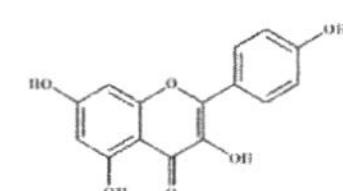	It exhibited an anti-inflammatory effect by • Decreasing iNOS, • Decreasing COX-2 levels • Inhibiting IL-6 release	Reduce inflammation	Kour et al. (2022)
3	*Cimicifuga heracleifolia* L.	Ferulic acid	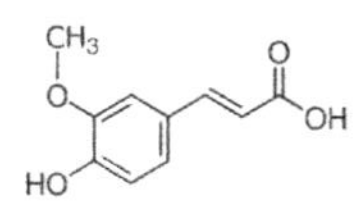	Ferulic acid has antiarthritic properties by • Blocking JAK-2 • Reducing TNF-α • Reducing TGF-β levels	It can control signal transduction pathways, gene expression, and vasodilation and ultimately reduce arthritis.	Zhu et al. (2020)
4	*Berberis vulgaris* L.	Berberine	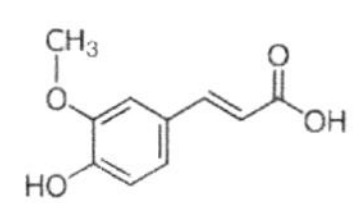	It reduced inflammation by • Decreasing the levels of iNOS • Downregulation of COX-2	Reduce inflammation of synovial fluid	Yang et al. (2007)
5	*Andrographis paniculate* (Burm.f.)	Andrographolide	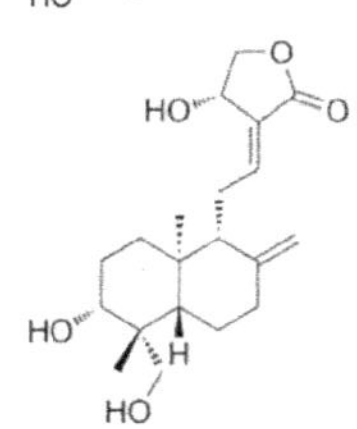	It is affected by • Downregulating PAD4 expression in the ankle joint • It also promotes neutrophil apoptosis by lowering the antiapoptotic component Bcl-2 and raising the proapoptotic protein Bax via blocking the ERK1/2 pathway	Produce antiarthritic effects	Gupta et al. (2020)

(Continued)

TABLE 9.1 (*Continued*)
Phytochemicals Targeting TGF-β and iNOS for the Treatment of Rheumatoid Arthritis

Sr. #	Plant Source	Phytoconstituent	Structure	Mechanism of Action	Effect	Reference
6	*Arctium lappa* L.	Arctigenin	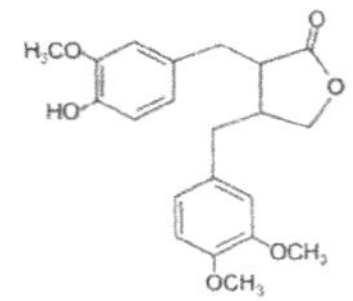	Inhibited the generation of NO in lipopolysaccharide (LPS) stimulated macrophages by • Downregulating iNOS expression, • Decreasing iNOS enzymatic activity • Decreasing proinflammatory cytokine release, including TNF-alpha and IL-6.	Treat the inflammatory problems linked to RA	Alam et al. (2014)
7	*Abies holophylla Maxi* L.	6-Shogaol	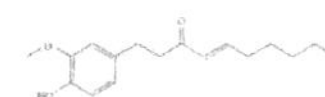	6-shogaol efficiently lowered • The LPS-induced nitric oxide production due to which there we observed a remarkable decrease in iNOS expression	Decrease cartilage damage and inflammation	Pan et al. (2008)
8	*Curcuma longa* L.	Curcumin	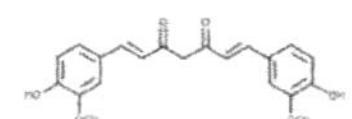	Suppressed the inflammatory cytokines and transcription factor NF-κb. which ultimately • Reduce the activity of TGF-β • Decrease iNOS production	Shows strong anti-inflammatory and antioxidant effect	Gandhi et al. (2022)
9	*Glycyrrhiza echinata* L.	Licochalcone A, C	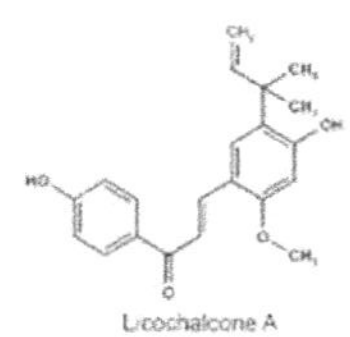 Licochalcone A 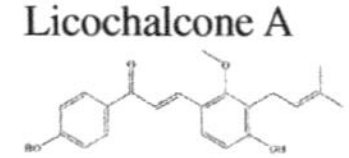 Licochalcone C	Showed its activity in attenuation of the LPS-induced inflammatory response by • Reducing the expression of iNOS • Decreasing the activity of iNOS • Regulating the NF-κB pathway.	Exhibit antiarthritic activity	Thiyagarajan et al. (2011)
10	*Moringa oleifera* L.	Chlorogenic acid	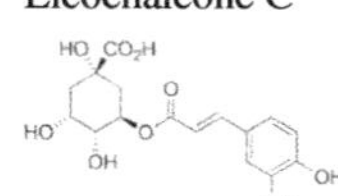	• Reduce cytokines produced by macrophage cells • Reduce the amount of autacoid (COX2 and PGE2) involved in inflammation • Reduced levels of iNOS.	Produce antiarthritic effects by reducing inflammation	Bagdas et al. (2020)

9.5.2 Kaempferol

Kaempferol is a dietary flavonoid, and it shows extensive biological and pharmacological activity. Major sources are broccoli, grapes, tea, strawberries, and apples. It shows anti-inflammatory effects by activating COX-2, decreasing iNOS, and inhibiting IL-6 release. It also regulates the NF-κB pathway to produce anti-inflammatory effects (Kour et al., 2022). Additionally, by decreasing COX-2 and IL-6, Kamfferol prevented the synthesis of NO and prostaglandin E2 (PGE2) that was generated by LPS (Lee et al., 2018).

9.5.3 Ferulic Acid

Ferulic acid is a phenolic compound that occurs naturally and has antioxidant effects. It also possesses neuroprotective, hepatoprotective, anti-inflammatory, antiallergic, antiarthritic, and antithrombotic effects. It can control signal transduction pathways, gene expression, and vasodilation (Kumar & Pruthi, 2014). Ferulic acid exhibited anti-arthritic activity in rats by increasing TGF-β and decreasing TNF-α and JAK-2 levels (Table 9.1). Ferulic acid has antiarthritic properties by influencing TNF-α and TGF-β levels and by blocking the JAK-2 gene (Zhu et al., 2020).

9.5.4 Berberine and Rutin

Berberis lycium belongs to the *Berberidaceae* family, which is widely present in the Pakistan northern region and the Indian northwest region at high altitudes. It is a very valuable medicinal herb and is widely used in treatments and therapies for multiple diseases (Sharma, Sharma, Kumar, & Padwad, 2020). Different compounds that show biological activity (phenolic and alkaloids) are present in Berberis liquid fruit extract (BLFE), which is observed in its characteristics in reducing LPS and Con-A-mediated oxidative inflammation exacerbation by inhibiting the activation of NF-κB. It was assessed *in vivo* that the effectiveness of Rutin and Berberine is found because different molecules that are bioactive are present in BLFE, which contribute to different medicinal effects. Studies have demonstrated that the stimulation of STAT-3 and PI3K/Akt signaling results in the activation of many mediators, including COX-2 and iNOS, that show their role in the progression of RA. A previous study showed that BLFE, rutin, and berberine reduce the levels of iNOS and COX-2 in CIA (collagen-induced arthritis) mice (Yang et al., 2007).

9.5.5 Andrographolide

Andrographis paniculata is a biologically active compound of Andrographolide and shows various therapeutic effects such as antidiabetes, anticancer, and anti-inflammation. In peritoneal macrophages of rats after the complete Freund's adjuvant-induced arthritis, treatment with andrographolide radically reduced the expression of NF-κB and the production of pro-inflammatory cytokines such as COX-2, iNOS, IL-1β, and IL-6 (Gupta, Mishra, Kumar, Singh, & Ganju, 2020). It also decreases the levels of myeloperoxidase and NO activity in the joints by increasing the effect of different enzymes including catalase, glutathione, and superoxide dismutase, thus decreasing proliferation and penetration of neutrophils in rats (Luo et al., 2020). It promotes neutrophil apoptosis by lowering the anti-apoptotic component Bcl-2 and raising the pro-apoptotic protein Bax via blocking the ERK1/2 pathway in RA mice (Table 9.1).

9.5.6 Arctigenin

Arctigenin is a lignan present in different plants and has anti-inflammatory properties. It is a phenyl propanoid dibenzyl butyrolactone. It inhibited the generation of NO in LPS-stimulated macrophages by decreasing iNOS expression. It also inhibited iNOS enzymatic activity and activation of pro-inflammatory cytokine release. It reduced the activation of MAPK, which results in

AP-1 (transcription factor) inactivation, and the generation of TNF-α is reduced. Additionally, the arctigenin reduced inflammation by preventing ROS-dependent STAT activation. Arctigenin might be utilized to treat inflammatory conditions such as RA (Alam, Rahman, & Khalil, 2014).

9.5.7 6-Shogaol

6-Shogoal displays its activity in two ways by suppressing LPS-induced expression of iNOS and by inhibiting COX-2 activities in murine macrophages. 6-Shogaol effectively suppressed mRNA and protein expression of iNOS and COX-2 in treated animals (Pan et al., 2008). It efficiently lowered the LPS-induced NO production followed by a remarkable decrease in iNOS expression, which is responsible for its antiarthritic activity.

9.5.8 Licochalcone A, C

Liquorice is from the family of Glycyrrhiza and its roots are commonly used in ancient China during the past centuries. A previous study reported its anti-inflammatory activity. Liquorice extract exhibited anti-inflammatory action by inhibiting LPS-stimulated nitric oxide, IL-6, and PGE2, and IL-1β production in murine macrophages (Thiyagarajan, Chandrasekaran, Deepak, & Agarwal, 2011). This anti-inflammatory action might be due to the presence of pharmacologically active compounds such as glabridin and isoliquiritigenin. Licochalcone C is in it. The glycyrrhizin and licochalcone A also exhibited an antiarthritic effect (Franceschelli et al., 2011; Yang, Or, Ho, & Lau, 2013).

9.5.9 Curcumin

Zingiberaceae family member *Curcuma longa* L., a popular Ayurvedic medication in Indian cooking, reduces knee soreness in osteoarthritic patients when compared to a placebo (Gandhi et al., 2022). Additionally, it has been observed that the *C. longa* extract reduces inflammatory biomarkers like IL-1 β and ROS. Furthermore, *C. longa* and its main phytochemicals have antiarthritic properties similar to those of NSAIDs. One of the main compounds found in the rhizome is curcumin, which shows anti-inflammatory and antioxidant potentials. Curcumin suppressed the inflammatory cytokines and transcription factor NF-κB, which ultimately reduced the activity of transforming growth factors and decreased the expression of iNOS that is responsible for its anti-arthritic and anti-inflammatory actions in treated animals (Kou et al., 2023).

9.5.10 Chlorogenic Acid

Chlorogenic acid obtained from *Moringa olifera*, which is a member of the Moringaceae family, has anti-inflammatory and antioxidant properties. Moringa leaves are extracted with ethanol and exhibited antiarthritic potential in treated animals. It exhibited anti-inflammatory properties at a dose of 250 mg/kg by significant reduction of TNF-alpha, IL-6, IL-1, PGE2, and COX2 (Lou et al., 2016; Bagdas et al., 2020) as described in Table 9.1.

9.6 CONCLUSION

A potential treatment strategy for RA is one that targets transforming growth factors and iNOS. It has been found that phytochemicals offer promising approaches to reduce inflammation and joint damage in RA patients. It is important to note that clinical trials and research are currently being conducted on these natural compounds in order to determine their safe and effective use. As a result of integrating phytonutrients with existing therapies, it may be possible to improve the control of disease and improve the possible outcome of patients suffering from arthritis. To fully understand how these phytochemicals act and to regulate their clinical application in the management of RA, we need to emphasize on conducting clinical trials.

REFERENCES

Alam, M. N., Rahman, M. M., & Khalil, M. I. (2014). Nutraceuticals in arthritis management: a contemporary prospect of dietary phytochemicals. *The Open Nutraceuticals Journal, 7*(1), 1–13.

Bagdas, D., Gul, Z., Meade, J. A., Cam, B., Cinkilic, N., & Gurun, M. S. (2020). Pharmacologic overview of chlorogenic acid and its metabolites in chronic pain and inflammation. *Current Neuropharmacology, 18*(3), 216–228.

Bao, Y., Sun, Y.-W., Ji, J., Gan, L., Zhang, C.-F., Wang, C.-Z., & Yuan, C.-S. (2019). Genkwanin ameliorates adjuvant-induced arthritis in rats through inhibiting JAK/STAT and NF-κB signaling pathways. *Phytomedicine, 63*, 153036.

Bierie, B., & Moses, H. L. (2010). Transforming growth factor beta (TGF-β) and inflammation in cancer. *Cytokine & Growth Factor Reviews, 21*(1), 49–59.

Carrier, Y., Yuan, J., Kuchroo, V. K., & Weiner, H. L. (2007). Th3 cells in peripheral tolerance. I. Induction of Foxp3-positive regulatory T cells by Th3 cells derived from TGF-β T cell-transgenic mice. *The Journal of Immunology, 178*(1), 179–185.

Carvajal Alegria, G., Cornec, D., Saraux, A., Devauchelle-Pensec, V., Jamin, C., Hillion, S., … Pochard, P. (2021). Abatacept promotes regulatory B cell functions, enhancing their ability to reduce the Th1 response in rheumatoid arthritis patients through the production of IL-10 and TGF-β. *The Journal of Immunology, 207*(2), 470–482.

Chuah, J. J., Hertzog, P. J., & Campbell, N. K. (2022). Immunoregulation by type I interferons in the peritoneal cavity. *Journal of Leukocyte Biology, 111*(2), 337–353.

Danielpour, D. (2005). Functions and regulation of transforming growth factor-beta (TGF-β) in the prostate. *European Journal of Cancer, 41*(6), 846–857.

Dobolyi, A., Vincze, C., Pál, G., & Lovas, G. (2012). The neuroprotective functions of transforming growth factor beta proteins. *International Journal of Molecular Sciences, 13*(7), 8219–8258.

Franceschelli, S., Pesce, M., Vinciguerra, I., Ferrone, A., Riccioni, G., Patruno, A., … Speranza, L. (2011). Licocalchone-C extracted from Glycyrrhiza glabra inhibits lipopolysaccharide-interferon-γ inflammation by improving antioxidant conditions and regulating inducible nitric oxide synthase expression. *Molecules, 16*(7), 5720–5734.

Gandhi, Y., Kumar, R., Grewal, J., Rawat, H., Mishra, S. K., Kumar, V., … Sharma, P. (2022). Advances in anti-inflammatory medicinal plants and phytochemicals in the management of arthritis: a comprehensive review. *Food Chemistry Advances, 1*, 100085.

Gonzalez-Gay, M. A., Llorca, J., Sanchez, E., Lopez-Nevot, M., Amoli, M., Garcia-Porrua, C., … Martin, J. (2004). Inducible but not endothelial nitric oxide synthase polymorphism is associated with susceptibility to rheumatoid arthritis in northwest Spain. *Rheumatology, 43*(9), 1182–1185.

Gonzalo-Gil, E., & Galindo-Izquierdo, M. (2014). Role of transforming growth factor-beta (TGF) beta in the physiopathology of rheumatoid arthritis. *Reumatología Clínica (English Edition), 10*(3), 174–179.

Groves, J. T., & Wang, C. C. (2000). Nitric oxide synthase: models and mechanisms. *Current Opinion in Chemical Biology, 4*(6), 687–695.

Guerby, P., Tasta, O., Swiader, A., Pont, F., Bujold, E., Parant, O., … Negre-Salvayre, A. (2021). Role of oxidative stress in the dysfunction of the placental endothelial nitric oxide synthase in preeclampsia. *Redox Biology, 40*, 101861.

Gupta, S., Mishra, K., Kumar, B., Singh, S., & Ganju, L. (2020). Andrographolide attenuates complete Freund's adjuvant induced arthritis via suppression of inflammatory mediators and pro-inflammatory cytokines. *Journal of Ethnopharmacology, 261*, 113022.

Huang, J.-B., Chen, Z.-R., Yang, S.-L., & Hong, F.-F. (2023). Nitric oxide synthases in rheumatoid arthritis. *Molecules, 28*(11), 4414.

Hung, C.-Y., Hsueh, T.-Y., Rethi, L., Lu, H.-T., & Chuang, A. E.-Y. (2025). Advancements in regenerative medicine: a comprehensive review of stem cell and growth factor therapies for osteoarthritis. *Journal of Materials Chemistry B, 13*, 4494–4526.

Javed, M., Saleem, A., & Akhtar, M. F. (2023). Diosgenin, a steroidal sapogenin, arrests arthritis through modulation of inflammatory cytokines and oxidative stress biomarkers in Wistar rats. *Inflammopharmacology, 31*(4), 1951–1966.

Kashfi, K., Kannikal, J., & Nath, N. (2021). Macrophage reprogramming and cancer therapeutics: role of iNOS-derived NO. *Cells, 10*(11), 3194.

Kim, S. K., Henen, M. A., & Hinck, A. P. (2019). Structural biology of betaglycan and endoglin, membrane-bound co-receptors of the TGF-beta family. *Experimental Biology and Medicine, 244*(17), 1547–1558.

Kou, H., Huang, L., Jin, M., He, Q., Zhang, R., & Ma, J. (2023). Effect of curcumin on rheumatoid arthritis: a systematic review and meta-analysis. *Frontiers in Immunology, 14*, 1121655.

Kour, G., Choudhary, R., Anjum, S., Bhagat, A., Bajaj, B. K., & Ahmed, Z. (2022). Phytochemicals targeting JAK/STAT pathway in the treatment of rheumatoid arthritis: is there a future? *Biochemical Pharmacology, 197*, 114929.

Kumar, N., & Pruthi, V. (2014). Potential applications of ferulic acid from natural sources. *Biotechnology Reports, 4*, 86–93.

Lee, S.-B., Shin, J.-S., Han, H.-S., Lee, H.-H., Park, J. C., & Lee, K.-T. (2018). Kaempferol 7-O-β-D-glucoside isolated from the leaves of Cudrania tricuspidata inhibits LPS-induced expression of pro-inflammatory mediators through inactivation of NF-κB, AP-1, and JAK-STAT in RAW 264.7 macrophages. *Chemico-Biological Interactions, 284*, 101–111.

Liu, L., Wang, Y., Yu, S., Liu, H., Li, Y., Hua, S., & Chen, Y. G. (2023). Transforming growth factor beta promotes inflammation and tumorigenesis in Smad4-deficient intestinal epithelium in a YAP-dependent manner. *Advanced Science, 10*(23), 2300708.

Lou, L., Zhou, J., Liu, Y., Wei, Y., Zhao, J., Deng, J., ... Yang, Y. (2016). Chlorogenic acid induces apoptosis to inhibit inflammatory proliferation of IL-6-induced fibroblast-like synoviocytes through modulating the activation of JAK/STAT and NF-κB signaling pathways. *Experimental and Therapeutic Medicine, 11*(5), 2054–2060.

Luo, S., Li, H., Liu, J., Xie, X., Wan, Z., Wang, Y., ... Yang, M. (2020). Andrographolide ameliorates oxidative stress, inflammation and histological outcome in complete Freund's adjuvant-induced arthritis. *Chemico-Biological Interactions, 319*, 108984.

Mao, X., Wu, W., Nan, Y., Sun, W., & Wang, Y. (2023). SMAD2 inhibits pyroptosis of fibroblast-like synoviocytes and secretion of inflammatory factors via the TGF-β pathway in rheumatoid arthritis. *Arthritis Research & Therapy, 25*(1), 144.

McInnes, I. B., & Schett, G. (2007). Cytokines in the pathogenesis of rheumatoid arthritis. *Nature Reviews Immunology, 7*(6), 429–442.

Nagy, G., Koncz, A., Telarico, T., Fernandez, D., Érsek, B., Buzás, E., & Perl, A. (2010). Central role of nitric oxide in the pathogenesis of rheumatoid arthritis and sysemic lupus erythematosus. *Arthritis Research & Therapy, 12*, 1–6.

Oliveira, F. R. M. B., Assreuy, J., & Sordi, R. (2022). The role of nitric oxide in sepsis-associated kidney injury. *Bioscience Reports, 42*(7), BSR20220093.

Pan, M. H., Hsieh, M. C., Hsu, P. C., Ho, S. Y., Lai, C. S., Wu, H., ... Ho, C. T. (2008). 6-Shogaol suppressed lipopolysaccharide-induced up-expression of iNOS and COX-2 in murine macrophages. *Molecular Nutrition & Food Research, 52*(12), 1467–1477.

Pap, T. (2004). Direct interaction of immunoglobulins with synovial fibroblasts: a missing link in the pathogenesis of rheumatoid arthritis? *Arthritis Research Therapy, 7*, 1–3.

Pap, T., Müller-Ladner, U., Gay, R. E., & Gay, S. (2000). Fibroblast biology: role of synovial fibroblasts in the pathogenesis of rheumatoid arthritis. *Arthritis Research & Therapy, 2*, 1–7.

Park, J. S., Chu, J. S., Tsou, A. D., Diop, R., Tang, Z., Wang, A., & Li, S. (2011). The effect of matrix stiffness on the differentiation of mesenchymal stem cells in response to TGF-β. *Biomaterials, 32*(16), 3921–3930.

Puppala, E. R., Prasad, N., Prakash, A. N., Abubakar, M., Syamprasad, N., Gangasani, J. K., & Naidu, V. (2024). Mesua assamica (King & Prain) kosterm. bark ethanolic extract attenuates rheumatoid arthritis via down-regulating TLR4/NF-κB/COX-2/iNOS and activation of Nrf2/HO-1 pathways: a comprehensive study on in-vitro and in-vivo models. *Journal of Ethnopharmacology, 335*, 118671.

Radu, A.-F., & Bungau, S. G. (2021). Management of rheumatoid arthritis: an overview. *Cells, 10*(11), 2857.

Rahman, M., Beg, S., Verma, A., Al Abbasi, F. A., Anwar, F., Saini, S., ... Kumar, V. (2017). Phytoconstituents as pharmacotherapeutics in rheumatoid arthritis: challenges and scope of nano/submicromedicine in its effective delivery. *Journal of Pharmacy and Pharmacology, 69*(1), 1–14.

Rico, M. C., Manns, J. M., Driban, J. B., Uknis, A. B., Kunapuli, S. P., & Cadena, R. A. D. (2008). Thrombospondin-1 and transforming growth factor beta are pro-inflammatory molecules in rheumatoid arthritis. *Translational Research, 152*(2), 95–98.

Saleem, A., Javed, M., Akhtar, M. F., Mobashar, A., & Khan, M. I. (2025). Assessment of the possible ameliorative effects of beta-caryophyllene oxide either alone or in combination with methotrexate in complete Freund's adjuvant-induced arthritis in Wistar rats. *Future Journal of Pharmaceutical Sciences, 11*(1), 1–16.

Saleem, A., Khalid, H., Akhtar, M. F., & Zeb, A. (2023). Appraisal of anti-arthritic potential of Coronopus didymus (L.) Sm. aqueous extract and its safety study in Wistar rats. *Inflammopharmacology, 31*(6), 3167–3182.

Sharma, A., Sharma, R., Kumar, D., & Padwad, Y. (2020). Berberis lycium Royle fruit extract mitigates oxi-inflammatory stress by suppressing NF-κB/MAPK signalling cascade in activated macrophages and Treg proliferation in splenic lymphocytes. *Inflammopharmacology, 28*, 1053–1072.

Sun, Y.-W., Bao, Y., Yu, H., Chen, Q.-J., Lu, F., Zhai, S., … Yuan, C.-S. (2020). Anti-rheumatoid arthritis effects of flavonoids from Daphne genkwa. *International Immunopharmacology, 83*, 106384.

Tahir, M., Saleem, A., & Akhtar, M. F. (2025). Diosgenin loaded-chitosan biodegradable nanoparticles ameliorate adjuvant-induced arthritis, pain, and peripheral neuropathy through moderation of inflammatory and oxidative stress biomarkers. *International Journal of Biological Macromolecules, 290*, 138926.

Thiyagarajan, P., Chandrasekaran, C., Deepak, H., & Agarwal, A. (2011). Modulation of lipopolysaccharide-induced pro-inflammatory mediators by an extract of Glycyrrhiza glabra and its phytoconstituents. *Inflammopharmacology, 19*, 235–241.

Tran, N., Garcia, T., Aniqa, M., Ali, S., Ally, A., & Nauli, S. M. (2022). Endothelial nitric oxide synthase (eNOS) and the cardiovascular system: in physiology and in disease states. *American Journal of Biomedical Science & Research, 15*(2), 153.

Tzavlaki, K., & Moustakas, A. (2020). TGF-β signaling. *Biomolecules, 10*(3), 487.

Yang, C. L., Or, T. C., Ho, M. H., & Lau, A. S. (2013). Scientific basis of botanical medicine as alternative remedies for rheumatoid arthritis. *Clinical Reviews in Allergy & Immunology, 44*, 284–300.

Yang, X. O., Panopoulos, A. D., Nurieva, R., Chang, S. H., Wang, D., Watowich, S. S., & Dong, C. (2007). STAT3 regulates cytokine-mediated generation of inflammatory helper T cells. *Journal of Biological Chemistry, 282*(13), 9358–9363.

Yin, Q., Wang, L., Yu, H., Chen, D., Zhu, W., & Sun, C. (2021). Pharmacological effects of polyphenol phytochemicals on the JAK-STAT signaling pathway. *Frontiers in Pharmacology, 12*, 716672.

Yoshimura, A., & Muto, G. (2011). TGF-β function in immune suppression. *Negative Co-Receptors and Ligands, 350*, 127–147.

Yoshimura, A., Wakabayashi, Y., & Mori, T. (2010). Cellular and molecular basis for the regulation of inflammation by TGF-β. *The Journal of Biochemistry, 147*(6), 781–792.

Zhu, L., Zhang, Z., Xia, N., Zhang, W., Wei, Y., Huang, J., … Yang, L. (2020). Anti-arthritic activity of ferulic acid in complete Freund's adjuvant (CFA)-induced arthritis in rats: JAK2 inhibition. *Inflammopharmacology, 28*, 463–473.

10 Phytochemicals Targeting Vascular Endothelial Growth Factor (VEGF) and RANKL Pathway in the Treatment of Rheumatoid Arthritis

Rida Siddique, Muhammad Asim Shehzad, Rabia Arshad, and Ammara Saleem

WHAT YOU WILL LEARN IN THIS CHAPTER?

In this chapter, we will explore various phytochemical therapies targeting vascular endothelial growth factor (VEGF) and receptor activator of nuclear factor-κB ligand (RANKL) pathways, specifically in contact with the treatment of rheumatoid arthritis (RA). We begin with a brief introduction of VEGF, its families, and roles in different organ systems of the body along with the RANKL pathway. Next, we delve into the role and significance of the VEGF and the RANKL pathways in mediating important inflammatory conditions such as RA, psoriasis, periodontal inflammatory diseases, and inflammatory bowel disease. Additionally, this chapter will provide an in-depth insight into their role in mediating the pathogenesis of RA and will go into an extensive detail of the preclinical studies done on phytochemical therapy targeting mainly VEGF and RANKL pathways. By the end of this chapter, the readers will have a comprehensive understanding of phytochemicals targeting VEGF and RANK pathways in the treatment of RA.

LIST OF ABBREVIATIONS

Bcl-2	B-cell lymphoma 2
CIA	Collagen-induced arthritis
COX-2	Cyclooxygenase-2
CRP	C-reactive protein
ESR	Erythrocyte sedimentation rate
IL-6	Interleukin-6
LTi	Lymphoid tissue inducer
MMP	Matrix metalloproteinase
NF-κB	Nuclear factor-kappa B
OA	Osteoarthritis
OPG	Osteoprotegerin
PKC	Protein kinase C
RA	Rheumatoid arthritis
RANKL	Receptor activator of nuclear factor-κB ligand
SLE	Systemic lupus erythematosus
SSC	Systemic sclerosis

DOI: 10.1201/9781003535201-12

TRAP Tartrate-resistant acid phosphatase
VEGF Vascular endothelial growth factor
VEGFR Vascular endothelial growth factor receptor

10.1 INTRODUCTION TO VEGF AND RANKL PATHWAYS

Vascular endothelial growth factor (VEGF) is an endothelial permeability factor, with its family containing seven members. All of these variants differ in their role, which in turn depends on the variations in their receptors, modes of expression, and their activity (Parveen et al., 2019). They are mostly expressed extracellularly, playing their role by binding to an extracellular receptor through the tyrosine kinase pathway. However, in some cases, they bind through the intracellular receptors, forming an intracranial signaling pathway (Wiszniak & Schwarz, 2021). Different cell types, including tumor cells, mesangial cells, keratinocytes, and macrophages, produce this factor (Duffy, Bouchier-Hayes, & Harmey, 2011). VEGF plays various roles, depending upon the system in which it is a part. VEGFs are mostly known for their proangiogenetic properties, but they have also been known to play an important part in many nonendothelial roles. Vascular endothelial growth factor receptor (VEGFR) 1 and VEGFR2 are expressed in many nonendothelial and cancer cell types (Wiszniak & Schwarz, 2021). VEGF is found at the infarction sites in the cardiovascular system due to its role in growth and repair. In the central nervous system, its role is mostly neuroprotective. It protects hippocampal neurons from ischemic injury. In hypoxic conditions, VEGFR-1 and VEGFR-2 increase, which in turn inhibits hypoxic preconditioning (Wick et al., 2002). During bone formation, hypertrophic chondrocytes in the growth plate produce VEGF, which modulates bone remodeling, angiogenesis, and growth. Similarly, VEGF is expressed in bone marrow and plays a central role in hematopoiesis. Cytokine stimulates the increased expression of VEGF. The lack of a single allele in VEGF resulted in embryonic lethality due to impaired angiogenesis and hematopoiesis (Burger et al., 2022).

Receptor activator of nuclear factor-κB ligand (RANKL)/receptor activator of nuclear factor-κB (RANK)/ osteoprotegerin (OPG) triad was discovered and identified in the early 1990s, an important mediator in T-cell receptor signaling and the function of dendritic cells. Receptor activator of NF-κB (RANK) and its ligand (RANKL) and OPG form this triad, interacting with one another to regulate the development and growth of osteoclasts and lymph node formation. Their role in bone metabolism and the immune system has been confirmed by later studies (Ono, Hayashi, Sasaki, & Nakashima, 2020).

In bone, osteoclasts and osteoblasts maintain homeostasis. During bone development, hypertrophic chondrocytes and osteoblasts, while in postgrowth phases, osteocytes are the predominant sources of RANKL. RANKL induces osteoclast differentiation, which results in bone matrix resorption. Most of the bone and joint diseases that are related to age, including rheumatoid arthritis (RA), involve the pivotal role of RANK–RAKL interaction (Nakashima et al., 2011; Ono et al., 2020). OPG acts as a RANKL decoy receptor produced by osteoblasts, which inhibits RANK–RANKL interaction and prevents osteoclast formation (Udagawa et al., 2021).

T and B lymphocytes, along with other hematopoietic stem cells, are found in bone marrow. In addition to that, erythropoiesis also takes place in the bone marrow. The RANKL pathway maintains the bone marrow capacity through the function of osteoclasts (Ono et al., 2020; Penna, Capo, Palagano, Sobacchi, & Villa, 2019).

In the development of the thymus, lymphoid tissue inducer (LTi) cells, T cells, and invariant natural killer T cells produce RANKL, which interacts with medullary thymic epithelial cells, inducing autoimmune regulator expression that is crucial for the establishment of self-tolerance. Similarly, in lymph nodes, RANKL from lymphoid tissue organizer cells stimulates LTi via RANK, which leads to macrophage recruitment (Akiyama, Shinzawa, & Akiyama, 2012; Ono et al., 2020). In the gastrointestinal system, RANKL regulates inflammation while promoting M-cell differentiation and facilitating antigen transfer to dendritic cells for IgA production through the interplay of group 3 innate lymphoid cells (ILC3s) (Ohno, 2016). In cases of CNS inflammation, cytokine production and motif chemokine ligand (CCL) 20 expression are influenced by RANKL signaling via astrocytes and microglial cells, which in turn mediates the recruitment of immune cells (Guerrini et al., 2015; Ono et al., 2020).

10.2 ROLE OF VEGF AND RANKL PATHWAYS IN INFLAMMATORY DISEASES

Both VEGF and RANKL have been implicated in many inflammatory conditions, including seronegative spondyloarthritis, systemic lupus erythematosus (SLE), systemic sclerosis (SSC), and RA. The seronegative spondyloarthritis group consists of a number of disorders, including psoriatic arthritis, Ankylosing spondylitis (AS), enteropathy arthropathy, reactive arthritis, and undifferentiated spondyloarthropathy. AS is considered the most common among all seronegative spondyloarthritis diseases (Le & Kwon, 2021). Osteoblast differentiation from synovial fibroblast might be regulated by VEGF and RANKL pathways, which, in turn, take part in the formation of new bone in AS (Liu, He, Tan, & Liao, 2015). VEGF also contributes to the stimulation of COX-2 pathways that take part in the ongoing inflammatory processes in AS. Their serum levels rise significantly, and the patient starts presenting with the signs of illness. The VEGF levels correlate with clinical signs that include morning joint stiffness, pain in joints, peripheral arthritis, enteritis, fatigue, and spinal pain. However, the extra-articular manifestations of the disease are unlikely to be affected by the level of VEGF in the blood. Therefore, to some extent, serum VEGF can be used as a marker of disease progression and activity in AS (Le & Kwon, 2021).

Another prevalent inflammatory and autoimmune disorder that might be mediated by VEGF and RANKL pathways includes SLE. The exact etiology of SLE has not been established yet, but it involves a complex interplay of genetics, ethnicity, and criminal and environmental factors. The progression of SLE can be halted, but it can pose some life-threatening complications (Justiz-Vaillant et al., 2024).

Like AS, the risk of SLE is also correlated with the serum VEGF levels. In the Chinese Han population, a correlation has been found between VEGFR1 genetic polymorphism and associated SLE risk (Yuan et al., 2020). VEGF also serves as an independent predictor of disease and can be used as an important biomarker in the case of SLE activity in pediatric patients and organ involvement, helping in the overall understanding of disease pathology and its monitoring. VEGF is also linked to various clinical manifestations of SLE, including oral ulceration and neuropsychiatric disorders (García-Ríos, Pecci-Lloret, & Oñate-Sánchez, 2022; Justiz-Vaillant et al., 2024). However, recent studies indicate significant differences in VEGF levels in pregnant SLE patients, lupus nephritis, and preeclampsia, suggesting its potential as a diagnostic marker (de Jesús et al., 2021).

Lupus nephritis, another clinically severe manifestation of SLE, is also related to elevated VEGF-A levels, which contribute to the pathogenesis of the disease by affecting endothelial and epithelial cellular interactions (Gómez-Bernal et al., 2022). VEGF levels are related to the development of kidney damage and are elevated in lupus nephritis in contrast to non-lupus nephritis. Treatment with immunosuppressant drugs such as mycophenolate mofetil decreased VEGF-A levels, and its long-term use reduced VEGFR-2 levels in SLE patients. This is indicative of VEGF's relation with disease activity (Le & Kwon, 2021).

Osteoarthritis (OA) is another inflammatory condition affecting the joints of mostly middle-aged and elderly individuals. All the joint components, including cartilage bone, synovium, ligaments, and muscles, are affected by this disease. Angiogenesis has been implicated in the pathogenesis of this disease by forming an endovascular network in the synovium, as in RA, called pannus, which is associated with osteophyte formation. The increased VEGF expression exacerbates rage disease symptoms by increasing osteoclastic bone degradation. This is done by promoting osteoclast differentiation, matrix metalloproteinase (MMP) activity, and chondrocyte apoptosis. The receptor expression in the joint space increases and leads to elevated VEGF levels in the joint space. The OA therapy response is monitored by VEGF because of its role in bone homeostasis and its ability to enhance osteoclast differentiation and RANKL secretion. The VEGF levels also correlate with pain scores and clinical manifestations in an osteoarthritic knee because they increase sensory nerve growth and promote pain hypersensitivity. VEGF also upregulates MMP impression. MMP-13 has been implicated in cartilage degeneration (Patil, Sable, & Kothari, 2012).

Hypoxia is also mediated by the VEGF pathway and can be an important contributor to temporomandibular joint OA by enhancing cartilage angiogenesis in the condyle via the hypoxia inducible factor-1-VEGF-Notch signaling pathway. In vivo studies showed encouraging outcomes with therapy targeting VEGF, such as administering bevacizumab directly into the joint in a rabbit model of OA. This approach resulted in reduced pain, cartilage degradation, and inflammation of the synovial membrane, all without adverse effects. Furthermore, administering an oral inhibitor targeting VEGF receptor-2 kinase slowed the progression of OA in a mouse model, indicating the potential efficacy of anti-VEGF therapy in managing OA (Le & Kwon, 2021).

SSC is another very rare autoimmune disorder that is marked by skin fibrosis and internal organ involvement, especially lungs, kidneys, and vascular system. It is caused by vascular changes and a large amount of collagen deposition brought about by changes in VEGF expression. There is an abnormal decrease in VEGF levels and impaired angiogenesis in early as well as established stages of SSC patients. VEGFR-2 can be used as an indicator for microangiopathic changes in SSc and can be used in establishing a proper diagnosis. VEGFR-1 may be decreased in skin lesions of SSC patients when compared with a normal person. The dysfunctional VEGF and Ang/Tie2 eventually becomes a leading pathway toward peripheral micro maculopathy in SSc (Le & Kwon, 2021; Moritz et al., 2017).

10.3 ROLE OF VEGF AND RANKL PATHWAYS IN RA

RA is one of the many inflammatory disorders that is mediated primarily by the VEGF and RANKL pathways. It is defined as a chronic autoimmune disorder affecting the synovial membrane of joints. It mainly affects females and has a prevalence of around 0.5%–1% in the general population (Javed, Saleem, & Akhtar, 2023; Saleem, Saleem, & Akhtar, 2020).

The pathophysiology of RA involves a complex interplay between innate and adaptive immunity and the process of angiogenesis, although the exact etiology is not established as of now. Immune cells, chondrocytes, fibroblasts, and dendritic cells mainly regulate it. It can lead to symmetrical involvement of joints that can progress to an inferior and poor quality of life by limiting mobility, increasing chances of disability, and leading to more comorbidities, including cardiovascular risks (Hansildaar et al., 2021).

Many similarities have been observed between the processes occurring in synovial membranes during the progression of RA and the development of atherosclerotic plaques along with elevated levels of proinflammatory cytokines, which lead to endothelial dysfunction.

It involves the progressive infiltration and proliferation of synovial fibroblasts. The inflammatory cellular infiltration leads to a hypoxic synovial environment, which ultimately leads to changes in the overall genes responsible for the expression of angiogenesis and cartilage degradation (Saleem, Saleem, Akhtar, Shahzad, & Jahan, 2020; Hansildaar et al., 2021). Angiogenesis, mediated mainly by VEGF, plays one of the major roles in the maintenance and progression of synovitis in RA (Kim, Kim, Kim, Cho, & Lee, 2015). There is endothelial proliferation and an overall increased number of blood vessels in the synovium. During very early stages, this endothelial activation and synovium hyperplasia can be detected through inflammatory cellular infiltration, which progressively increases during the course of the disease (Le & Kwon, 2021; Szekanecz & Koch, 2009).

The overall serum and synovial concentrations of VEGF and rheumatoid factor (RF) are relatively higher in RA when compared to a normal person or OA. Thus, the VEGF levels reflect the progression of disease with the correlation to ESR, C-reactive protein (CRP), and RF (Lee et al., 2001). The VEGF-A and its receptors increase their overall expression in the synovial tissue during the pathogenesis of RA. Not only this, VEGF-NRP induces anti-apoptotic protein, Bcl-2 expression, and translocation of Bax that prevents synoviocyte apoptosis, which leads to synovial hyperplasia or pannus that is the most prominent characteristic feature of RA (Kim et al., 2006; Le & Kwon, 2021).

The overall bone destruction is a direct function of osteoclastic differentiation, and the significance of VEGF's role cannot be denied in this. It is a direct regulator of osteoclastic differentiation from monocytes and leads to the induction of fibroblast-secreted RANKL, which acts as a pivotal cytokine in osteoclast differentiation (Kim et al., 2015). In addition to this, genetic polymorphism of VEGF-C can also contribute to variations in susceptibility, therapeutic targets, and possible diagnostic markers in patients with RA (Cheng et al., 2020).

There is increased RANKL expression in synovial tissues of patients with RA. It has also been reported that in response to interleukin-6 (IL-6), the RANKL expression was induced. Tumor necrosis factor (TNF)-α, IL-17, or IL-1β lead to stimulation of IL-6 production in synovial fibroblasts, confirming the role of inflammatory cytokine involvement in synovial fibroblastic RANKL production (Tanaka, 2019). The cellular interaction between fibroblasts and immune cells is a major cause of the imbalance between regulatory T (Treg) cells and helper T (Th)17 cells, which in turn speeds up the inflammatory processes and bone destruction. This is done mainly by upregulating RANKL expression in synovial fibroblasts (Komatsu & Takayanagi, 2022). Role of RANKL and VEGF in RA is described in Figure 10.1.

The production of RANKL and OPG by osteoblasts under the influence, either direct or indirect, of hormones, including parathyroid hormone, estrogens, calcitriol, prolactin, glucocorticoids, growth factor, inflammatory cytokines, and inflammatory cells (Geusens, 2012). TNF-α and IL-6 enhance RANKL expression in synovial fibroblasts. Transcription factor (NF-kappa B) induces these inflammatory cytokines and causes inflammation (Siddique et al., 2024).

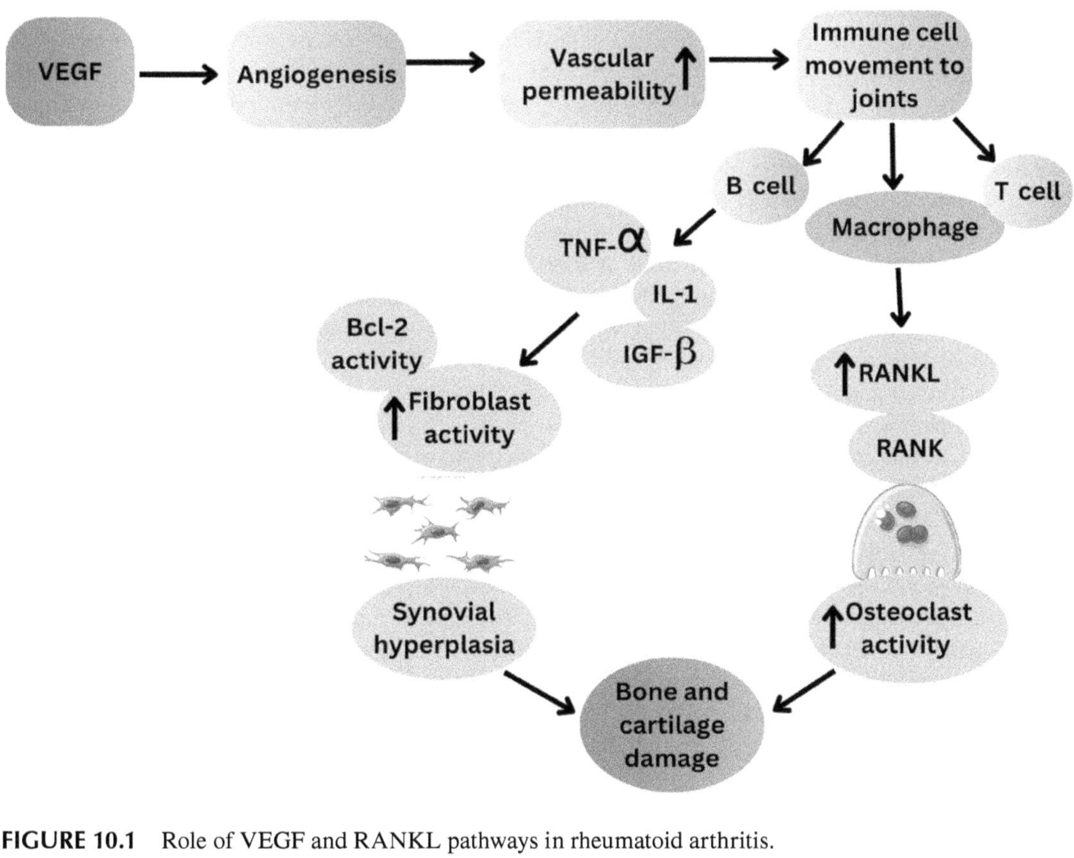

FIGURE 10.1 Role of VEGF and RANKL pathways in rheumatoid arthritis.

Dickkopf (DKK)-1 is a potent inhibitor of the Wnt pathway and is crucial for bone formation. TNF-α induces DKK1 and suppresses osteogenesis in RA, which in turn is further suppressed by the environment created by inflammatory cellular infiltrate. Since the canonical Wnt pathway regulates OPG production, the elevated DKK1 promotes bone resorption. Resultantly, there is a decrease in osteogenesis while bone resorption increases in RA. This process disrupts the coupling between bone formation and resorption, leading to decreased bone mineralization, which is another important feature of RA (Maeda et al., 2022). Research studies indicated that the baseline serum RANKL/OPG ratio of early untreated RA patients can predict bone damage at 5 and 11 years of follow-up, regardless of other predictors such as C-terminal cross-linked telopeptide type II collagen (CTX-II), erythrocyte sedimentation rate (ESR), and joint damage. Patients presenting with a high baseline RANKL/OPG ratio and ESR showed the greatest long-term radiographic progression of joint damage. All these findings suggest that the interplay between inflammation and bone destruction may be a fundamental factor influencing disease progression in RA (Geusens, 2012).

10.4 ROLE OF VEGF AND RANKL IN PATHOPHYSIOLOGY OF RA

The vascularity of the synovium is increased as the pannus invades the underlying cartilage and bone and the synovial lining of joints expands in RA, resulting in increased nutrient demand. Therefore, angiogenesis is the prime mechanism that results in pannus formation in RA. VEGF has the prime involvement in angiogenesis. It increases vascular permeability resulting in edematous swelling in RA. Thus, targeting angiogenesis ameliorates the swelling in RA resulting in improvement in symptoms and disease reversal.

In an earlier study, the role of VEGF in osteoclast genesis was determined. Real-time PCR, ELISA, and luciferase assays were used to determine the expression of VEGF-induced RANKL in synovial fibroblasts in RA. The number of positive multinucleated cells and tartrate-resistant acid phosphatase (TRAP) determines osteoclast genesis in VEGF-cultured peripheral blood monocytes. VEGF concentration was compared with synovial fluid RANKL in patients with RA, as RANKL expression in synovial fibroblasts of RA was stimulated by VEGF. The inhibition of protein kinase C (PKC), p38 MAPK, Src, and VEGF receptors 1 and 2 reduced the expression of RANKL- and VEGF-induced differentiation of osteoclasts. Thus, inhibition of VEGF and src pathways results in the inhibition of differentiation of monocytes into osteoclasts, causing the inhibition of osteoclast genesis. Therefore, the inhibition of VEGF causes inhibition of osteoclast genesis as direct induction from precursors is inhibited, as well as RANKL production is inhibited via inhibition of Src and PKC pathways. Thus, targeting RANKL production ameliorates the disease (Kim et al., 2015).

Osteoclast genesis is inhibited by OPG, which is a decoy receptor and prevents RANKL-mediated osteoclast differentiation by competing and binding with RANKL. RANKL levels were raised when RA treated and a decline in the expression of OPG was noted; thus, the ratio of RANKL to OPG expression was increased.

The inflammatory cytokines in osteoclast genesis cause destruction of bone by activating osteoclasts. Osteoclasts are present at cartilage destruction sites in RA patients. Factors essential for osteoclast genesis include RANK and RANKL. IL-6 and soluble interleukin (sIL)-6R induced RANKL expression in RA-fibroblast-like synoviocytes (FLS). On the other hand, RANKL expression wasn't induced by TNF-α and IL-17. Although, both are growth stimulators and are responsible for IL-6 production. Interestingly, RANKL expression is induced in the presence of TNF-α, sIL-6R, and IL-17 (Yokota et al., 2021). In a coculture of osteoclast precursor cells, sIL-6R and IL-6 induced NFATc1 and TRAP5b mRNA expression in the osteoclast precursor cells. Osteoclast genesis is directly induced by IL-6/sIL-6R by inducing RANKL expression in RA-FLS. Moreover, bone resorption is stimulated by IL-6 in the presence of SIL-6R, which was decreased by osteoclast inhibitors; thus, it reveals that IL-6 signaling results in osteoclast genesis induced by the interaction of osteoblast and osteoclast. Thus, it is established that IL-6 induces osteoclast formation by inducing RANKL in RA-FLS as well as osteoblast.

10.5 PHYTOCHEMICALS TARGETING VEGF AND RANK PATHWAYS IN THE TREATMENT OF RA

The overall knowledge of VEGF and RANKL pathways has helped a lot in increasing the overall treatment horizon for RA. Many drugs that target VEGF pathways have been indicated in many diseases, including inflammatory conditions, cancer, and retinopathy (Le & Kwon, 2021).

The current treatment of RA is mainly done through disease-modifying antirheumatic drugs, which target the ongoing inflammatory processes of bone erosion (Cohen et al., 2008). There is a limitation of research that studies the effect of treatments that specifically target VEGF and RANKL pathways.

There is a very large percentage of patients with RA who use herbal methods for the management of RA symptoms, including joint pain, inflammation, and swelling. Phytochemicals are the herbal products that can be included in this category. Over the past few decades, there has been an increased inclination of researchers toward conducting clinical trials to test the efficiency of natural phytochemical agents in alleviating symptoms of RA. The phytochemical groups that show an anti-inflammatory role against RA include flavonoids, alkaloids, steroids, polyphenols, and terpenoids. However, their properties, including a very high molecular weight, poor permeability, and water solubility, limit their role to some extent. Researchers have been trying to use nanotechnology to improve the pharmacokinetic properties and the overall ability of these drugs (Pimple, Shah, & Singh, 2024).

The first group of phytochemicals that are discussed is Parthenolide (*Tanacetum parthenium*), a sesquiterpene lactone found in the Asteraceae family of medicinal plants. In both in vitro and in vivo models, it showed anti-inflammatory and antibone restorative characteristics. In in vitro trials, parthenolide showed inhibitory effects on IL-1β and TNF-α on chondrocytes, leading to the prevention of osteoclast formation and bone formation. Similarly, the in vivo studies showed that parthenolide (0.5 and 1 mg/kg) blocked osteolysis and prevented bone loss in murine calvarial models. Similarly, within the collagen-induced arthritis (CIA) rat model, parthenolide (1 mg/kg) reduced pannus formation and inhibited the inflammatory processes (Williams et al., 2020).

Genistein (GEN: 4′, 57-trihydroxy isoflavone) is a natural isoflavone compound, mainly found in leguminous plants. This herbal medicine has been of great interest due to its potential effects in treating cardiovascular diseases, bone destruction, RA, and osteoporosis (Li et al., 2013). Again, a CIA rat model was used in which genistein exerted anti-inflammatory and anti-angiogenic properties by targeting the expression of VEGF, IL-6, and TNF-α. MH7A cells were deployed to check the inhibitory effects of GEN on VEGF. EA.hy926 cells were used to verify the inhibition of cellular migration and tube formation in vascular endothelial cells under an inflammatory circumstance (Cheng et al., 2020). It has been proven that excessive activation of STAT3 enhances VEGF expression which leads to angiogenesis, which in turn contributes to inflammatory processes in RA (Chen & Han, 2008). Therefore, inhibition of VEGF and IL-6 by GEN via IL-6/JAK/STAT3/VEGF pathway showed its great therapeutic potential in suppressing angiogenesis in RA (Cheng et al., 2020).

Another important flavonoid used in the treatment of RA is quercetin, a phytochemical found in fruits and vegetables with significant anti-inflammatory and antioxidant effects. Studies have been done to assess the immune-regulated function of quercetin in IL-17-induced osteoclastogenesis (OCG) in RA. The RA-FLS were stimulated by IL-17, and the increased expression of RANKL was detected in the in vitro study (Tang et al., 2023).

CD-14 monocytes were exposed to IL-17 and RANKL, with or without quercetin. These monocytes act as precursors for osteoclasts. TRAP activity was measured to evaluate the osteoclastic differentiation. To further explore, IL-17-stimulated RA-FLS and Th17 cells were cultured with monocytes. In addition to this, CD4+ T cells were cultured in combination with quercetin, and their differentiation into TH17 and Treg cells was assessed under specific conditions. The results showed that IL-17 caused RA-FLS to generate RANKL while, on the other hand, quercetin mediated/mitigated IL-17-induced RANKL expression along with attenuation of IL-17-induced activation of mammalian target of rapamycin, extracellular signal-regulated kinase, and inhibitor of kappa B-alpha. Quercetin reduced the formation of mature osteoclasts due to stimulation of either IL-17 or RANKL. Quercetin also suppressed Th17-differentiation and IL-17 production under Th17-differentiation conditions. The study suggested quercetin inhibited IL-17-induced RANKL production in RA-FLS and IL-17-induced osteoclast formation, thus potentially serving as an additional therapeutic avenue for managing bone-destructive processes in RA (Kim et al., 2019).

Adenosine deaminase (ADA) is considered one of the important inflammatory markers in RA and acts as a checkpoint for control of the overall immune system through the regulation of adenosine pathways (Nalesnik, Nikolić, & Jandrić, 2011). A study was conducted in a rat model to assess the inhibitory effect of quercetin on the ADA enzyme in silico and in vivo. Quercetin showed significant inhibition of the overall joint ADA, leading to reduced inflammation and RA-associated cytokines. In addition to this, it enhances the overall efficacy of methotrexate and reduces its toxicity, thus proving its effective role in the overall treatment of RA (Nalesnik et al., 2011).

Icariin (ICA), extracted from *Epimedium brevicornum Maxim,* is another important phytochemical that reveals a promising therapeutic role in the treatment of RA. Although the exact mechanism is not known (Wei, Ping, You, Qiang, & Tao, 2016), studies have shown ICA's role as an important immunoregulatory and anti-inflammatory agent (Shao et al., 2015). In an earlier study, RA-FLS were treated with different concentrations of ICA, and the effects were observed on the overall levels of inflammatory cytokines. The results showed a significant reduction in TNF-α, IL-1β, and IL-6 and an increase in the apoptosis of RA-FLS (Wu et al., 2024).

In another study, antigen-induced arthritis was developed in rabbits and followed by treatment with ICA at different concentrations. The expression of RANKL and OPG in joint tissues was modulated. The histological analysis showed very minute chondrocyte damage in the cartilage of the group treated with ICA. The results of micro CT showed higher bone mineral density in the ICA-treated rabbits as compared to the disease control group. The investigations also showed decreased RANKL and increased OPG levels in these rabbits compared to the disease control group (Wei et al., 2016).

Curcumin, a natural polyphenol and the main compound of Turmeric genus (*Curcuma longa*), is widely used for medical processes, including pain reduction and improvement of inflammatory conditions (Pourhabibi-Zarandi, Shojaei-Zarghani, & Rafraf, 2021). Curcumin acts as a therapeutic agent by inhibiting the NF-κB signaling pathway and promoting macrophage apoptosis (Wang et al., 2019). An investigation was done to assess the effects of curcumin in RANKL-related molecular signaling in osteoclast precursor cells (OCPs) and to investigate the significance of RANK–TRAF6 signaling in curcumin-treated OCG (Ke et al., 2024).

In a study, mice were used to assess the effects of curcumin on OCG and RANKL-mediated bone loss. In addition to this, the significance of the JNK-BCL2-Beclin1 pathway in the curcumin-regulated OCG pathway was explored through rescue assays and BCL2 phosphorylation detection. It turned out that curcumin inhibited RANKL signaling in OCPs, which resulted in the repression of osteoclast differentiation and autophagy in mainly $RANK^+$ OCPs. These changes were recovered by TRAF6 overexpression. Additionally, a decrease in bone mass and increased osteoclastic formation and autophagy in $RANK^+$ OCPs were also prevented by curcumin in treated mice. Furthermore, curcumin inhibited BCL2 phosphorylation at Ser70 and promoted the protein interaction between BCL2 and Beclin1 in OCPs (Ke et al., 2024).

Another systematic review showed that curcumin can improve ESR and CRP in patients with RA. The findings suggested more benefit from curcumin in the experimental group than the patients in the control group with no curcumin (Kou et al., 2023).

Resveratrol (RES) is another natural antioxidant found in plants (*Cassia obtusifolia*). It shows immense therapeutic potential in treating inflammatory conditions, including RA. Many clinical studies have demonstrated RES's ability to improve RA symptoms and reduce the level of inflammatory markers. Sirtulin (SIRT) 1 is a key regulator in RA pathology. RES activated SIRT1 and inhibited the NF-κB pathway and promoted synovial cell apoptosis, leading to reduced synovial hyperplasia and inflammation (Deng et al., 2019). The RES inhibited TNF-β-induced inflammation in chondrocytes and reduced IL-1β and MMP-3 production in RA-FLS (Wang et al., 2020). In CIA models, RES also inhibited various angiogenetic and signaling pathways in synovial tissue, leading to a progressive halt in overall RA progression (Li et al., 2022).

A randomized controlled trial was done to investigate the RES therapeutic potential in treating RA patients. Patients received RES capsules alongside standard treatment or standard treatment alone for three months. Results showed that the RES group had reduced joint swelling, tenderness, and disease activity scores compared to the control group. In addition to that, the level of inflammatory markers was significantly reduced in RES-treated patients, thus promising to be a beneficial therapeutic agent in the overall treatment of RA (Khojah, Ahmed, Abdel-Rahman, & Elhakeim, 2018).

Green tea is widely regarded as an important healing substance in the treatment of different chronic diseases. Its health benefits are mostly attributed to the healing nature of its components, which include epigallocatechin gallate (EGCG) and polyphenol. EGCG inhibits the activation of different signaling pathways, including NF-κB and MAPKs. EGCG inhibits osteoclastic development through suppression of mRNA expression of various osteoclast-related molecules. These include β3 integrins, calcitonin receptor, and RANKL (Lin et al., 2020). The downregulation of RANKL-induced NF-ATc1, which is a potent transcription factor involved in osteoclastic differentiation, suggests the therapeutic potential of EGCG in reducing the activity of osteoclasts in RA joints (Chacko, Thambi, Kuttan, & Nishigaki, 2010).

The bone regulatory mechanisms of EGCG have been proven by various studies, which have provided considerable evidence of the apoptotic induction ability in osteosarcoma cells by green tea polyphenols. This induction of apoptosis is done by the activation of caspase-3 and NF-κB p65 regulation. EGCG also inhibits monocyte chemotactic protein-1 (MCP-1)/CCL2 synthesis, which is induced by oncostatin-M. In response to various stimuli, IL-6 stimulates the activation of osteoclast formation and the resulting bone resorption. EGCG results in the inhibition of fibroblast growth factor-2 induced IL-6 in MC3T3-E1 cells through suppression of ERK1/2 and p38 MAP kinase pathways. In addition to all this, EGCG blocks the activation of RANKL-induced activation of NF-κB and of c-Jun-N-terminal kinase pathways. This results in the suppression of NFATc1 and c-Fos expression in osteoclast precursor cells and reduces osteoclast formation and differentiation (Ahmed, 2010).

Tripterygium wilfordii Hook F (TWHF) is a Chinese herb that has been widely used in different inflammatory conditions for its therapeutic potential. Recent studies that included placebo-controlled clinical trials have proved the efficacy of TWHF in the treatment of RA. The TWHF extract consists of over 70 different compounds. Among these compounds, diterpene trioxide is an important triptolide responsible for TWHF's major therapeutic properties. TWHF inhibits NF-AT, NF-κB, and AP-1, the important transcription factors in the pathogenesis of RA. In the experiments conducted with CIA mice, triptolide was given orally in various concentrations of 8, 16, and 32 μg/kg/day. This resulted in the reduction of inflammatory cells and NF-κB-p65 transcripts in synovial cells. The 32 μg/kg/day concentration proved to be the most effective as it resulted in the inhibition of NF-κB and its translocation (Zhou et al., 2018).

Studies have also shown that the combination of TWHF with methotrexate (MTX) has far superior therapeutic effects in the treatment of RA as compared to MTX alone. The concentration of

TWHF and MTX combined therapy, with TWHF being at 30–60 mg/d combined with MTX at ~10 mg/w, is considered optimal (Luo et al., 2023).

Another comprehensive survey of different in vitro and in vivo trials has tried exploring guggulsterone's therapeutic potential in the management of RA and inflammation by targeting different molecular pathways, including NF-κB, VEGF, JAK/STAT, COX-2, and inducible nitric oxide synthase. The results showed significant alleviation of arthritis and inflammation by guggulsterone. The study confirmed its therapeutic potential in the treatment of RA and other inflammatory conditions.

Table 10.1 summarizes the role of some of the phytochemicals discussed above as well as some other important phytochemicals along with their class and mechanism of action in the treatment of RA.

TABLE 10.1
Phytochemicals Targeting VEGF and RANKL Pathways in the Treatment of Rheumatoid Arthritis

Phytochemical	Class	Mechanism of Action	References
Parthenolide	Sesquiterpene lactones	NF-κB inhibition, suppression of pro-inflammatory cytokines including TNF-α, IL-1β, and IL-6, and induction of apoptosis in synovial cells.	Darwish et al. (2019)
Genistein	Flavonoid	Suppression of STAT3 activation, inhibition of VEGF and IL-6 through IL-6/JAK/STAT3/VEGF pathway.	George, Shyni, and Raghu (2020)
Quercetin	Flavonoids	Downregulation of NF-κB, MAPK, and PI3K/Ak signaling pathways and inhibition of the production of reactive oxygen species (ROS).	Liu et al. (2020)
Icariin	Flavonoids	Suppression of the NF-κB pathway, reduction of the production of pro-inflammatory cytokines and decrease in proliferation and invasion of FLSs.	Li and Zhang (2020)
Curcumin	Polyphenol	Inhibition of NF-κB signaling pathway and promotion of macrophage apoptosis	Cho et al. (2006)
Resveratrol	Polyphenol	Activation of SIRT1, inhibition of NF-κB pathway, promotion of synovial cell apoptosis, inhibition of TNF-β-induced inflammation in chondrocytes, and reduction in IL-1β and MMP-3 production in RA-FLS.	Behl et al. (2022)
Epigallocatechin gallate	Flavonoids (Catechins)	Inhibits the activation of different signaling pathways, including NF-κB and MAPKs, and inhibits monocyte chemotactic protein-1 (MCP-1)/CCL2 synthesis	Ding et al. (2023)
Tripterygium wilfordii Hook F (TWHF)	Flavonoids	Inhibits NF-AT, NF-κB, and AP-1 pathways	Zhang et al. (2021)
Guggulsterone	Plant sterol	Suppresses IKK and blocks NF-κB in FLS	Patidar et al. (2022)
Hesperidin	Flavonoid	Reduces TNF-α levels in RA joints and prevents joint destruction.	Gandhi et al. (2021)

(*Continued*)

TABLE 10.1 (*Continued*)
Phytochemicals Targeting VEGF and RANKL Pathways in the Treatment of Rheumatoid Arthritis

Phytochemical	Class	Mechanism of Action	References
Liquiritin	Flavonoid	Inhibits the production of TNF-α, IL-1β, and IL-6, reduces oxidative stress by neutralizing reactive oxygen species (ROS), and interferes with the NF-κB pathway	Patidar et al. (2022)
Berberine	Alkaloid	Inhibits the proliferation and migration of RA-FLS, inhibits TNF-α, IL-6, and IL-1β, and suppresses NF-κB pathway	Ma and Jiang (2016)
Oroxylin A	Flavonoid	Reduces IL-1β, IL-6, TNF-α, and IL-17, increases Treg cells, inhibits NF-κB pathway, and reduces MMPs.	Sajeev et al. (2022)
Boswellic acids	Triterpenoid	Inhibits 5-LOX, decreases the levels of cytokines like TNF-α and IL-1β, and inhibits NF-κB	Sharma, Chaubey, and Suvarna (2021)
Kaempferol	Flavonoid	Inhibits FLS migration and TNF-α expression	Ren et al. (2019)
Sinomenine	Alkaloid	Inhibits NF-κB and JAK/STAT pathways	Feng et al. (2019)
Baicalin	Flavonoid	Acts as an antioxidant and reduces ROS and NF-κB activity	Feng et al. (2019)
Emodin	Anthraquinone	Inhibits NF-κB and MAPK pathways and reduces cytokines.	Lü et al. (2015)

10.6 KEY TAKEAWAYS

- VEGF and its various subtypes play an important role in angiogenesis by interacting with different key receptors, influencing processes like cardiovascular repair, neuroprotection, and, most importantly, bone development.
- The RANKL/RANK/OPG triad regulates OCG and immune functions. Dysregulation in this pathway contributes to inflammatory bone disorders and immune dysfunctions.
- Both VEGF and RANKL pathways are implicated in inflammatory conditions like RA, SLE, and SSC. Understanding their roles can offer insights into disease mechanisms and potential therapeutic targets.
- Phytochemicals targeting VEGF and RANKL pathways show a promising role in managing inflammatory diseases. Compounds like quercetin, genistein, curcumin, and RES exhibit anti-inflammatory and bone-protective effects, offering new avenues for treatment.
- Monitoring VEGF and RANKL levels may serve as diagnostic markers for disease activity and progression in RA and SLE. Targeting these pathways presents opportunities for developing novel therapies and improving patient outcomes.

10.7 FURTHER READING

- Exploring the mechanism of celastrol in the treatment of RA based on systems pharmacology and multi-omics.
- Resveratrol as an effective adjuvant therapy in the management of RA: a clinical study.
- Quercetin, a plant polyphenol, has potential for the prevention of bone destruction in RA.
- Effect of curcumin on RA: a systematic review and meta-analysis.

- Molecular docking and interactions between VEGF receptors and phytochemicals: an in-silico study.
- Multiple targets directed multiple ligands: an in silico and in vitro approach to evaluating the effect of triphala on angiogenesis.

REFERENCES

Ahmed, S. (2010). Green tea polyphenol epigallocatechin 3-gallate in arthritis: progress and promise. *Arthritis Research & Therapy, 12*(2), 208. https://doi.org/10.1186/ar2982

Akiyama, T., Shinzawa, M., & Akiyama, N. (2012). RANKL-RANK interaction in immune regulatory systems. *World Journal of Orthopedics, 3*(9), 142–150. https://doi.org/10.5312/wjo.v3.i9.142

Behl, T., Mehta, K., Sehgal, A., Singh, S., Sharma, N., Ahmadi, A., … Nutrition. (2022). Exploring the role of polyphenols in rheumatoid arthritis. *Critical Reviews in Food Science and Nutrition, 62*(19), 5372–5393.

Burger, M. G., Grosso, A., Briquez, P. S., Born, G. M., Lunger, A., Schrenk, F., … Schaefer, D. J. (2022). Robust coupling of angiogenesis and osteogenesis by VEGF-decorated matrices for bone regeneration. *Acta Biomaterialia, 149*, 111–125.

Chacko, S. M., Thambi, P. T., Kuttan, R., & Nishigaki, I. (2010). Beneficial effects of green tea: a literature review. *China Medical Journal, 5*, 13. https://doi.org/10.1186/1749-8546-5-13

Chen, Z., & Han, Z. C. (2008). STAT3: a critical transcription activator in angiogenesis. *Medical Research Reviews, 28*(2), 185–200. https://doi.org/10.1002/med.20101

Cheng, W. X., Huang, H., Chen, J. H., Zhang, T. T., Zhu, G. Y., Zheng, Z. T., … Zhang, P. (2020). Genistein inhibits angiogenesis developed during rheumatoid arthritis through the IL-6/JAK2/STAT3/VEGF signalling pathway. *Journal of Orthopedic Translation, 22*, 92–100. https://doi.org/10.1016/j.jot.2019.07.007

Cho, M.-L., Jung, Y. O., Moon, Y.-M., Min, S.-Y., Yoon, C.-H., Lee, S.-H., … Kim, H.-Y. J. I. l. (2006). Interleukin-18 induces the production of vascular endothelial growth factor (VEGF) in rheumatoid arthritis synovial fibroblasts via AP-1-dependent pathways. *Immunology Letters, 103*(2), 159–166.

Cohen, S. B., Dore, R. K., Lane, N. E., Ory, P. A., Peterfy, C. G., Sharp, J. T., … Newmark, R. (2008). Denosumab treatment effects on structural damage, bone mineral density, and bone turnover in rheumatoid arthritis: a twelve-month, multicenter, randomized, double-blind, placebo-controlled, phase II clinical trial. *Arthritis & Rheumatism, 58*(5), 1299–1309. https://doi.org/10.1002/art.23417

Darwish, N. H., Sudha, T., Godugu, K., Bharali, D. J., Elbaz, O., El-ghaffar, H. A. A., … Mousa, S. A. J. M. (2019). Novel targeted nano-parthenolide molecule against NF-kB in acute myeloid leukemia. *Molecule, 24*(11), 2103.

de Jesús, G. R., Lacerda, M. I., Rodrigues, B. C., Dos Santos, F. C., do Nascimento, A. P., Cristóvão Porto, L., … Klumb, E. M. (2021). Soluble Flt-1, placental growth factor, and vascular endothelial growth factor serum levels to differentiate between active lupus nephritis during pregnancy and preeclampsia. *Arthritis Care Research, 73*(5), 717–721. https://doi.org/10.1002/acr.24360

Deng, Z., Li, Y., Liu, H., Xiao, S., Li, L., Tian, J., … Zhang, F. (2019). The role of sirtuin 1 and its activator, resveratrol in osteoarthritis. *Bioscience Reports, 39*(5), BSR20190189.

Ding, Q., Hu, W., Wang, R., Yang, Q., Zhu, M., Li, M., … therapy, t. (2023). Signaling pathways in rheumatoid arthritis: implications for targeted therapy. *Signal Transduction and Targeted Therapy, 8*(1), 68.

Duffy, A., Bouchier-Hayes, D., & Harmey, J. (2011). *Vascular Endothelial Growth Factor (VEGF) and Its Role in Non-Endothelial Cells: Autocrine Signalling by VEGF* (pp. 133–144). Madame Curie Bioscience Database.

Feng, Z.-T., Yang, T., Hou, X.-Q., Wu, H.-Y., Feng, J.-T., Ou, B.-J., … Mei, Z. G. (2019). Sinomenine mitigates collagen-induced arthritis mice by inhibiting angiogenesis. *Biomedicine & Pharmacotherapy, 113*, 108759.

Gandhi, G. R., Jothi, G., Mohana, T., Vasconcelos, A. B. S., Montalvão, M. M., Hariharan, G., … Li, H.-B. J. P. (2021). Anti-inflammatory natural products as potential therapeutic agents of rheumatoid arthritis: a systematic review. *Phytomedicine, 93*, 153766.

García-Ríos, P., Pecci-Lloret, M. P., & Oñate-Sánchez, R. E. (2022). Oral manifestations of systemic lupus erythematosus: a systematic review. *International Journal of Environmental Research and Public Health, 19*(19). https://doi.org/10.3390/ijerph191911910

George, G., Shyni, G., & Raghu, K. J. I. (2020). Current and novel therapeutic targets in the treatment of rheumatoid arthritis. *Inflammopharmacology, 28*(6), 1457–1476.

Geusens, P. (2012). The role of RANK ligand/osteoprotegerin in rheumatoid arthritis. *Therapeutic Advances in Musculoskelet Disease, 4*(4), 225–233. https://doi.org/10.1177/1759720x12438080

Gómez-Bernal, F., Fernández-Cladera, Y., Quevedo-Abeledo, J. C., García-González, M., González-Rivero, A. F., de Vera-González, A., … Ferraz-Amaro, I. (2022). Vascular endothelial growth factor and its soluble receptor in systemic lupus erythematosus patients. *Biomolecules, 12*(12). https://doi.org/10.3390/biom12121884

Guerrini, M. M., Okamoto, K., Komatsu, N., Sawa, S., Danks, L., Penninger, J. M., … Takayanagi, H. (2015). Inhibition of the TNF family cytokine RANKL prevents autoimmune inflammation in the central nervous system. *Immunity, 43*(6), 1174–1185. https://doi.org/10.1016/j.immuni.2015.10.017

Hansildaar, R., Vedder, D., Baniaamam, M., Tausche, A. K., Gerritsen, M., & Nurmohamed, M. T. (2021). Cardiovascular risk in inflammatory arthritis: rheumatoid arthritis and gout. *Lancet Rheumatology, 3*(1), e58–e70. https://doi.org/10.1016/s2665-9913(20)30221-6

Javed, M., Saleem, A., & Akhtar, M. F. (2023). Diosgenin, a steroidal sapogenin, arrests arthritis through modulation of inflammatory cytokines and oxidative stress biomarkers in Wistar rats. *Inflammopharmacology, 31*(4), 1951–1966.

Justiz-Vaillant, A. A., Gopaul, D., Soodeen, S., Arozarena-Fundora, R., Barbosa, O. A., Unakal, C., … Akpaka, P. E. (2024). Neuropsychiatric systemic lupus erythematosus: molecules involved in its imunopathogenesis, clinical features, and treatment. *Molecules, 29*(4). https://doi.org/10.3390/molecules29040747

Ke, D., Xu, H., Han, J., Dai, H., Wang, X., Luo, J., … Xu, J. (2024). Curcumin suppresses RANKL-induced osteoclast precursor autophagy in osteoclastogenesis by inhibiting RANK signaling and downstream JNK-BCL2-Beclin1 pathway. *Biomedical Journal, 47*(1), 100605. https://doi.org/10.1016/j.bj.2023.100605

Khojah, H. M., Ahmed, S., Abdel-Rahman, M. S., & Elhakeim, E. H. (2018). Resveratrol as an effective adjuvant therapy in the management of rheumatoid arthritis: a clinical study. *Clinical Rheumatology, 37*(8), 2035–2042. https://doi.org/10.1007/s10067-018-4080-8

Kim, H. R., Kim, B. M., Won, J. Y., Lee, K. A., Ko, H. M., Kang, Y. S., … Kim, K. W. (2019). Quercetin, a plant polyphenol, has potential for the prevention of bone destruction in rheumatoid arthritis. *Journal of Medical Food, 22*(2), 152–161. https://doi.org/10.1089/jmf.2018.4259

Kim, H. R., Kim, K. W., Kim, B. M., Cho, M. L., & Lee, S. H. (2015). The effect of vascular endothelial growth factor on osteoclastogenesis in rheumatoid arthritis. *PLoS One, 10*(4), e0124909. https://doi.org/10.1371/journal.pone.0124909

Kim, W. U., Kang, S. S., Yoo, S. A., Hong, K. H., Bae, D. G., Lee, M. S., … Cho, C. S. (2006). Interaction of vascular endothelial growth factor 165 with neuropilin-1 protects rheumatoid synoviocytes from apoptotic death by regulating Bcl-2 expression and Bax translocation. *Journal of Immunology, 177*(8), 5727–5735. https://doi.org/10.4049/jimmunol.177.8.5727

Komatsu, N., & Takayanagi, H. (2022). Mechanisms of joint destruction in rheumatoid arthritis – immune cell-fibroblast-bone interactions. *Nature Reviews Rheumatology, 18*(7), 415–429. https://doi.org/10.1038/s41584-022-00793-5

Kou, H., Huang, L., Jin, M., He, Q., Zhang, R., & Ma, J. (2023). Effect of curcumin on rheumatoid arthritis: a systematic review and meta-analysis. *Frontiers in Immunology, 14*, 1121655. https://doi.org/10.3389/fimmu.2023.1121655

Le, T. H. V., & Kwon, S. M. (2021). Vascular endothelial growth factor biology and its potential as a therapeutic target in rheumatic diseases. *International Journal of Molecuar Science, 22*(10). https://doi.org/10.3390/ijms22105387

Lee, S. S., Joo, Y. S., Kim, W. U., Min, D. J., Min, J. K., Park, S. H., … Kim, H. Y. (2001). Vascular endothelial growth factor levels in the serum and synovial fluid of patients with rheumatoid arthritis. *Clinical and Experimental Rheumatology, 19*(3), 321–324.

Li, J., Gang, D., Yu, X., Hu, Y., Yue, Y., Cheng, W., … Zhang, P. (2013). Genistein: the potential for efficacy in rheumatoid arthritis. *Clinical Rheumatology, 32*(5), 535–540. https://doi.org/10.1007/s10067-012-2148-4

Li, W., Wang, K., Liu, Y., Wu, H., He, Y., Li, C., … Su, W. (2022). A novel drug combination of mangiferin and cinnamic acid alleviates rheumatoid arthritis by inhibiting TLR4/NFκB/NLRP3 activation-induced pyroptosis. *Frontiers in Immunology, 13*, 912933.

Li, X.-Z., & Zhang, S.-N. J. C. M. (2020). Recent advance in treatment of osteoarthritis by bioactive components from herbal medicine. Chinese Medicine, *15*(1), 80.

Lin, S.-Y., Kan, J. Y., Lu, C.-C., Huang, H. H., Cheng, T.-L., Huang, H.-T., … Lin, Y.-S. (2020). Green tea catechin (-)-Epigallocatechin-3-Gallate (EGCG) facilitates fracture healing. *Biomolecules, 10*(4), 620.

Liu, C., He, L., Wang, J., Wang, Q., Sun, C., Li, Y., … Ming, R. J. J. O. E. (2020). Anti-angiogenic effect of Shikonin in rheumatoid arthritis by downregulating PI3K/AKT and MAPKs signaling pathways. *Journal of Ethnopharmacology*, *260*, 113039.

Liu, K. G., He, Q. H., Tan, J. W., & Liao, G. J. (2015). Expression of TNF-α, VEGF, and MMP-3 mRNAs in synovial tissues and their roles in fibroblast-mediated osteogenesis in ankylosing spondylitis. *Genetics and Molecular Research, 14*(2), 6852–6858.

Lü, S., Wang, Q., Li, G., Sun, S., Guo, Y., & Kuang, H. (2015). The treatment of rheumatoid arthritis using Chinese medicinal plants: from pharmacology to potential molecular mechanisms. *Journal of Ethnopharmacology, 176*, 177–206.

Luo, Y., Hou, X., Xi, A., Luo, M., Wang, K., & Xu, Z. (2023). *Tripterygium wilfordii* Hook F combination therapy with methotrexate for rheumatoid arthritis: an updated meta-analysis. *Journal of Ethnopharmacology, 307*, 116211. https://doi.org/10.1016/j.jep.2023.116211

Ma, Q., & Jiang, J.-G. J. C. D. T. (2016). Functional components from nature-derived drugs for the treatment of rheumatoid arthritis. *Current Drug Targets, 17*(14), 1673–1686.

Maeda, K., Yoshida, K., Nishizawa, T., Otani, K., Yamashita, Y., Okabe, H., … Saito, M. (2022). Inflammation and bone metabolism in rheumatoid arthritis: molecular mechanisms of joint destruction and pharmacological treatments. *International Journal of Molecular Science, 23*(5). https://doi.org/10.3390/ijms23052871

Moritz, F., Schniering, J., Distler, J. H. W., Gay, R. E., Gay, S., Distler, O., & Maurer, B. (2017). Tie2 as a novel key factor of microangiopathy in systemic sclerosis. *Arthritis Research Therap, 19*(1), 105. https://doi.org/10.1186/s13075-017-1304-2

Nakashima, T., Hayashi, M., Fukunaga, T., Kurata, K., Oh-Hora, M., Feng, J. Q., … Takayanagi, H. (2011). Evidence for osteocyte regulation of bone homeostasis through RANKL expression. *Nature Medicine, 17*(10), 1231–1234. https://doi.org/10.1038/nm.2452

Nalesnik, M., Nikolić, J. M., & Jandrić, S. (2011). Adenosine deaminase and C-reactive protein in diagnosing and monitoring of rheumatoid arthritis. *Medicinski Glasnik (Zenica), 8*(1), 163–168.

Ohno, H. (2016). Intestinal M cells. *Journal of Biochemistry, 159*(2), 151–160. https://doi.org/10.1093/jb/mvv121

Ono, T., Hayashi, M., Sasaki, F., & Nakashima, T. (2020). RANKL biology: bone metabolism, the immune system, and beyond. *Inflammation Regeneration, 40*, 2. https://doi.org/10.1186/s41232-019-0111-3

Parveen, A., Subedi, L., Kim, H. W., Khan, Z., Zahra, Z., Farooqi, M. Q., & Kim, S. Y. (2019). Phytochemicals targeting VEGF and VEGF-related multifactors as anticancer therapy. *Journal of Clinical Medicine, 8*(3). https://doi.org/10.3390/jcm8030350

Patidar, V., Shah, S., Kumar, R., Singh, P. K., Singh, S. B., & Khatri, D. K. (2022). A molecular insight of inflammatory cascades in rheumatoid arthritis and anti-arthritic potential of phytoconstituents. *Molecular Biology Reports, 5*, 1–17.

Patil, A., Sable, R., & Kothari, R. (2012). Genetic expression of MMP-Matrix-mettalo-proteinases (MMP-1 and MMP-13) as a function of anterior mandibular repositioning appliance on the growth of mandibular condylar cartilage with and without administration of Insulin like growth factor (IGF-1) and Transforming growth factor-B (TGF-β). *The Angle Orthodontist, 82*(6), 1053–1059. https://doi.org/10.2319/122011-780.1

Penna, S., Capo, V., Palagano, E., Sobacchi, C., & Villa, A. (2019). One disease, many genes: implications for the treatment of osteopetroses. *Frontiers in Endocrinology, 10*, 85. https://doi.org/10.3389/fendo.2019.00085

Pimple, P., Shah, J., & Singh, P. (2024). Emerging phytochemical formulations for management of rheumatoid arthritis: a review. *Current Drug Delivery*. https://doi.org/10.2174/0115672018270434240105110330

Pourhabibi-Zarandi, F., Shojaei-Zarghani, S., & Rafraf, M. (2021). Curcumin and rheumatoid arthritis: a systematic review of literature. *International Journal of Clinical Practice, 75*(10), e14280. https://doi.org/10.1111/ijcp.14280

Ren, J., Lu, Y., Qian, Y., Chen, B., Wu, T., & Ji, G. J. E. (2019). Recent progress regarding kaempferol for the treatment of various diseases. *Experimental and Therapeutic Medicine, 18*(4), 2759–2776.

Sajeev, A., Hegde, M., Daimary, U. D., Kumar, A., Girisa, S., Sethi, G., & Kunnumakkara, A. B. J. P. (2022). Modulation of diverse oncogenic signaling pathways by oroxylin A: an important strategy for both cancer prevention and treatment. *Phytomedicine, 105*, 154369.

Saleem, A., Saleem, M., & Akhtar, M. F. (2020). Antioxidant, anti-inflammatory and antiarthritic potential of Moringa oleifera Lam: an ethnomedicinal plant of Moringaceae family. *South African Journal of Botany, 128*, 246–256.

Saleem, A., Saleem, M., Akhtar, M. F., Shahzad, M., & Jahan, S. (2020). Moringa rivae leaf extracts attenuate Complete Freund's adjuvant-induced arthritis in Wistar rats via modulation of inflammatory and oxidative stress biomarkers. *Inflammopharmacology, 28*, 139–151.

Shao, H., Shen, J., Wang, M., Cui, J., Wang, Y., Zhu, S., … Geng, D. (2015). Icariin protects against titanium particle-induced osteolysis and inflammatory response in a mouse calvarial model. *Biomaterials, 60*, 92–99. https://doi.org/10.1016/j.biomaterials.2015.04.048

Sharma, D., Chaubey, P., & Suvarna, V. J. J. O. F. B. (2021). Role of natural products in alleviation of rheumatoid arthritis—a review. *Journal of Food Biochemistry*, *45*(4), e13673.

Siddique, R., Muhammad, F., Faisal, M. N., Akhtar, B., Saleem, A., Kousar, S., … Muhammad, S. (2024). Gingerol nanoparticles attenuate complete Freund adjuvant-induced arthritis in rats via targeting the RANKL/OPG signaling pathway. *Inflammopharmacology, 32*(5), 3311–3326.

Szekanecz, Z., & Koch, A. E. (2009). Angiogenesis and its targeting in rheumatoid arthritis. *Vascular Pharmacology, 51*(1), 1–7. https://doi.org/10.1016/j.vph.2009.02.002

Tanaka, S. (2019). RANKL is a therapeutic target of bone destruction in rheumatoid arthritis. *F1000Res, 8*. https://doi.org/10.12688/f1000research.17296.1

Tang, M., Zeng, Y., Peng, W., Xie, X., Yang, Y., Ji, B., & Li, F. (2023). Pharmacological aspects of natural quercetin in rheumatoid arthritis. *Drug Design, Development and Therapy*, *29*(16), 2043–2053.

Udagawa, N., Koide, M., Nakamura, M., Nakamichi, Y., Yamashita, T., Uehara, S., … Tsuda, E. (2021). Osteoclast differentiation by RANKL and OPG signaling pathways. *Journal of Bone and Mineral Metabolism, 39*(1), 19–26. https://doi.org/10.1007/s00774-020-01162-6

Wang, G., Xie, X., Yuan, l., Qiu, J., Duan, W., Xu, B., & Chen, X. (2020). Resveratrol ameliorates rheumatoid arthritis via activation of SIRT1-Nrf2 signaling pathway. *Biofactors, 46*(3), 441–453.

Wang, Q., Ye, C., Sun, S., Li, R., Shi, X., Wang, S., … Liu, R. (2019). Curcumin attenuates collagen-induced rat arthritis via anti-inflammatory and apoptotic effects. *International Immunopharmacology, 72*, 292–300. https://doi.org/10.1016/j.intimp.2019.04.027

Wei, C. C., Ping, D. Q., You, F. T., Qiang, C. Y., & Tao, C. (2016). Icariin prevents cartilage and bone degradation in experimental models of arthritis. *Mediators Inflammation, 2016*, 9529630. https://doi.org/10.1155/2016/9529630

Wick, A., Wick, W., Waltenberger, J., Weller, M., Dichgans, J., & Schulz, J. B. (2002). Neuroprotection by hypoxic preconditioning requires sequential activation of vascular endothelial growth factor receptor and Akt. *Journal of Neuroscience, 22*(15), 6401–6407. https://doi.org/10.1523/jneurosci.22-15-06401.2002

Williams, B., Lees, F., Tsangari, H., Hutchinson, M. R., Perilli, E., & Crotti, T. N. (2020). Assessing the effects of parthenolide on inflammation, bone loss, and glial cells within a collagen antibody-induced arthritis mouse model. *Mediators Inflammation, 2020*, 6245798. https://doi.org/10.1155/2020/6245798

Wiszniak, S., & Schwarz, Q. (2021). Exploring the intracrine functions of VEGF-A. *Biomolecules, 11*(1). https://doi.org/10.3390/biom11010128

Wu, Z., Liu, Q., Cao, Z., Li, H., Zhou, Y., & Zhang, P. (2024). Icariin decreases cell proliferation and inflammation of rheumatoid arthritis-fibroblast like synoviocytes via GAREM1/MAPK signaling pathway. *Immunopharmacology Immunotoxicology, 46*(1), 86–92. https://doi.org/10.1080/08923973.2023.2253990

Yokota, K., Sato, K., Miyazaki, T., Aizaki, Y., Tanaka, S., Sekikawa, M., … Mimura, T. (2021). Characterization and function of tumor necrosis factor and interleukin-6–induced osteoclasts in rheumatoid arthritis. *Arthritis & Rheumatology, 73*(7), 1145–1154.

Yuan, Z. C., Xu, W. D., Wang, J. M., Wu, Q., Zhou, J., & Huang, A. F. (2020). Gene polymorphisms and serum levels of sVEGFR-1 in patients with systemic lupus erythematosus. *Science Reports, 10*(1), 15031. https://doi.org/10.1038/s41598-020-72020-8

Zhang, Y., Mao, X., Li, W., Chen, W., Wang, X., Ma, Z., & Lin, N. J. M. R. R. (2021). *Tripterygium wilfordii*: an inspiring resource for rheumatoid arthritis treatment. *Medicinal research reviews*, *41*(3), 1337–1374.

Zhou, Y. Y., Xia, X., Peng, W. K., Wang, Q. H., Peng, J. H., Li, Y. L., … Huang, Q. C. (2018). The effectiveness and safety of *Tripterygium wilfordii* Hook. F extracts in rheumatoid arthritis: a systematic review and meta-analysis. *Frontiers in Pharmacology, 9*, 356. https://doi.org/10.3389/fphar.2018.00356

11 Clinical Studies about Phytomedicines for the Management of Rheumatoid Arthritis

Atika Afzal, Ayesha Mahmood,
Shaneel Kousar, Sajid Ali, and Ammara Saleem

WHAT WILL YOU LEARN?

In this chapter, the main focus will be on the naturally occurring compounds present in plants, known as phytochemicals, that represent crucial reservoirs for developing novel pharmaceuticals. In this chapter, clinical studies on herbs and phytocompounds will be provided.

LIST OF ABBREVIATIONS

BSE	Boswellia serrata extract
BSO	Black seed oil
cAMP	Cyclic adenosine monophosphate levels
CBD	Cannabidiol
CRP	C-reactive protein
EBV	Epstein-Barr virus
ERK	Extracellular signal-regulated kinase
ESR	Erythrocyte sedimentation rate
FOX3a	Forkhead box class O 3a
HLA	Human leukocyte antigen
IKK-beta kinase	Inhibitory Kappa B kinase beta
NO	Nitric oxide
PGs	Prostaglandins
TNF	Tumor necrosis factor
VAS	Visual analog scale
WBE	Willow bark extract

11.1 INTRODUCTION

Rheumatoid arthritis (RA) is a chronic autoimmune condition that impacts nearly a quarter of the population in the United States (Lindler, Long, Taylor, & Lei, 2020). RA is a long-lasting systemic condition marked by inflammation, primarily affecting the joints, feet, and hands. It affects all joints, mainly small joints of the hand and foot, with erosive alterations. In autoimmune diseases like this, the body's immune system will mistakenly attack its tissues, weakening the defense mechanisms and potentially increasing susceptibility to various infections (Javed, Saleem, & Akhtar, 2023).

 DOI: 10.1201/9781003535201-13

For individuals with autoimmune conditions, these cells and antibodies primarily focus on the body's tissues, causing inflammation around the joints (Sailaja, Rani, & Swetha, 2020). RA can develop at any age but is more frequent in women than in men (Qindeel, Ullah, & Ahmed, 2020). It is the most prevalent systemic autoimmune disease globally, with a prevalence of 0.5%–1% (Neidhart & Neidhart, 2019).

11.2 SIGNS AND SYMPTOMS

Symptoms of RA can vary widely among individuals, depending on the extent of inflammation and the specific body tissues affected by the inflammation. Severity can range from mild to severe (Banerjee, Nara, Chakraborty, Chowdhury, & Haldar, 2022).

11.3 RISK FACTORS FOR DEVELOPING RA

Several factors significantly contribute to the onset and progression of the current disease. The immune response is triggered by several factors, particularly a combination of sex hormones, genetic predisposition, and environmental factors (Singh, Singh, Mahajan, & Dhiman, 2020).

11.4 PATHOPHYSIOLOGICAL MECHANISM

RA is identified by synovitis, which involves the thickening of the synovial lining and inflammatory cell infiltration. Angiogenic cytokines are responsible for stimulating the growth of new blood vessels in the synovium, leading to synovial proliferation and the formation of a tumor-like mass known as pannus. Pannus then damages the articular cartilage and subchondral bone, ultimately causing bony erosions. The pathophysiology involves an immune system imbalance, with increased activation and infiltration of neutrophils, macrophages, lymphocytes, and dendritic cells, accompanied by synovial membrane cell proliferation. This process generates autoantibodies within adaptive immune cells, recognizing various modified proteins, including post-translational modifications. Integrin, a critical transmembrane protein, plays a vital role in regulating these immune cell actions (Mehmood, Saleem, Akhtar, & Mobashar, 2025; Radu & Bungau, 2021).

11.5 MANAGEMENT OF RA USING PHYTOMEDICINES

Phytochemicals are invaluable for developing new pharmaceuticals and therapies for certain conditions such as cancer, diabetes mellitus, fatty liver associated with hyperlipidemia, and chronic inflammatory disorders like osteoarthritis and RA (Choudhari, Mandave, Deshpande, Ranjekar, & Prakash, 2020; Tahir, Saleem, & Akhtar, 2025). The importance of phytomedicines in managing chronic joint inflammatory diseases, particularly RA, has been substantial (Lippert & Renner, 2022). Plant-based treatments are opted for their nontoxic nature, affordability, and fewer side effects compared to modern chemical drugs (Choudhari et al., 2020).

The World Health Organization (WHO) has documented around 21,000 plant species as medicinally active plants (Sharangi, n.d.). Significant progress has been made in the pharmacological appraisal of medicinal plants traditionally used to treat chronic diseases. For this reason, a large number of the population has started to depend on herbal practitioners as their main healthcare providers in developing countries (Sholikhah et al., 2020).

Phytochemicals have therapeutic potential, providing both natural and holistic approaches in order to manage a wide range of health conditions, particularly chronic inflammatory diseases. The combination of traditional knowledge with modern scientific research highlights the pivotal role of medicinal plants in the global healthcare market. Despite widespread use of herbal and traditional medicines by a significant number of population in many developing countries, there are still chances that herbal preparations may pose serious health risks, thus emphasizing the importance of evaluating safety, efficacy, and quality of herbal remedies and detecting possible adverse effects associated with them (Niyomchan et al., 2023).

A major challenge in incorporating herbal formulations into modern medical practices is the limited availability of preclinical data. For the development of traditional medicines, both preclinical and clinical testing are necessary. Preclinical testing includes toxicity studies, which are conducted in animal models in order to gather data regarding the safety and efficacy of herbal medicines that are being used for a longer period of time for the treatment of many ailments (Sholikhah et al., 2020). Oriental herbology involves the practice of blending various herbs to achieve maximum therapeutic results for individual herbs (Ishtiaq et al., 2017). Naturally occurring compounds present in plants are known as phytochemicals that represent crucial reservoirs for developing novel pharmaceuticals and serve as potential sources for different therapies, such as cancer, diabetes mellitus, hyperlipidemia, and chronic inflammatory disorders such as osteoarthritis and RA (Choudhari et al., 2020).

Many polyherbal formulations have been reported to treat many ailments from ancient times. NawaTab is a tablet formulation derived from the Nawametho polyherbal formula, which has been traditionally utilized in Surat Thani province, Southern Thailand, for managing hyperlipidemia. Nawametho is an herbal preparation that blends nine medicinal herbs (Niyomchan et al., 2023). The Committee on Herbal Medicinal Products, which operates under the European Medicines Agency, produces monographs on herbal drugs and plant-based products. These monographs gather evidence for both "traditional" and "well-established" uses. The latter necessitates scientific proof of beneficial effects, a minimum of ten years of practical application, and safety publications. These monographs serve as reference points during the assessment of market authorization applications. Products recognized with "traditional use" or "well-established use" status undergo an expedited approval process (Lippert & Renner, 2022).

Due to side effects, safety, and costs associated with conventional arthritis treatment and a chance of relapse, there has been interest in treating RA with medicinal plants. Difficulty in chronic pain management associated with RA has made researchers investigate herbal extracts having potentially active phytochemicals used in different herbal remedies. Herbs may provide a safe and effective alternative approach when used as a single or in combination form for treatment (Crowson, Rahman, & Matteson, 2009). Recently, studies have been reported in which herbal treatments using herbs have been examined in vitro, in vivo, and in human clinical studies.

11.6 CLINICAL TRIALS ON SINGLE HERB/CLINICAL FORMULATIONS OF HERBAL PLANTS

Several herbal extracts have shown positive clinical outcomes (Dragos et al., 2017). The main aim of RA treatment is to control and halt the progression of the disease. The overall purpose of RA treatment is to minimize pain and inflammation (Crowson et al., 2009). Therapy using herbal medicine is one of the basic aspects of complementary and alternative medicine. Continuous use of synthetic drugs may lead to many side effects, resulting in limited use. Traditionally used herbs contain various active phytoconstituents possessing antioxidant and anti-inflammatory actions, which tend to be an effective approach for managing and treating RA (Sharangi, n.d.; Singh et al., 2020). These herbs have been evaluated by conducting clinical and experimental studies for their potential role in the inhibition of inflammatory mediators, and clinical data have been provided in research studies.

11.6.1 *Withania somnifera*

Withania somnifera (WS) is commonly known as ashwagandha and exhibits analgesic, anti-inflammatory, and anti-arthritic properties. It belongs to the Solanaceae family and is a renowned Ayurvedic medicine. WS roots and leaves contain withanolides, the main compounds that mimic the action of steroids and are believed to be responsible for the plant's therapeutic effects. Treatment with doses of 125 and 250 mg of *W. somnifera* taken twice daily for 12 weeks, with the earliest efficacy observed at 4 weeks, significantly reduced redness, swelling, and deformity. The anti-inflammatory activity is attributed to Withaferin A, which is bound to the enzyme *IKK-beta kinase*, thereby inhibiting NF-κβ signaling and causing suppression of the production of inflammatory cytokines such as COX-2, IL-1, IL-6, and tumor necrosis factor alpha (TNF-α). Additionally, it caused inhibition of FOX3a and the pro-apoptotic protein (Gupta & Singh, 2014; Ramakanth, Kumar, Kishan, & Usharani, 2016).

11.6.2 *Tripterygium wilfordii* Hook F (TwHF)

It is a Chinese traditional herb, which has been extensively utilized in autoimmune and inflammatory conditions, particularly in the treatment of RA. Studies have indicated that the TwHF root extract inhibited the expression of mediators, pro-inflammatory cytokines, and histamines (Bao & Dai, 2011). TwHF extracts were administered at 360 and 180 mg/day for 20 weeks and exhibited immunosuppressive, cartilage-protective, and anti-inflammatory effects, making them a potential treatment for RA patients who are nonresponsive to conventional methods of treatment (Lindler et al., 2020; Tang & Zuo, 2012).

11.6.3 *Paeonia lactiflora*

It belongs to the family Ranunculaceae and is widely used in Chinese traditional medicine for the treatment of numerous inflammatory disorders, including RA. It exhibits anti-inflammatory action by decreasing prostaglandins (PGs), and nitric oxide (NO) synthesis (Lindler et al., 2020; Zhang & Dai, 2012).

11.6.4 *Matricaria chamomilla* L.

It is also called Chamomile and is traditionally used for the treatment of various inflammatory disorders, mostly in the form of tea, and it alleviates joint and rheumatic pain. It is available in herbal preparations, as FDA has approved chamomile as "generally recognized as safe". It contains apigenin, luteolin, quercetin, and glycosides that are responsible for its anti-inflammatory effects. These phenolic compounds reduce the production of PGE2 and cytokines that play a key role in RA pathogenesis. A clinical study reported that consuming 6 g of chamomile tea per day causes a decrease in erythrocyte sedimentation rate (ESR) and decreased tender joints in RA patients (Pirouzpanah, Mahboob, Sanayei, Hajaliloo, & Safaeiyan, 2017; Shoara et al., 2015).

11.6.5 *Eremostachys laciniata*

It belongs to the family Lamiaceae and is available in many dosage formulations, such as decoctions and herbal teas, consisting of root and flower extracts. It is effective in treating inflammatory disorders, including osteoarthritis, RA, and septic arthritis (Delazar et al., 2013). In a single-blinded, randomized clinical trial, it has been shown that applying 0.5% *E. laciniata* ointment to affected joints for 2 weeks alleviates pain in arthritic patients (Crowson et al., 2009).

11.6.5.1 *Curcuma* species

Curcuma longa, known as turmeric, is a member of the Zingiberaceae family. This plant contains curcumin, a crucial ingredient renowned for having anti-inflammatory potential and antioxidant properties. It has been widely utilized in Chinese traditional and Ayurvedic medicine.

Scientific studies have demonstrated that consumption of turmeric extracts having the composition of 1000 mg/day of curcumin, for 8–12 weeks, can alleviate arthritic symptoms such as pain and inflammation. Curcumin exerts its effects by inhibiting COX-2, causing a reduction in the synthesis of PGE2 (Daily, Yang, & Park, 2016). Moreover, it provides a synergistic effect in herbal formulations with other medicines by inhibiting 5-lipoxygenase (Karlapudi, Prasad Mungara, Sengupta, Davis, & Raychaudhuri, 2018). Additionally, it suppressed the activation of the NF-KB pathway (Rajguru, Rajput, Sonawane, & Agrawal, 2023). BCM-95® is an available formulation of curcumin (Mahajan, Mali, & Mehta, 2007).

11.6.6 *Boswellia* Species

It is also known as Indian frankincense or Salai guggal. This plant is commonly used in traditional Ayurvedic medicine for treating patients with arthritis (Crowson et al., 2009). *Boswellia serrata,* belonging to the family Burseraceae, is commonly known as Indian Olibanum. It is native to India (Majeed, Majeed, Narayanan, & Nagabhushanam, 2019). This herb is thought to exert its anti-inflammatory effects due to the abundance of active phytoconstituents such as terpenoids, β-boswellic acid, and β-pineneis, inhibiting PGE2, reducing NO and (COX)-2, and preventing the synthesis of leukotriene (Majeed et al., 2024). A double-blind, placebo-controlled human trial was conducted to evaluate the safety and efficacy of *Boswellia serrata* extract (BSE). These include oral supplementation of BSE *(Boswellin®)* for eight weeks or more than four months, which significantly reduced pain and stiffness and also improved physical function. *Boswellia Carterii B.* (oliban oil) was applied topically and alleviated symptoms in arthritic patients (Crowson et al., 2009).

11.6.7 *Equisetum arvense*

This herb is commonly known as horsetail, belonging to the family *Equisetaceae.* It has a long history as an anti-inflammatory remedy in European ethnomedicine. Kynurenic acid was the main constituent responsible for its anti-inflammatory potential. A decrease in TNF-α was considered one of the contributing mechanisms that resulted in a decrease in morning stiffness and pain scores in RA patients (Parada-Turska, Zgrajka, & Majdan, 2013).

11.6.8 *Arnica montana*

In traditional herbal medicine, *Arnica montana* has been employed for centuries to address strain, trauma, and inflammation-related problems of the locomotor system. It stands as one of the most prevalent folk remedies for rheumatologic conditions. Belonging to the family Asteraceae, *Arnica montana*'s anti-inflammatory effect is attributed to the inhibition of inflammatory mediator production, which arises from the combined effects of its phenolic and flavonoid compounds (Sharma, Arif, Nirala, Gupta, & Thakur, 2016).

11.6.9 *Harpagophytum procumbens*

It is commonly referred to as devil's claw, is a member of the Pedaliaceae family, and contains harpagoside as the main constituent, which exhibits anti-inflammatory properties. The root extract of *Harpagophytum procumbens* inhibits the production of NO, inflammatory cytokines such as IL-6, IL-1β, and TNF-α, as well as PGE2. Human clinical studies have demonstrated that daily administration of *Harpagophytum procumbens* tuber extracts, equivalent to 50–60 mg of harpagoside, for 8–16 weeks, significantly reduced the severity of symptoms such as pain and inflammation (Mariano et al., 2020).

11.6.10 *Cannabis sativum*

This herb is locally known as marijuana, belonging to the family Cannabaceae. Cannabidiol (CBD) is the main active component showing therapeutic effects by inhibiting inflammatory mediators. CBD is administered orally at 10 and 25 mg/kg. It decreased inflammatory cytokines and mediators. Oil extracted from *Cannabis sativa* seeds has significant potential in alleviating pain and inflammation (Singh et al., 2020).

11.6.11 Ribes nigrum

Blackcurrant (*Ribes nigrum*) is a renowned plant with therapeutic properties. Blackcurrant oil is extracted from seeds and is a rich source of polyunsaturated fatty acids. A clinical trial was conducted on RA patients, and it was found that blackcurrant seed oil treatment after 24 weeks improved symptoms of RA significantly by suppressing proinflammatory cytokines in peripheral blood monocytes. It also caused a decrease in morning stiffness in treated patients (Yarnell, 2017).

11.6.12 *Borago officinalis*

It belongs to the Boraginaceae family. This plant is native to Europe. Borage (*Borago officinalis*) is a valued plant having a high percentage of gamma-linolenic acid, which is important in suppressing inflammatory pathways and acts as a potent suppressor. This plant caused a significant reduction in cyclic adenosine monophosphate (cAMP) levels and an increase in PGE2. In a clinical study, it has been investigated that borage seed oil in doses of 1.4 and 2.8 g/day for a period of 6 months exhibited antiarthritic activity, making borage oil an effective remedy for treating RA (Singh et al., 2020).

11.6.13 *Berberis orthobotrys*

Zereshk (barberry) belongs to the family Berberidaceae. This plant contains alkaloids as main active constituents, among them berberine is the most active phytochemical constituent that has anti-inflammatory effects by inhibiting inflammatory mediators such as interleukin (IL)-1, IL-6, and TNF-α, IFN-α, and platelet-derived growth factor. Synovial hyperplasia has been significantly improved because of the inhibitory effects produced by zereshk (Singh et al., 2020).

11.6.14 Salix Species

Salix (willow) exists in various species and has been used in traditional medicines to relieve pain and inflammation. Salix species is one of the members belonging to the family Salicaceae. A double-blind, randomized, placebo-controlled study was conducted to assess the anti-inflammatory effects of willow bark extract (WBE). A 240mg salicin/day dose of WBE was administered for 2 and 6 weeks. The findings of the study showed inflammation-suppressing effects of WBE by inhibiting COX-2 enzymes, pro-inflammatory cytokines (TNF-α), and NF-κB (Saller, Melzer, & Felder, 2008).

11.6.15 *Aloe barbadensis*

Aloe vera is another commonly used name. This plant is a member of the Asphodelaceae family. Aloe contains anthraquinone glycosides as the main active chemical constituents. It also includes anthracene, anthranilic acid, and cinnamic acid, which are responsible for producing anti-inflammatory effects. Aloe exhibits antiarthritic activity by inhibiting signaling pathways such as nuclear factor-kappa B (NF-κB), p38, janus kinases, and ERK activated by lipopolysaccharide. *Aloe vera* is valued because of its ability to regulate the immune system and provide relief from muscle and joint pain (Singh et al., 2020).

11.6.16 *Nigella sativa*

Black cumin or Cumin is another name for this plant. Cumin belongs to the Ranunculaceae family. *N. sativa* seed oil or black seed oil (BSO) contains thymoquinone, involved in the reduction of oxidative stress in RA patients when administered at a dose of 500 mg, two times a day for 8 weeks. It acts by inhibiting cyclooxygenase and 5-lipoxygenase. BSO can be recommended to use topically and orally (Mahboubi, Kashani, & Mahboubi, 2018; Zielińska, Dereń, Polak-Szczybyło, & Stępień, 2021).

11.6.17 *Zingiber officinalis*

It is also known as ginger, belonging to the family Zingiberaceae, a common spice and traditional remedy used in ethnomedicine for joint disorders. Ginger extracts can be administered at 30, 250, and 500 mg daily, showing an improvement in pain scores in different clinical trials. In another study, it was reported that daily intake of two tablets having a dose of 500 mg of ginger improved pain. This was due to a reduction in the production of PGE2, NO, IL-1β, IL-12, and TNF-α (Crowson et al., 2009). Its marketed formulation contains ginger in powder form (1 g/day) that reduced NO and inflammatory markers when given for three months of therapy. Ginger extract has been used as an anti-inflammatory agent in treating RA (Rondanelli et al., 2020).

11.6.18 *Moringa oleifera*

This plant belongs to the Moringaceae family and possesses anti-inflammatory and anti-oxidant properties. Various parts of the plant, particularly the leaves and seeds, are rich sources of chemically active constituents such as flavonoids, phenolics (such as quinic acid and chlorogenic acid), vitamins, carotenoids, minerals, sterols (such as β-sitosterol), amino acids, alkaloids, and glycosides, which are responsible for these activities. Extracts of moringa leaves having a 250 mg/kg dose inhibited NO production by macrophage cells and reduced cytokine (IL-1, IL-6, and TNF-α) levels, resulting in the improvement of inflammation and pain (Mahajan et al., 2007).

11.6.19 *Piper nigrum*

It is commonly known as black pepper, which belongs to the Piperaceae family. Piperine is an active constituent known for its ability to inhibit the enzymes that are responsible for the biosynthesis of leukotriene (Singh et al., 2020).

11.7 CLINICAL TRIALS ON HERBAL COMBINATION THERAPIES

Combining different herbs can have synergistic effects, enhancing both therapeutic outcomes and safety as compared to a single herb (Table 11.1).

11.8 CLINICAL STUDIES ON PHYTOCHEMICALS AND THEIR FORMULATIONS

11.8.1 Curcumin

The first pilot randomized clinical trial was conducted by Chandran and his colleagues in order to evaluate the efficacy of curcumin in RA patients against a standard NSAID (diclofenac sodium). RA patients were divided into three groups receiving diclofenac sodium and curcumin alone and in combination.

TABLE 11.1
Summary of Herbal Remedies and Clinical Formulations Used to Treat Rheumatoid Arthritis

Clinical Formulation	Dosage (Per Day)	Treatment Time (Day)	Mechanism of Action	Clinical Implications	References
LI73014F2 (novel herbal formulation containing *Terminalia chebula* fruit, *Curcuma longa* rhizome, and *Boswellia serrata* gum resin)	200–400 mg	90 days but shows efficacy at 14 days	5-Lipoxygenase (5-LOX) inhibition	Management of joint discomfort, pain relief, improved quality of life with better physical function	Karlapudi et al. (2018)
Urtica dioica and Oenothera biennis (evening primrose oil) (EPO)		Three-month follow-up	Decrease inflammatory factors	Improve symptoms of RA patients	Abd-Nikfarjam et al. (2022)
Withania somnifera (Ashwashila Amulya joint care capsules)	250 and 125 mg twice daily	12 weeks, but showed the earliest efficacy at 4 weeks	It binds to the IKK-beta kinase, effectively hindering NF-κβ signaling and thereby suppressing TNF-α, IL-1, IL-6, and COX-2. Additionally, it inhibits FOX3a and the pro-apoptotic protein BIM	Protecting Joint destruction	Ramakanth et al. (2016)
ZO preparation (*Zingiber officinalis* extract and glucosamine)	(340 mg EV.EXT 35 *Zingiber officinalis* extract) and glucosamine (1,000 mg) orally daily	4 weeks	Inflammatory markers decreased	Arthritic pain decreased during standing and moving	Majeed et al. (2019)

The primary end point was a decline in disease activity score (DAS). The secondary end points included tenderness and swelling of joint scores. All three treatment groups exhibited improvement in DAS, but the curcumin treatment group showed a relatively higher level improvement in DASs without any side effects.

Another clinical study was conducted by Jobi and his colleagues in order to investigate the efficacy of CuroWhite (hydrogenated curcuminoid formulation) in RA patients. Three months' treatment with high-dose CuroWhite considerably reduced ESR, C-reactive protein (CRP), and rheumatoid factor (RF) in RA patients. Comparison between placebo control and treated patients was made, and Visual Analog Scale (VAS) and DAS were found to be improved in the treatment group. Additionally, CuroWhite also improved joint tenderness and swelling. These results clearly suggested the promising role of this formulation in RA. Hemmati and his co-workers also conducted another randomized clinical trial. During this clinical-based study, curcumex was administered to the RA patients for eight weeks, which was the mixture of ginger and curcumin. This mixture considerably attenuated the disease severity by improving DAS, tenderness, and swelling of joints.

During another randomized, double-blind placebo-controlled study, nanomicelles of curcumin (120 mg/day) were administered to the patients for 12 weeks. This formulation improved DAS, tenderness, and swelling of joints. More interestingly, no adverse effect was observed in these nanomicelle-treated RA patients (Javadi, Khadem Haghighian, Goodarzy, Abbasi, & Nassiri-Asl, 2019). Almaraj et al. conducted a clinical trial in RA patients. The results of this double-blind, placebo-controlled study clearly suggested that turmeric matrix formulation (consisting of curcuminoids) significantly improved pathological manifestations of RA. This formulation, which was subjected in the trial, when compared to the placebo-controlled group, was found to improve the values of VAS, CRP, ESR, and RF (Augustine Amalraj et al., 2017).

11.8.2 Resveratrol

Resveratrol belongs to the polyphenol group of phytochemicals. It is found abundantly in grape seeds, skin, and red wine. It has antioxidant and anti-inflammatory activities. Khoja et al. constructed a randomized controlled clinical trial in order to evaluate the therapeutic effects of resveratrol in patients suffering from different stages of RA. Hundred patients with RA (32 men, 68 women) were randomly enrolled. Patients received 1 g capsule of resveratrol for three months along with conventional treatment. When the blood sample of these patients was evaluated for different biomarkers, there was a significantly reduced level of CRP, ESR, TNF-α, IL-6, and MMP3 as compared to those patients who received conventional therapy alone (Khojah, Ahmed, Abdel-Rahman, & Elhakeim, 2018).

Preclinical research has explored various medicinal plants and phytochemicals using cellular and animal models for treating RA, although human clinical trials have not been conducted in detail. The underlying mechanisms in preclinical trials provide valuable insights into the therapeutic potential of these herbal remedies for the treatment of RA.

Various herbs such as angelica, cinnamon, saposhnikovia, and glycyrrhiza and phytochemicals such as curcumin, boswellic acid, and resveratrol have exhibited anti-inflammatory potential and pain-reducing effects (Amirah et al., 2025). Research studies indicate that these herbs and phytocompounds have been contributing to cartilage structure and texture improvement and bone remodeling regulation and inhibiting osteoclastogenesis in order to support their role in treating arthritis. Further clinical trials are required for the assessment of the safety and efficacy of these herbs and phytocompounds so that they could be beneficial for treating arthritis in humans (Crowson et al., 2009).

11.9 CONCLUSION

Phytomedicines offer a promising and viable alternative for managing RA as compared to conventional therapies. Their natural origins, lower toxicity, cost-effectiveness, and minimal side effects make them a safer and more accessible option for many patients. The therapeutic properties of phytochemicals, particularly anti-inflammatory and immunomodulatory, supported by extensive traditional use and modern research validation, provide advantages in managing chronic inflammation and alleviating pain associated with RA.

The worldwide recognition of phytomedicines highlights their potential to improve the quality of life of patients suffering from RA. By integrating traditional knowledge and modern scientific research, herbal therapies offer sustainable and patient-friendly treatment options for chronic inflammatory joint diseases.

REFERENCES

Abd-Nikfarjam, B., Abbasi, M., Memarzadeh, M., Farzam, S.-A., Jamshidian, A., & Dolati-Somarin, A. (2022). Therapeutic efficacy of *Urtica dioica* and evening primrose in patients with rheumatoid arthritis: a randomized double-blind, placebo-controlled clinical trial. *Journal of Herbal Medicine, 32*, 100556.

Amirah, S., Abdurrahman, M. F., Sudiasa, I. N. S., Duta, T. F., Fakri, F., & Iqhrammullah, M. (2025). Efficacy of Phytopharmaca (Cinnamon, Turmeric, Ginger and Garlic) as an adjuvant in rheumatoid arthritis management: a systematic review and meta-analysis. *Indian Journal of Rheumatology, 20*(1), 60–72.

Augustine Amalraj, A. A., Karthik Varma, K. V., Jacob, J., Chandradhara Divya, C. D., Kunnumakkara, A., Stohs, S., & Sreeraj Gopi, S. G. (2017). A novel highly bioavailable curcumin formulation improves symptoms and diagnostic indicators in rheumatoid arthritis patients: a randomized, double-blind, placebo-controlled, two-dose, three-arm, and parallel-group study. *Journal of Medicinal Food*, *2*(10), 1–13

Banerjee, S., Nara, R., Chakraborty, S., Chowdhury, D., & Haldar, S. (2022). Integrin regulated autoimmune disorders: understanding the role of mechanical force in autoimmunity. *Frontiers in Cell and Developmental Biology, 10*, 852878.

Bao, J., & Dai, S.-M. (2011). A Chinese herb *Tripterygium wilfordii* Hook F in the treatment of rheumatoid arthritis: mechanism, efficacy, and safety. *Rheumatology international, 31*, 1123–1129.

Choudhari, A. S., Mandave, P. C., Deshpande, M., Ranjekar, P., & Prakash, O. (2020). Phytochemicals in cancer treatment: from preclinical studies to clinical practice. *Frontiers in Pharmacology, 10*, 497776.

Crowson, C. S., Rahman, M. U., & Matteson, E. L. (2009). Which measure of inflammation to use? A comparison of erythrocyte sedimentation rate and C-reactive protein measurements from randomized clinical trials of golimumab in rheumatoid arthritis. *The Journal of Rheumatology, 36*(8), 1606–1610.

Daily, J. W., Yang, M., & Park, S. (2016). Efficacy of turmeric extracts and curcumin for alleviating the symptoms of joint arthritis: a systematic review and meta-analysis of randomized clinical trials. *Journal of Medicinal Food, 19*(8), 717–729.

Delazar, A., Sarker, S. D., Nahar, L., Jalali, S. B., Modaresi, M., Hamedeyazdan, S., … Moghadam, S. B. (2013). Rhizomes of *Eremostachys laciniata*: isolation and structure elucidation of chemical constituents and a clinical trial on inflammatory diseases. *Advanced Pharmaceutical Bulletin, 3*(2), 385.

Dragos, D., Gilca, M., Gaman, L., Vlad, A., Iosif, L., Stoian, I., & Lupescu, O. (2017). Phytomedicine in joint disorders. *Nutrients, 9*(1), 70.

Gupta, A., & Singh, S. (2014). Evaluation of anti-inflammatory effect of *Withania somnifera* root on collagen-induced arthritis in rats. *Pharmaceutical biology, 52*(3), 308–320.

Ishtiaq, S., Akram, M., Kamran, S. H., Hanif, U., Afridi, M. S. K., Afzal, A., … Akbar, S. (2017). Acute and sub-acute toxicity study of a Pakistani polyherbal formulation. *BMC Complementary and Alternative Medicine, 17*, 1–13.

Javadi, M., Khadem Haghighian, H., Goodarzy, S., Abbasi, M., & Nassiri-Asl, M. (2019). Effect of curcumin nanomicelle on the clinical symptoms of patients with rheumatoid arthritis: a randomized, double-blind, controlled trial. *International Journal of Rheumatic Diseases, 22*(10), 1857–1862.

Javed, M., Saleem, A., & Akhtar, M. F. (2023). Diosgenin, a steroidal sapogenin, arrests arthritis through modulation of inflammatory cytokines and oxidative stress biomarkers in Wistar rats. *Inflammopharmacology, 31*(4), 1951–1966.

Karlapudi, V., Prasad Mungara, A. V. V., Sengupta, K., Davis, B. A., & Raychaudhuri, S. P. (2018). A placebo-controlled double-blind study demonstrates the clinical efficacy of a novel herbal formulation for relieving joint discomfort in human subjects with osteoarthritis of knee. *Journal of Medicinal Food, 21*(5), 511–520.

Khojah, H. M., Ahmed, S., Abdel-Rahman, M. S., & Elhakeim, E. H. (2018). Resveratrol as an effective adjuvant therapy in the management of rheumatoid arthritis: a clinical study. *Clinical Rheumatology, 37*, 2035–2042.

Lindler, B. N., Long, K. E., Taylor, N. A., & Lei, W. (2020). Use of herbal medications for treatment of osteoarthritis and rheumatoid arthritis. *Medicines, 7*(11), 67.

Lippert, A., & Renner, B. (2022). Herb–drug interaction in inflammatory diseases: review of phytomedicine and herbal supplements. *Journal of Clinical Medicine, 11*(6), 1567.

Mahajan, S. G., Mali, R. G., & Mehta, A. A. (2007). Protective effect of ethanolic extract of seeds of *Moringa oleifera* Lam. against inflammation associated with development of arthritis in rats. *Journal of Immunotoxicology, 4*(1), 39–47.

Mahboubi, M., Kashani, L. M. T., & Mahboubi, M. (2018). Nigella sativa fixed oil as alternative treatment in management of pain in arthritis rheumatoid. *Phytomedicine, 46*, 69–77.

Majeed, A., Majeed, S., Satish, G., Manjunatha, R., Rabbani, S. N., Patil, N. V., & Mundkur, L. (2024). A standardized *Boswellia serrata* extract shows improvements in knee osteoarthritis within five days-a double-blind, randomized, three-arm, parallel-group, multi-center, placebo-controlled trial. *Frontiers in Pharmacology, 15*, 1428440.

Majeed, M., Majeed, S., Narayanan, N. K., & Nagabhushanam, K. (2019). A pilot, randomized, double-blind, placebo-controlled trial to assess the safety and efficacy of a novel *Boswellia serrata* extract in the management of osteoarthritis of the knee. *Phytotherapy Research, 33*(5), 1457–1468.

Mariano, A., Di Sotto, A., Leopizzi, M., Garzoli, S., Di Maio, V., Gullì, M., … Scotto d'Abusco, A. (2020). Antiarthritic effects of a root extract from *Harpagophytum procumbens* DC: novel insights into the molecular mechanisms and possible bioactive phytochemicals. *Nutrients, 12*(9), 2545.

Mehmood, H., Saleem, A., Akhtar, M. F., & Mobashar, A. (2025). Amelioration of Freund's adjuvant-induced polyarthritis by Diosmin-loaded nanoparticles in Wistar rats; a mechanistic study. *Inflammopharmacology, 33*(6), 3099–3117.

Neidhart, S., & Neidhart, M. (2019). Rheumatoid arthritis and the concept of autoimmune disease. *International Journal of Clinical Rheumatology, 14*(2), 75–79.

Niyomchan, A., Chatgat, W., Chatawatee, B., Keereekoch, T., Issuriya, A., Jaisamut, P., … Kunworarath, N. (2023). Safety evaluation of the polyherbal formulation NawaTab: acute and subacute oral toxicity studies in rats. *Evidence-Based Complementary and Alternative Medicine, 2023,* 1–11.

Parada-Turska, J., Zgrajka, W., & Majdan, M. (2013). Kynurenic acid in synovial fluid and serum of patients with rheumatoid arthritis, spondyloarthropathy, and osteoarthritis. *The Journal of Rheumatology, 40*(6), 903–909.

Pirouzpanah, S., Mahboob, S., Sanayei, M., Hajaliloo, M., & Safaeiyan, A. (2017). The effect of chamomile tea consumption on inflammation among rheumatoid arthritis patients: randomized clinical trial. *Progress in Nutrition, 19*(1-S), 27–33.

Qindeel, M., Ullah, M. H., & Ahmed, N. (2020). Recent trends, challenges and future outlook of transdermal drug delivery systems for rheumatoid arthritis therapy. *Journal of Controlled Release, 327*, 595–615.

Radu, A.-F., & Bungau, S. G. (2021). Management of rheumatoid arthritis: an overview. *Cells, 10*(11), 2857.

Rajguru, N., Rajput, N., Sonawane, M., & Agrawal, S. (2023). Use of herbal medications for treatment of osteoarthritis and rheumatoid arthritis. *International Journal of Pharmaceutical Sciences, 1*(12), 1–1.

Ramakanth, G., Kumar, C. U., Kishan, P., & Usharani, P. (2016). A randomized, double blind placebo controlled study of efficacy and tolerability of *Withaina somnifera* extracts in knee joint pain. *Journal of Ayurveda and Integrative Medicine, 7*(3), 151–157.

Rondanelli, M., Riva, A., Allegrini, P., Faliva, M. A., Naso, M., Peroni, G., … Iannello, G. (2020). The use of a new food-grade lecithin formulation of highly standardized ginger (*Zingiber officinale*) and *Acmella oleracea* extracts for the treatment of pain and inflammation in a group of subjects with moderate knee osteoarthritis. *Journal of Pain Research, 12*, 761–770.

Sailaja, A. K., Rani, E. V., & Swetha, V. (2020). Herbal medicine for the treatment of rheumatoid arthritis. *Journal of Physics & Optics Sciences, 114*, 3.

Saller, R., Melzer, J., & Felder, M. (2008). Pain relief with a proprietary extract of Willow bark in rheumatology. An open trial. *Schweizerische Zeitschrift Für Ganzheitsmedizin/Swiss Journal of Integrative Medicine, 20*(3), 156–162.

Sharma, S., Arif, M., Nirala, R. K., Gupta, R., & Thakur, S. C. (2016). Cumulative therapeutic effects of phytochemicals in *Arnica montana* flower extract alleviated collagen-induced arthritis: inhibition of both pro-inflammatory mediators and oxidative stress. *Journal of the Science of Food and Agriculture, 96*(5), 1500–1510.

Shoara, R., Hashempur, M. H., Ashraf, A., Salehi, A., Dehshahri, S., & Habibagahi, Z. (2015). Efficacy and safety of topical *Matricaria chamomilla* L.(chamomile) oil for knee osteoarthritis: a randomized controlled clinical trial. *Complementary Therapies in Clinical Practice, 21*(3), 181–187.

Sholikhah, E., Mustofa, M., Nugrahaningsih, D., Yuliani, F., Purwono, S., Sugiyono, S., … Santosa, D. (2020). Acute and subchronic oral toxicity study of polyherbal formulation containing *Allium sativum* L., terminalia bellirica (Gaertn.) roxb., curcuma aeruginosa roxb., and amomum compactum sol. ex. maton in rats. *BioMed Research International, 2020,* 1–9.

Singh, S., Singh, T. G., Mahajan, K., & Dhiman, S. (2020). Medicinal plants used against various inflammatory biomarkers for the management of rheumatoid arthritis. *Journal of Pharmacy and Pharmacology, 72*(10), 1306–1327.

Tahir, M., Saleem, A., & Akhtar, M. F. (2025). Diosgenin loaded-chitosan biodegradable nanoparticles ameliorate adjuvant-induced arthritis, pain, and peripheral neuropathy through moderation of inflammatory and oxidative stress biomarkers. *International Journal of Biological Macromolecules, 290*, 138926.

Tang, W., & Zuo, J.-P. (2012). Immunosuppressant discovery from *Tripterygium wilfordii* Hook F: the novel triptolide analog (5R)-5-hydroxytriptolide (LLDT-8). *Acta Pharmacologica Sinica, 33*(9), 1112–1118.

Yarnell, E. (2017). Herbs for rheumatoid arthritis. *Alternative and Complementary Therapies, 23*(4), 149–156.

Zhang, W., & Dai, S.-M. (2012). Mechanisms involved in the therapeutic effects of *Paeonia lactiflora* Pallas in rheumatoid arthritis. *International Immunopharmacology, 14*(1), 27–31.

Zielińska, M., Dereń, K., Polak-Szczybyło, E., & Stępień, A. E. (2021). The role of bioactive compounds of *Nigella sativa* in rheumatoid arthritis therapy—current reports. *Nutrients, 13*(10), 3369.

Section C

Nutraceuticals and Toxicity of Phytochemicals

12 Phytochemicals as Nutraceuticals in the Management of Rheumatoid Arthritis

Shakila Sabir, Muhammad Naeem, Ammara Saleem, Ghulam Md Ashraf, and Muhammad Furqan Akhtar

WHAT WILL YOU LEARN FROM THIS CHAPTER?

This chapter provides you with an in-depth exploration of potential phytochemicals for the treatment of rheumatoid arthritis (RA). By the end of this chapter, you will learn that which phytonutrients are used in the treatment of RA with special emphasis on the molecular mechanisms that are responsible for exerting their immunomodulatory and anti-inflammatory effects. An opportunistic approach to future directions toward the effective use of these phytochemicals, in addition to a thorough analysis based on preclinical and clinical studies.

LIST OF ABBREVIATIONS

COX-2	Cyclooxygenase-2
ERK-1/2	Extracellular signal-regulated kinase ½
H1F-1α	Hypoxia-inducible factor 1-alpha
IL-1	Interleukin-1
JAK2/STAT3	Janus kinase/signal transducer and activator of transcription
JNK	Jun-N terminal kinase
MAPK	Mitogen-activated protein kinase
MMP-1	Matrix metalloproteinase-1
NF-κB	Nuclear factor kappa beta
PGE2	Prostaglandin E2
ROS	Reactive oxygen species
TH	Helper T cells
TNF	Tumor necrosis factor
Treg cell	Regulatory T cells
VEGF	Vascular endothelial growth factor
VEGFR	Vascular endothelial growth factor receptor

12.1 NUTRACEUTICALS

Biologically active substances found in the food are frequently referred to as nutraceuticals (Puri et al., 2022), because these substances have the characteristics of both nutrients and drugs. In addition to their nutritional values, these biologically active substances also have the ability to promote health, along with the prevention and treatment of diseases (Chopra et al.,2022). Nutraceuticals include

DOI: 10.1201/9781003535201-15

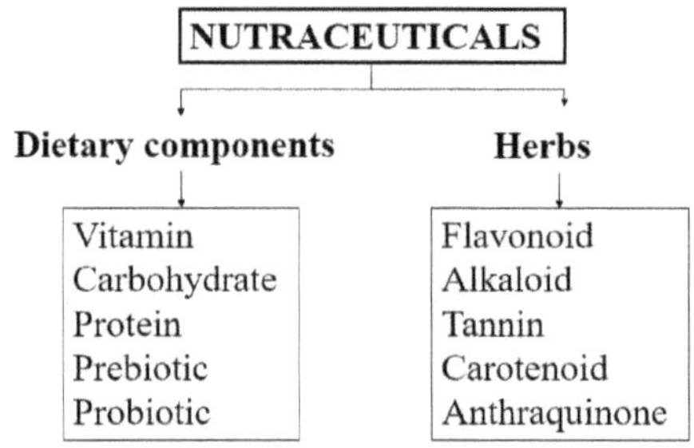

FIGURE 12.1 Types of nutraceuticals.

carbohydrates, lipids, proteins, minerals, vitamins, or some other essential nutrients. Furthermore, antioxidants, dietary fibers, prebiotics, probiotics, mineral supplements, and herbs also come under the category of nutraceuticals. Nutraceuticals have shown promise as therapeutic options for a number of illnesses, including arthritis, atherosclerosis, diabetes, hypertension, cancer, neurological, and cardiovascular disorders. The World Health Organization maintains the Global Health Expenditure Database that shows that the United States has invested 17% of its GDP in the health sector (Sachdeva, Roy, & Bharadvaja, 2020). The massive burden of healthcare expenditure has directed the interest of research and development sectors of numerous multinational companies toward nutraceuticals with promising health benefits (Hussain, Panjagari, Singh, & Patil, 2015). The nutraceuticals business is still in its early stages, despite its rapid expansion, given the quantity of marketed products and its contribution to nation's GDP. A minute quantity of bioactive substances is found in the food items including fruits, vegetables, and whole grains. Beyond the fundamental nutritional worth, nutraceuticals have numerous health benefits and potential pharmaceutical uses (Banwo et al., 2021). Nutraceuticals are categorized under dietary supplements and herbal sources based on where they come from. For centuries, special attention has been given to herbs because of the wide range of flavors and aromas that herbs have to offer, and they have been used in cooking for millennia in addition to being used medicinally. The bioactive substances obtained from the herbs include carotenoids, flavonoids, coumarins, lignans, sterols, phthalates, polyphenols, sulfides, saponins, and terpenoids (Roy & Bharadvaja, 2018; Sen & Samanta, 2015; Sharma & Sarkar, 2013) (Figure 12.1). These nutraceuticals are preferable over the synthetic drugs in the context of managing rheumatoid arthritis (RA) because nutraceuticals have fewer side effects when used within the recommended dose range, while synthetic drugs like NSAIDs and corticosteroids have various side effects upon persistent use like immunosuppression, hepatotoxicity, and gastrointestinal side effects. Additionally, nutraceuticals have multiple beneficial effects at the same dose level like curcumin, and they have a good anti-inflammatory effect along with promising antioxidant properties. If nutraceuticals are used as adjuvant therapy, they are helpful to reduce the required doses of synthetic drugs with reduced toxicity (Sharma, Chaubey, & Suvarna, 2021).

Nutraceuticals play a crucial part in promoting health and improving the quality of life. There are some other alternative terms for nutraceuticals, common in the United States and Europe, including pharmaceutical food/pharma food, functional food, food additives, therapeutic food, preventive substances, or phytochemicals (Sawicka, Ziarati, Krochmal-Marczak, & Skiba, 2019). Further investigation through studies performed *in vitro* and *in vivo* is required for the evaluation of the potential mechanism of action, efficacy, and toxicity profile of these bioactive substances in order to develop the commercial nutraceutical products.

12.2 CURCUMIN

The perennial herb turmeric is a member of the Zingiberaceae family. An important bioactive constituent of turmeric is curcumin. It is well documented that curcumin has strong antioxidant, anti-inflammatory, anti-angiogenic, and anti-cancer activities with mild adverse effects (Hewlings

& Kalman, 2017). Literature has shown that curcumin has a promising role in the treatment of multiple diseases including diabetes, asthma, obesity, depression, anxiety, cancer, and cardiovascular disorders (Shah et al., 2022). The expression of interleukin (IL)-1β, IL-6, IL-8, IL-12, and TNF (tumor necrosis factor)-α as pro-inflammatory biomarkers as well as monocyte chemoattractant protein-1 (MCP-1) is strongly prevented by curcumin, as traditionally evident from the literature, in different immune cells and various animal models (Gorabi et al., 2021; Rahardjo, Widjajanto, Sujuti, & Keman, 2014).

Lee and his colleagues treated synovial fibroblast (SF) cells stimulated by horbol-12-myristate acetate with, [2,6-bis[2,5-dimethoxybenzylidence] cyclohexanone], a curcumin analog. This curcumin analog acted in a dose-dependent manner and targeted signaling pathway involving nuclear factor kappa B (NF-κB) and caused a significant reduction in the exaggerated expression of IL-6 as well as COX-2 (Lee et al., 2015). Another *in vitro* study exhibited that curcumin also causes significant downregulation of signaling pathway, involving TLR4/ NF-κB, which in turn reduced the production of TNF-α as well as IL-1β primarily in a dose-dependent pattern (Li et al., 2019).

An animal-based study conducted on a mouse psoriasis model exhibited that curcumin considerably suppressed the proliferation of TH (Helper T)17 cells up to 50% and reduced the expression of IL-17, IL-22, and TNF-α as inflammatory cytokines by 60%. In animals treated with curcumin, a marked reduction was also seen in the level of some other cytokines including IL-22, IL-23, and IFN-γ (Kang et al., 2016). More interestingly, Handono et al. suggested that inflammatory reactions induced by TH17 are significantly prevented by curcumin, which inhibits the differentiation of naïve CD4+ cells into TH17 subset of helper T cells. Significant improvement in autoimmune disorders was found to be associated with these advanced findings (Dolati et al., 2018; Handono, Pratama, Endharti, & Kalim, 2015). Handono et al also recommended that pronounced depression in the percentage of TH17 cells was seen by exposure of curcumin to naïve CD4+ cells obtained from systemic lupus erythematosus (SLE) patients (Handono et al., 2015) (Table 12.1).

TABLE 12.1
Potential Phytochemicals Used to Treat Rheumatoid Arthritis

Compound	Source	Dose/Duration	Animal Model/ Cell Line	Mechanism	Reference
Curcumin	*Curcuma longs*, *Curcuma longa*	200 mg/kg for 3 weeks	Rats	↓IL-1β ↓TNFα ↓MMP-1 ↓MMP-3	Dai, Zhou, Xu, and Song (2018)
Apigenin	*Petroselinum crsipum*, *Matricaria recutia*	20 mg/kg for 50 days	Mice	↓VEGF and VEGFR ↓Synovial hyperplasia ↓Angiogenesis ↓Osteoclastofenesis	
Ferulic acid		25, 50, 100 μM for 24 hours	AA FLS	↓NF-κB, ↑Osteoprotegerin ↓osteoclast cells	
Gallic acid	*Quercus infectoria*, *Ananas comosus*	40 and mg/kg for 21 days	Mice	↓ Arthritis score ↓ Hand paw volume ↓TH17 cells ↑Treg cells ↓MMPs level	Liu et al. (2023)

(*Continued*)

TABLE 12.1 (*Continued*)
Potential Phytochemicals Used to Treat Rheumatoid Arthritis

Compound	Source	Dose/Duration	Animal Model/ Cell Line	Mechanism	Reference
Genistein	*Genista tinctoria*	25 μM/L for 24 hours	MH7A cells	↓VEGF through JAK2/STAT3 pathway	Cheng et al. (2020)
EGCG	*Camellia sinensis*	5-20 μM for 24 hours	RA SF	↓IL-6, IL-8 , MMP-2 ↓NF-κB, p38	Fechtner, Singh, Chourasia, and Ahmed (2017)
Kaempferol	*Vitis infectoria, Ginko biloba, Solanum lycopersicum*	100 μM for 2 days	Synovial tissues	↓MMP1 ↓MMP-3 ↓COX-2 ↓PGE2 ↓Phosphorylation of ERK-1/2, JNK, and p38	Yoon et al. (2013)
Resveratrol	*Polygonum cuspidatum,* veratrum gradiflorum, *Arachis hypogaea*	200 and 400 mg/bw for 21 days	Rats	↓H1F-1α ↓MAPK ↓Synovial hyperplasia ↓ROS	Yang et al. (2018)

IL-1: interleukin-1; TNF: tumor necrosis factor; MMP-1: matrix metalloproteinase; VEGF: vascular endothelial growth factor; VEGFR: vascular endothelial growth factor receptor; NF-κB: nuclear factor kappa beta; TH: helper T cells; Treg cell: regulatory T cells; JAK2/STAT3: janus kinase/signal transducer and activator of transcription; COX-2: cyclooxygenase-2; PGE2: prostaglandin E2; ERK-1/2: extracellular signal-regulated kinase ½; JNK: Jun-N terminal kinase; H1F-1α: hypoxia-inducible factor 1-alpha; MAPK: mitogen-activated protein kinase; ROS: reactive oxygen species.

Another *in vitro* study showed that when T cells obtained from mice suffering from psoriasis were exposed to curcumin (10 μM) 60% reduction in the expression of IL-17, IL-22, and TNF-α was seen. Furthermore, curcumin at a dose of 100 μM was found to cause up to 50% reduction in the proliferation of TH17 cells (Kang et al., 2016).

Larmonier et al. conducted an *in vivo* study based on the colitis animal model. The study suggested that curcumin would be a promising active phyto-constituent in order to attenuate the progression and severity of the disorder by reducing the production of related cytokines, especially IFN-γ, primarily by virtue of its ability to halt the proliferation of TH cells (Larmonier et al., 2008). The results of the aforementioned study were supported by another animal model-based *in vivo* study on encephalomyelitis. Curcumin treatment attenuated the symptoms of encephalomyelitis by reducing the TH17 cells differentiation and function. Furthermore, in dendritic cells, the level of Heme Oxygenase-1 (HO-1) and mRNA was considerably reduced, thus subsequently increasing the phosphorylation of STAT3. The consequent inhibition of immune response mediated by TH17 was found to be associated with the enhanced production of anti-inflammatory dendritic cells in addition to markedly reduced expression of IL-12 (Brück et al., 2017; Kanakasabai et al., 2012).

Rogers and his co-workers conducted a study in order to investigate the effects of curcumin on Treg cells. The study exhibited that curcumin caused the induction of immature dendritic cells, thus mediated the immunomodulatory effects. The immature state of dendritic cells showed reduced expression of crucial co-stimulatory biomarkers including CD80, CD86, and CD40 and also affected the process of T cell priming by causing a marked reduction in the expression of IL-12. Therefore, there was an enhanced production of Treg cells and a reduction in the differentiation and activation of naïve T cells (Mahnke, Johnson, Ring, Enk, 2007; Raker, Domogalla, & Steinbrink, 2015). It is well documented that the imbalance of TH17/Treg contributed to the immunological intolerance in RA and SLE. Thus, there is a need to identify active constituents with the potential to maintain TH17/Treg balance in order to successfully treat autoimmune disorders. By keeping in view this crucial mechanistic approach, Handono et al. designed the treatment of the CD4+T cell with low-dose curcumin, where the cell culture was obtained from SLE patients and the dose of curcumin was 0.1 and 1 µg/ml. Curcumin efficiently restored the TH17/Treg balance without any adverse effect. Curcumin treatment decreased the TH17 cells, on the other hand, the population of Treg cells as well as CD4+ TGF-1ß producer cells has been found considerably increased (Handono et al., 2015).

Wang et al. projected this evidence that the differentiation of B cells into their respective subsets, i.e., B10 was induced by curcumin. B10 subset suppressed the immune system in the animal model of myasthenia gravis. Furthermore, curcumin elevated the level of anti-inflammatory mediator, namely IgG1 antibody, and reduced the level of pathogenic mediator, namely IgG2b antibody (Wang et al., 2016).

Curcumin downregulated the production of certain biomarkers belonging to TNF family, which are involved in activation of B cells (B-cell activating factor: a cytokine, crucially important for development, process of auto-antibody production, and survival of B lymphocytes) with subsequent suppression of autoreactive B cells's activation and reduction in the development of various auto-antibodies, which are produced in RA (de la Torre et al., 2010; Huang et al., 2013). An *in vitro* study based on curcumin's immunomodulatory properties revealed that it can prevent B cells from activating and proliferating as well as from producing several antibodies, including IgG2a (Sharma, Chopra, Kulkarni, & Agrewala, 2007). Jacob and his colleagues reported that curcumin treatment reduced the complement activation. The ratio of CD19/CD3 cells and the overall number of CD19+ cells were found to be reduced after treatment with curcumin, with no significant change in the population of T cells (Jacob et al., 2013).

A randomized clinical trial on phytocompounds was conducted in which one group of RA patients was treated with curcumin, the second group with diclofenac sodium, and the third group was treated with a combination of both. The findings indicated more improvement in the curcumin-treated groups as compared to the other two groups with reduced side effects (Chandran & Goel, 2012). In another clinical study, three months' treatment with high-dose CuroWhite (curcuminoid formulation) considerably reduced biochemical parameters including C-reactve protein and erythrocyte sedimentation rats in RA patients in contrast to placebo control (Jacob et al., 2019). In a previous randomized clinical study, curcumex (a mixture of ginger and curcumin) was provided to RA patients for two months. The findings revealed improvement in disease severity and extra-articular manifestations (Hemmati et al., 2016).

In a randomized clinical study, nano-micelles of curcumin were administered to RA patients for 12 weeks. The findings exhibited reduced swelling and pain in arthritics with no adverse reactions (Javadi, Khadem Haghighian, Goodarzy, Abbasi, & Nassiri-Asl, 2019). In another clinical study, a curcuminoid-rich turmeric formulation was administered in a placebo-controlled double-blind study, which revealed improvement in RA patients (Amalraj et al., 2017).

12.3 APIGENIN

Apigenin is a 4′,5,7-trihydroxy flavone. It is a conjugate acid of an apigenin-7-olate. Apigenin is considered one of the most widely distributed flavonoids in the plant kingdom. It is among those phenolic compounds, which have been studied the most. Apigenin is mostly present in the glycosylated form in a considerable amount of vegetables (celery, parsley, and onion), fruits (orange), herbs (chamomile, oregano, basil, and thyme), and plant-based beverages (wine, beer, and tea) (Salehi et al., 2019).

Apigenin, in a dose-dependent manner, reduced the production of nitric oxide (NO) stimulated by lipopolysaccharide in macrophage cells and reduced the expression of cyclooxygenase-2 (COX-2). In LPS-stimulated mouse macrophages and human monocytes, the expression of inflammatory cytokines was also inhibited by apigenin. NF-κB activity by apigenin is regulated by inactivation of LPS-stimulated IKK complex and phosphorylation of serine (Ser) 536 located in p65 subunit (Nicholas et al., 2007).

The results of the aforementioned study were also supported by another *in vivo* study, in which apigenin caused a considerable reduction in NF-κB and TNF-α. The marked decline in transcriptional activity of GAL4-NF-κB p65 fusion protein with apigenin was also found. Additionally, reduction in janus kinases (JNKs) activation, CXCL1/KC (keratinocyte chemoattractant), and CCL2/MCP-1, induced by TNF-α, was seen with apigenin, but it showed no impact on p38 and extracellular signal-regulated kinase (ERK) pathway (Funakoshi, Nakamura, Tago, Mashino, & Kasahara, 2011). In RA-fibroblast-like synoviocytes (FLS) cells, it also exerted a profound inhibitory effect on the expression of TNF and matrix metalloproteinase (MMP)-1 and MMP-3. Furthermore, in arthritic rats, the severity of arthritis was significantly ameliorated by apigenin-4'-O-α-L-rhamnoside (Cao et al., 2022).

Maslikah et al. conducted a molecular docking study and reported that apigenin interacted with the active site in COX-2 enzymes similar to rofecoxib, as shown by the presence of Ile 517 amino acid residues. These residues bind to hydrogen bond and hydrophobic type interaction with amino acids including Val 523, Val 349, and Leu 352.

12.4 FERULIC ACID

Ferulic acid is a polyphenol and is abundant in citrus fruits, vegetables, grains, and coffee. The promising antioxidant, anticancer, antidiabetic, and anti-inflammatory activities of ferulic acid are evident from the literature.

In adjuvant-induced arthritis rats, ferulic acid profoundly reduced the expression of transmembrane receptors stimulated by IL-17. Ferulic acid also found to inhibit the signaling pathway involving IL-17/IL-17RA/STAT-3, which in turn caused hindrance in the expression of IL-23, TLR-3, Cyr61, and GM-CSF in FLS. Furthermore, ferulic acid also reduced the production of osteoclast cells and resorption of bones. Regulation of the signaling pathway based on IL-17/IL-17RA/STAT-3 by ferulic acid resulted in the enhancement of osteoprotegerin expression as well as a reduction in the expression of IL-17-induced receptor activation of nuclear factor-κB ligand (RANKL). Ganesan and his co-worker also conducted docking analysis, which strongly suggested that ferulic acid manifested ligand efficiency toward various proteins including STAT-3, IL-23, IL-17RA, and RANKL (Ganesan & Rasool, 2019). Ferulic acid also exhibited the anti-arthritic potential in the CFA-induced arthritis model via modulation of two key cytokines including TNFα and TGF-β, with subsequent inhibition of the JAK2 level (Zhu et al., 2020). In an earlier study, a phenolic substance-rich fraction consisting of ferulic acid was administered to adjuvant-induced arthritis animals and it was found that COX-2 expression was significantly inhibited by ferulic acid dimer (Hirata et al., 2005).

Qin and his colleagues reported that sodium ferulate has anti-inflammatory and antiapoptotic effects on the IL-1β-induced osteoarthritic model. Inhibition of the signal transduction pathway based on IKK/NF-κB and apoptosis mediated by the caspase-dependent pathway by sodium ferulate is thought to be responsible for the aforementioned effects (Qin et al., 2013). In another study, sodium ferulate effectively attenuated acetic acid-induced inflammatory reactions in an animal model. Anti-inflammatory action was reflected from reduced myeloperoxidase, SOD, MDA, NO, and PGE2 levels in these animals (Dong et al., 2003).

Sodium ferulate also exhibited the reversal of the degradation process of cartilage. The expression of BAX and MMP-1 in a rat osteoporosis model was also significantly inhibited. Sodium ferulate also prevented chondrocyte apoptosis and NO synthesis and attenuated the MMP-1/tissue inhibitor of MMP-1 in order to inhibit the degradation of the extracellular matrix in chondrocytes (Shang et al., 2009).

12.5 GALLIC ACID

Gallic acid is a hydroxybenzoic acid. Gallic acid is abundantly found in processed beverages like green tea and red wine. Gallic acid is widely distributed in the plant kingdom and represents the large family of secondary metabolites of plants (Rosas, Correa, das Graças Henriques, et al., 2019).

Literature exhibited that gallic acid induced apoptosis in RA-FLS via regulating the pro-apoptotic protein expression. It regulated the expression of p53, Bax, Bcl-2, and pAkt as well as induction of caspase-3 at the doses of 0.1 and 1M. In addition to MMP-9 and COX-2, it also suppressed IL-1 and IL-6 and CCL-7/MCP-3 and CCL-2/MCP-1 by suppressing the level of mRNA expression of proinflammatory cytokines and chemokines, respectively (Yoon et al., 2013).

Gallic acid treatment hindered the overexpression of MMPs and modified the dysregulated inflammatory cytokines. Furthermore, gallic acid also reduced the population of TH17 cells and elevated the population of Treg cells (Liu, Li, Feng, 2023). Bai et al. reported that gallic acid stimulated cAMP-mediated AMPK/SIRT1/NFκB signaling pathway via suppression of phosphodiesterase (PDE) 4 activity. These modifications brought huge improvement in local inflammatory responses (Bai et al., 2021). It is well documented that the pathogenesis of RA resembles cancer in a few aspects like malignant proliferation and gallic acid has the potential to hinder the aberrant proliferation (Cheng et al., 2020).

12.6 GENISTEIN

Genistein is an isoflavone that belongs to the flavonoid family. It is regarded as a phytoestrogen as it shares the structural similarities with mammalian estrogen. It is mainly obtained from legumes including lupine, fava bean, soya bean, willd, kudzu, and psoralea (Goh, Jalil, Lam, Husain, & Premakumar, 2022). Genistein has received special attention from researchers over the last decade because of its promising therapeutic effects in degenerative diseases like cardiovascular diseases, hormone-related cancers, and osteoporosis. Genistein shows weak estrogenic effects and well-documented nonspecific tyrosine kinase inhibitory activity at therapeutic doses (Chakraborty, Gupta, Biswas, 2021; Li et al., 2013). Genistein induced anti-inflammatory and antiangiogenic responses in a collagen-induced RA model by hindering the expression of IL-6, vascular endothelial growth factor (VEGF), and TNF-α (Hu et al., 2016). Literature has reported that overactivation of STAT3 may considerably stimulate the production of VEGF and promote angiogenesis. These alterations can be alleviated by suppressing the STAT3 activation (Chen & Han, 2008; Wei et al., 2003). Furthermore, genistein induced inhibition of VEGF expression in RA by IL-6-mediated JAK-STAT3/VEGF signaling cascade (Cheng et al., 2020). Additionally, genistein may inhibit the growth of RA-FLS under the stimulation of IL-1β and TNFα. This inhibitory effect resulted in reduced cell number owing to cell-cycle arrest, particularly at the G1 restriction point. Genistein also reduced colony number under anchorage-independent conditions (Zhang et al., 2012).

Genistein modulated TH1-predominant immune response with considerable suppression of IFN-γ secretion in a collagen-induced RA model. Furthermore, genistein also elevated IL-4 production and maintained TH1/TH2 balance (Wang et al., 2008). Another study reported genistein-mediated reduction in the expression of IL-6, formerly stimulated by the LPS, TNFα and activation of NF-κB in the respective macrophage cell line (Ji et al., 2012). Hamalainen et al. reported a marked reduction in the level of inducible nitric oxide synthase (iNOS) and thus subsequent reduction was found in a dose-dependent manner in the levels of NO with the use of genistein along with the suppression of STAT-1 and NF-κB activation (Hämäläinen, Nieminen, Vuorela, Heinonen, & Moilanen, 2007). Genistein also attenuated TNFα-mediated production of chemokines and adhesion molecules (Jia et al., 2013). Another study suggested that genistein showed an antiarthritic effect in RA-FLS via inhibition of MMP-9 and FLS proliferation at a dose of 10 mg/mL (Zhang et al., 2012).

12.7 EPIGALLOCATECHIN-3-GALLATE (EGCG)

It makes 50%–80% catechin content in green tea or 200–300 mg per brewed cup of green tea. EGCG is an ester of gallic acid and epigallocatechin. EGCG has a wide range of therapeutic effects, such as antidiabetic, antioxidant, anti-Parkinson, stroke, and antiobesity (Sharifi-Rad et al., 2020). Literature has shown that EGCG attenuated collagen and IL-1β-mediated inflammatory responses. For the first time, Haqqi et al. reported an antiarthritic effect of EGCG when EGCG-rich green tea was administered to a collagen-induced arthritic (CIA) animal model. The antiarthritic effect produced by the inhibition of TNFα, COX-2, and INFγ is associated with the consequent inhibition of inflammatory mediators. Moreover, EGCG treatment also causes an overall reduction of IgG antibodies as well as type II collagen-specific IgG in both serum and joints (Haqqi et al., 1999). When arthritic cartilage-derived chondrocytes were incubated with EGCG, there was a marked inhibition of iNOS and COX-2 expression with subsequently reduced NO and prostaglandin (PG) E2 synthesis (Singh, Ahmed, Islam, Goldberg, & Haqqi, 2002).

Lin et al. reported an antiarthritic activity of EGCG. The findings of the study suggested that EGCG suppressed macrophage infiltration. EGCG also reduced MCP-1/CCL2-synthesizing osteoblasts (Lin et al., 2008).

Various studies based on *in vitro* and *in vivo* models have highlighted the potential role of EGCG as an antiarthritic and anti-inflammatory agent. The expression of various chemokines and cytokines, such as MMPs, reactive oxygen species (ROS), aggrecanase, COX-2, PGE-2, and NO, is significantly regulated by EGCG as they play a pivotal role in the pathogenesis of RA and OA (Singh, Akhtar, & Haqqi, 2010). Additionally, the bone-preserving and SF regulatory effects of EGCG have also been reported (Ahmed, 2010). The literature has reported that EGCG significantly reduced ROS-induced cytotoxicity, particularly in chondrocytes and some other models of oxidative stress (Meng, Velalar, & Ruan, 2008).

Nrf-2 activated antioxidant gene transcription while increasing HO-1 expression. HO-1 has an important role in the prevention of joint damage and the regulation of various crucial cytokines. Nrf-2 deficiency resulted in augmented levels of TNFα and IL-6 (Maicas et al., 2011). A study has reported that there was an increased expression of Nrf-2 and HO-1 in the joints of the EGCG-treated group (Karatas et al., 2020).

12.8 KAEMPFEROL

Kaempferol belongs to the flavonol class of flavonoids. It is found in different commonly used herbs and vegetables like tomatoes, grapes, broccoli, tea, and ginkgo biloba leaves. A wide range and variety of pharmacological properties, i.e., anticancer, antidiabetic, antimicrobial, anti-inflammatory, and antioxidant, have been found to be possessed by kaempferol (Alam, Khan, Shah, Cauli, & Saso, 2020). The literature has suggested that kaempferol reduced the invasion, migration, and MMP expression in RA-FLS. Furthermore, kaempferol in arthritic animals ameliorated the disease severity of RA (Pan et al., 2018).

It has been found that in the case of RS (basic fibroblast growth factor [bFGF]) the level of synovial fluid basic FGF (bFGF) was considerably higher in comparison to the patients of osteoarthritis. Activation of bFGF-FGF receptor 3(FGFR3)-ribosomal S6 kinase 2 signaling axis causes bFGF-stimulated migration and proliferation of FLS. Lee and his co-worker have recently conducted a molecular docking study on kaempferol. The study suggested that kaempferol has binding affinity with the FGFR3 kinase domain and it inhibited FGFR3 activity by attaching and blocking its active pockets. Kaempferol made a hydrogen bond with both Ala558, which is the FGFR3 backbone oxygen and Glu555 as well as with a side chain of Lys508. Blocking of signaling pathway based on bFGF-FGFR3-RSK2 by kaempferol resulted in reduced migration and proliferation of RA-FLS and also suppressed the release of TNF-α, IL-17, and IL-21, which are inflammatory cytokines mediated by T-cell activation (Lee et al., 2018). It is evident from the literature review that kaempferol inhibits the proliferation of RA SFs. Phosphorylation of JNK, ERK-1/2, and p38 and activation of NF-κB were also hindered by kaempferol (Yoon et al., 2013).

Recently, a metabolomic study was conducted that revealed another interesting pathway for anti-arthritic effects of kaempferol. Kaempferol helped rebalance the microbial metabolism and gut flora. It helped to replenish the metabolism of tryptophan, fatty acids, and bile acids in the gut of CIA mice. It is a well-documented bile acid that is strongly linked to immune inflammation. In the presence of gut flora, the metabolic conversion of primary bile acids is done into secondary bile acids. The level of cholic and deoxycholic acid is increased, while the level of glycolic acid is decreased in RA. Deoxycholic acid has been recognized as a secondary bile acid primarily metabolized by gut microbiota. Deoxycholic acid is a crucial metabolite in the development of inflammation in arthritis. Kaempferol reduced the production of deoxycholic acid with minor effects on cholic acid and glycolic acid (Aa et al., 2020).

Xian and his co-workers have reported the suppression of p16 and p21 by kaempferol, which are senescence-related markers. Kaempferol also reduced the expression of IL-6. The level of β-galactosidase in RAW264.7 cells was found to be promoted by former treatment with LPS. This destructive alteration was reversed by kaempferol treatment. Thus, kaempferol prevented the premature aging of immune cells by regulating the expression of senescence-related genes, i.e., p16/p21, IL-1β, IL-6, and TNF-α (Xian et al., 2023). In an earlier study, Lee and his colleagues suggested that phosphorylation of p38 is mediated by RANKL. Treatment with kaempferol prevented the bone loss by reducing the formation of osteoclasts. Prevention of bone destruction was a key contributing factor for designing novel treatment protocols of inflammatory arthritis (Lee, Lee, Sung, & Yoo, 2014).

12.9 RESVERATROL

Resveratrol belongs to the polyphenol group of phytochemicals. It is found abundantly in grape seeds, skin, and red wine. It has a well-documented potent, antioxidant, anti-inflammatory, and anticancer activity. Literature has shown that it can inhibit NF-κB, MAPK, and some other signal transduction pathways. Thus, it has the potential to inhibit the production as well as the release of proinflammatory cytokines, ILs, and MMPs. Resveratrol can also induce the apoptosis of FLS, MH7A, and other inflammatory cells, which play a crucial role in RA. It has the potential to regulate the function and differentiation of macrophages. It normalized the mitochondrial redox kinetics. The recent literature has suggested that resveratrol may alleviate synovial inflammation, proliferation, and cartilage destruction. More interestingly, resveratrol also has beneficial effects in complications associated with RA like periodontitis (Dudics et al., 2018; Sheng et al., 2022) (Table 12.1).

12.10 CONCLUSION

A number of above-mentioned studies have clearly shown that nutraceuticals have great potential to treat RA. They have excellent anti-inflammatory and immune-modulatory activity and mechanisms behind these effects have been studied extensively. But data regarding the clinical perspective is limited. There is an intensive need to conduct clinical trials and validate their effective and toxic doses. There is also a great need to develop appropriate dosage forms with enhanced bioavailability of these nutraceuticals.

CONFLICT OF INTEREST

The authors declare no conflict of interest.

REFERENCES

Aa, L.-X., Fei, F., Qi, Q., Sun, R.-B., Gu, S.-H., Di, Z.-Z., … Liu, C.-X. (2020). Rebalancing of the gut flora and microbial metabolism is responsible for the anti-arthritis effect of kaempferol. *Acta Pharmacologica Sinica, 41*(1), 73–81. https://doi.org/10.1038/s41401-019-0279-8

Ahmed, S. (2010). Green tea polyphenol epigallocatechin 3-gallate in arthritis: progress and promise. *Arthritis Research Therapy, 12*(2), 208. https://doi.org/10.1186/ar2982

Alam, W., Khan, H., Shah, M. A., Cauli, O., & Saso, L. (2020). Kaempferol as a dietary anti-inflammatory agent: current therapeutic standing. *Molecules, 25*(18), 4073.

Amalraj, A., Varma, K., Jacob, J., Divya, C., Kunnumakkara, A. B., Stohs, S. J., & Gopi, S. (2017). A novel highly bioavailable curcumin formulation improves symptoms and diagnostic indicators in RA patients: a randomized, double-blind, placebo-controlled, two-dose, three-arm, and parallel-group study. *Journal of Medical Food, 20*(10), 1022–1030. https://doi.org/10.1089/jmf.2017.3930

Bai, J., Zhang, Y., Tang, C., Hou, Y., Ai, X., Chen, X., … Meng, X. (2021). Gallic acid: pharmacological activities and molecular mechanisms involved in inflammation-related diseases. *Biomedicine & Pharmacotherapy, 133*, 110985. https://doi.org/10.1016/j.biopha.2020.110985

Banwo, K., Olojede, A. O., Adesulu-Dahunsi, A. T., Verma, D. K., Thakur, M., Tripathy, S., … Utama, G. L. (2021). Functional importance of bioactive compounds of foods with Potential Health Benefits: a review on recent trends. *Food Bioscience, 43*, 101320.

Brück, J., Holstein, J., Glocova, I., Seidel, U., Geisel, J., Kanno, T., … Ghoreschi, K. (2017). Nutritional control of IL-23/Th17-mediated autoimmune disease through HO-1/STAT3 activation. *Science Reports, 7*, 44482. https://doi.org/10.1038/srep44482

Cao, D., Fan, Q., Li, Z., Chen, M., Jiang, Y., Lin, R., … Zhao, C. (2022). Transcriptomic profiling revealed the role of apigenin-4′-O-α-L-rhamnoside in inhibiting the activation of rheumatoid arthritis fibroblast-like synoviocytes via MAPK signaling pathway. *Phytomedicine, 102*, 154201. https://doi.org/10.1016/j.phymed.2022.154201

Chakraborty, D., Gupta, K., & Biswas, S. (2021). A mechanistic insight of phytoestrogens used for Rheumatoid arthritis: an evidence-based review. *Biomedicine & Pharmacotherapy, 133*, 111039. https://doi.org/10.1016/j.biopha.2020.111039

Chandran, B., & Goel, A. (2012). A randomized, pilot study to assess the efficacy and safety of curcumin in patients with active rheumatoid arthritis. *Phytotherapy Research, 26*(11), 1719–1725. https://doi.org/10.1002/ptr.4639

Chen, Z., & Han, Z. C. (2008). STAT3: a critical transcription activator in angiogenesis. *Medicinal Research Reviews, 28*(2), 185–200.

Cheng, W.-X., Huang, H., Chen, J.-H., Zhang, T.-T., Zhu, G.-Y., Zheng, Z.-T., … Bai, X.-L. (2020). Genistein inhibits angiogenesis developed during rheumatoid arthritis through the IL-6/JAK2/STAT3/VEGF signalling pathway. *Journal of Orthopaedic Translation, 22*, 92–100.

Chopra, A. S., Lordan, R., Horbańczuk, O. K., Atanasov, A. G., Chopra, I., Horbańczuk, J. O., … & Arkells, N. (2022). The current use and evolving landscape of nutraceuticals. *Pharmacological Research, 175*, 106001.

Dai, Q., Zhou, D., Xu, L., & Song, X. (2018). Curcumin alleviates rheumatoid arthritis-induced inflammation and synovial hyperplasia by targeting mTOR pathway in rats. *Drug Design, Development and Therapy, 12*(null), 4095–4105. https://doi.org/10.2147/DDDT.S175763

de la Torre, I., Moura, R. A., Leandro, M. J., Edwards, J., & Cambridge, G. (2010). B-cell-activating factor receptor expression on naive and memory B cells: relationship with relapse in patients with rheumatoid arthritis following B-cell depletion therapy. *Annals of Rheumatic Disease, 69*(12), 2181–2188. https://doi.org/10.1136/ard.2010.131326

Dolati, S., Ahmadi, M., Rikhtegar, R., Babaloo, Z., Ayromlou, H., Aghebati-Maleki, L., … Yousefi, M. (2018). Changes in Th17 cells function after nanocurcumin use to treat multiple sclerosis. *International Immunopharmacology, 61*, 74–81. https://doi.org/10.1016/j.intimp.2018.05.018

Dong, W. G., Liu, S. P., Yu, B. P., Wu, D. F., Luo, H. S., & Yu, J. P. (2003). Ameliorative effects of sodium ferulate on experimental colitis and their mechanisms in rats. *World Journal of Gastroenterology, 9*(11), 2533–2538. https://doi.org/10.3748/wjg.v9.i11.2533

Dudics, S., Langan, D., Meka, R. R., Venkatesha, S. H., Berman, B. M., Che, C.-T., & Moudgil, K. D. (2018). Natural products for the treatment of autoimmune arthritis: their mechanisms of action, targeted delivery, and interplay with the host microbiome. *International Journal of Molecular Science, 19*(9), 2508.

Fechtner, S., Singh, A., Chourasia, M., & Ahmed, S. (2017). Molecular insights into the differences in anti-inflammatory activities of green tea catechins on IL-1β signaling in rheumatoid arthritis synovial fibroblasts. *Toxicology and Applied Pharmacology, 329*, 112–120. https://doi.org/10.1016/j.taap.2017.05.016

Funakoshi-Tago, M., Nakamura, K., Tago, K., Mashino, T., & Kasahara, T. (2011). Anti-inflammatory activity of structurally related flavonoids, Apigenin, Luteolin and Fisetin. *International Immunopharmacology, 11*(9), 1150–1159. https://doi.org/10.1016/j.intimp.2011.03.012

Ganesan, R., & Rasool, M. (2019). Ferulic acid inhibits interleukin 17-dependent expression of nodal pathogenic mediators in fibroblast-like synoviocytes of rheumatoid arthritis. *Journal of Cellular Biochemistry, 120*(2), 1878–1893. https://doi.org/10.1002/jcb.27502

Goh, Y. X., Jalil, J., Lam, K. W., Husain, K., & Premakumar, C. M. (2022). Genistein: a review on its anti-inflammatory properties. *Frontiers in Pharmacology, 13*, 820969.

Gorabi, A. M., Razi, B., Aslani, S., Abbasifard, M., Imani, D., Sathyapalan, T., & Sahebkar, A. (2021). Effect of curcumin on proinflammatory cytokines: a meta-analysis of randomized controlled trials. *Cytokine, 143*, 155541.

Hämäläinen, M., Nieminen, R., Vuorela, P., Heinonen, M., & Moilanen, E. (2007). Anti-inflammatory effects of flavonoids: genistein, kaempferol, quercetin, and daidzein inhibit STAT-1 and NF-kappaB activations, whereas flavone, isorhamnetin, naringenin, and pelargonidin inhibit only NF-kappaB activation along with their inhibitory effect on iNOS expression and NO production in activated macrophages. *Mediators Inflammation, 2007*, 45673. https://doi.org/10.1155/2007/45673

Handono, K., Pratama, M. Z., Endharti, A. T., & Kalim, H. (2015). Treatment of low doses curcumin could modulate Th17/Treg balance specifically on CD4+ T cell cultures of systemic lupus erythematosus patients. *Central European Journal of Immunology, 40*(4), 461–469. https://doi.org/10.5114/ceji.2015.56970

Haqqi, T. M., Anthony, D. D., Gupta, S., Ahmad, N., Lee, M. S., Kumar, G. K., & Mukhtar, H. (1999). Prevention of collagen-induced arthritis in mice by a polyphenolic fraction from green tea. *Proceedings of the National Academy of Sciences of the United States of America, 96*(8), 4524–4529. https://doi.org/10.1073/pnas.96.8.4524

Hemmati, A. A., Rajaee, E., Houshmand, G., Fakhroddin, M., Dargahi-MalAmir, M., Hesam, S., & Maram, N. (2016). Study the effects of anti-inflammatory curcumex capsules containing three plants (ginger, curcumin and black pepper) in patients with active rheumatoid arthritis. *IIOAB J, 7*, 389–392.

Hewlings, S., & Kalman, D. (2017). Curcumin: a review of its' effects on human health. *Foods, 6*(10), 92.

Hirata, A., Murakami, Y., Atsumi, T., Shoji, M., Ogiwara, T., Shibuya, K., ... Fujisawa, S. (2005). Ferulic acid dimer inhibits lipopolysaccharide-stimulated cyclooxygenase-2 expression in macrophages. *In Vivo, 19*(5), 849–853.

Hu, Y., Li, J., Qin, L., Cheng, W., Lai, Y., Yue, Y., ... Zhang, P. (2016). Study in treatment of collagen-induced arthritis in DBA/1 mice model by genistein. *Current Pharmaceutical Design, 22*(46), 6975–6981.

Huang, G., Xu, Z., Huang, Y., Duan, X., Gong, W., Zhang, Y., ... He, F. (2013). Curcumin protects against collagen-induced arthritis via suppression of BAFF production. *Journal of Clinical Immunology, 33*(3), 550–557. https://doi.org/10.1007/s10875-012-9839-0

Hussain, S. A., Panjagari, N. R., Singh, R. R., & Patil, G. R. (2015). Potential herbs and herbal nutraceuticals: food applications and their interactions with food components. *Critical Reviews in Food Science and Nutrition, 55*(1), 94–122. https://doi.org/10.1080/10408398.2011.649148

Jacob, A., Chaves, L., Eadon, M. T., Chang, A., Quigg, R. J., & Alexander, J. J. (2013). Curcumin alleviates immune-complex-mediated glomerulonephritis in factor-H-deficient mice. *Immunology, 139*(3), 328–337. https://doi.org/10.1111/imm.12079

Jacob, J., Amalraj, A., Raj, K. K. J., Divya, C., Kunnumakkara, A. B., & Gopi, S. (2019). A novel bioavailable hydrogenated curcuminoids formulation (CuroWhite™) improves symptoms and diagnostic indicators in rheumatoid arthritis patients - a randomized, double blind and placebo controlled study. *Journal of Traditional Complementary Medicine, 9*(4), 346–352. https://doi.org/10.1016/j.jtcme.2018.06.001

Javadi, M., Khadem Haghighian, H., Goodarzy, S., Abbasi, M., & Nassiri-Asl, M. (2019). Effect of curcumin nanomicelle on the clinical symptoms of patients with rheumatoid arthritis: a randomized, double-blind, controlled trial. *International Journal of Rheumatic Diseases, 22*(10), 1857–1862.

Ji, G., Zhang, Y., Yang, Q., Cheng, S., Hao, J., Zhao, X., & Jiang, Z. (2012). Genistein suppresses LPS-induced inflammatory response through inhibiting NF-κB following AMP kinase activation in RAW 264.7 macrophages. *PLoS One, 7*(12), e53101. https://doi.org/10.1371/journal.pone.0053101

Jia, Z., Babu, P. V., Si, H., Nallasamy, P., Zhu, H., Zhen, W., ... Liu, D. (2013). Genistein inhibits TNF-α-induced endothelial inflammation through the protein kinase pathway A and improves vascular inflammation in C57BL/6 mice. *International Journal of Cardiology, 168*(3), 2637–2645. https://doi.org/10.1016/j.ijcard.2013.03.035

Kanakasabai, S., Casalini, E., Walline, C. C., Mo, C., Chearwae, W., & Bright, J. J. (2012). Differential regulation of CD4(+) T helper cell responses by curcumin in experimental autoimmune encephalomyelitis. *Journal of Nutritional Biochemistry, 23*(11), 1498–1507. https://doi.org/10.1016/j.jnutbio.2011.10.002

Kang, D., Li, B., Luo, L., Jiang, W., Lu, Q., Rong, M., & Lai, R. (2016). Curcumin shows excellent therapeutic effect on psoriasis in mouse model. *Biochimie, 123*, 73–80. https://doi.org/10.1016/j.biochi.2016.01.013

Karatas, A., Dagli, A. F., Orhan, C., Gencoglu, H., Ozgen, M., Sahin, N., … Koca, S. S. (2020). Epigallocatechin 3-gallate attenuates arthritis by regulating Nrf2, HO-1, and cytokine levels in an experimental arthritis model. *Biotechnology and Applied Biochemistry, 67*(3), 317–322.

Larmonier, C. B., Uno, J. K., Lee, K. M., Karrasch, T., Laubitz, D., Thurston, R., … Kiela, P. R. (2008). Limited effects of dietary curcumin on Th-1 driven colitis in IL-10 deficient mice suggest an IL-10-dependent mechanism of protection. *Amerian Journal of Physiology-Gastrointestinal and Liver Physiology, 295*(5), G1079–G1091. https://doi.org/10.1152/ajpgi.90365.2008

Lee, C.-J., Moon, S.-J., Jeong, J.-H., Lee, S., Lee, M.-H., Yoo, S.-M., … Cho, Y.-Y. (2018). Kaempferol targeting on the fibroblast growth factor receptor 3-ribosomal S6 kinase 2 signaling axis prevents the development of rheumatoid arthritis. *Cell Death & Disease, 9*(3), 401. https://doi.org/10.1038/s41419-018-0433-0

Lee, K. H., Abas, F., Mohamed Alitheen, N. B., Shaari, K., Lajis, N. H., Israf, D. A., & Syahida, A. (2015). Chemopreventive effects of a curcumin-like diarylpentanoid [2, 6-bis (2, 5-dimethoxybenzylidene) cyclohexanone] in cellular targets of rheumatoid arthritis in vitro. *International Journal of Rheumatic Diseases, 18*(6), 616–627.

Lee, W.-S., Lee, E.-G., Sung, M.-S., & Yoo, W.-H. (2014). Kaempferol inhibits IL-1β-stimulated, RANKL-mediated osteoclastogenesis via downregulation of MAPKs, c-Fos, and NFATc1. *Inflammation, 37*(4), 1221–1230. https://doi.org/10.1007/s10753-014-9849-6

Li, J., Gang, D., Yu, X., Hu, Y., Yue, Y., Cheng, W., … Zhang, P. (2013). Genistein: the potential for efficacy in rheumatoid arthritis. *Clinical Rheumatology, 32*(5), 535–540. https://doi.org/10.1007/s10067-012-2148-4

Li, X., Xu, D. Q., Sun, D. Y., Zhang, T., He, X., & Xiao, D. M. (2019). Curcumin ameliorates monosodium urate-induced gouty arthritis through Nod-like receptor 3 inflammasome mediation via inhibiting nuclear factor-kappa B signaling. *Journal of Cellular Biochemistry, 120*(4), 6718–6728.

Lin, S. K., Chang, H. H., Chen, Y. J., Wang, C. C., Galson, D. L., Hong, C. Y., & Kok, S. H. (2008). Epigallocatechin-3-gallate diminishes CCL2 expression in human osteoblastic cells via up-regulation of phosphatidylinositol 3-Kinase/Akt/Raf-1 interaction: a potential therapeutic benefit for arthritis. *Arthritis & Rheumatism, 58*(10), 3145–3156. https://doi.org/10.1002/art.23937

Liu, S., Li, J., & Feng, L. H. (2023). Gallic acid regulates immune response in a mouse model of rheumatoid arthritis. *Immunity, Inflammation and Disease, 11*(2), e782.

Mahnke, K., Johnson, T. S., Ring, S., & Enk, A. H. (2007). Tolerogenic dendritic cells and regulatory T cells: a two-way relationship. *Journal of Dermatological Science, 46*(3), 159–167. https://doi.org/10.1016/j.jdermsci.2007.03.002

Maicas, N., Ferrándiz, M. L., Brines, R., Ibáñez, L., Cuadrado, A., Koenders, M. I., … Alcaraz, M. J. (2011). Deficiency of Nrf2 accelerates the effector phase of arthritis and aggravates joint disease. *Antioxidants Redox Signalling, 15*(4), 889–901. https://doi.org/10.1089/ars.2010.3835

Meng, Q., Velalar, C. N., & Ruan, R. (2008). Effects of epigallocatechin-3-gallate on mitochondrial integrity and antioxidative enzyme activity in the aging process of human fibroblast. *Free Radicals and Biology Medicine, 44*(6), 1032–1041. https://doi.org/10.1016/j.freeradbiomed.2007.11.023

Nicholas, C., Batra, S., Vargo, M. A., Voss, O. H., Gavrilin, M. A., Wewers, M. D., … Doseff, A. I. (2007). Apigenin blocks lipopolysaccharide-induced lethality in vivo and proinflammatory cytokines expression by inactivating NF-kappaB through the suppression of p65 phosphorylation. *Journal of Immunology, 179*(10), 7121–7127. https://doi.org/10.4049/jimmunol.179.10.7121

Pan, D., Li, N., Liu, Y., Xu, Q., Liu, Q., You, Y., … Lin, C. (2018). Kaempferol inhibits the migration and invasion of rheumatoid arthritis fibroblast-like synoviocytes by blocking activation of the MAPK pathway. *International Immunopharmacology, 55*, 174–182. https://doi.org/10.1016/j.intimp.2017.12.011

Puri, V., Nagpal, M., Singh, I., Singh, M., Dhingra, G. A., Huanbutta, K., … & Sangnim, T. (2022). A comprehensive review on nutraceuticals: therapy support and formulation challenges. *Nutrients, 14*(21), 4637.

Qin, J., Shang, L., Ping, A. S., Li, J., Li, X. J., Yu, H., … Wang, H. (2013). Response to 'TNF/TNFR signal transduction pathway-mediated anti-apoptosis and anti-inflammatory effects of sodium ferulate on IL-1β-induced rat osteoarthritis chondrocytes in vitro'--authors' reply. *Arthritis Research Therapy, 15*(3), 409. https://doi.org/10.1186/ar4227

Rahardjo, B., Widjajanto, E., Sujuti, H., & Keman, K. (2014). Curcumin decreased level of proinflammatory cytokines in monocyte cultures exposed to preeclamptic plasma by affecting the transcription factors NF-κB and PPAR-γ. *Biomarkers and Genomic Medicine, 6*(3), 105–115.

Raker, V. K., Domogalla, M. P., & Steinbrink, K. (2015). Tolerogenic dendritic cells for regulatory T cell induction in man. *Frontiers in Immunology, 6*, 569. https://doi.org/10.3389/fimmu.2015.00569

Rosas, E. C., Correa, L. B., & das Graças Henriques, M. (2019). Chapter 28- Antiinflammatory Properties of Schinus terebinthifolius and Its Use in Arthritic Conditions. In R. R. Watson & V. R. Preedy (Eds.), *Bioactive Food as Dietary Interventions for Arthritis and Related Inflammatory Diseases (Second Edition)* (pp. 489–505). Academic Press.

Roy, A., & Bharadvaja, N. (2018). Biotechnological approaches for the production of pharmaceutically important compound: plumbagin. *Current Pharmaceutical Biotechnology, 19*(5), 372–381. https://doi.org/10.2174/1389201019666180629143842

Sachdeva, V., Roy, A., & Bharadvaja, N. (2020). Current prospects of nutraceuticals: a review. *Current Pharmaceutical Biotechnology, 21*(10), 884–896.

Salehi, B., Venditti, A., Sharifi-Rad, M., Kręgiel, D., Sharifi-Rad, J., Durazzo, A., … Martins, N. (2019). The therapeutic potential of apigenin. *International Journal Molecular Science, 20*(6). https://doi.org/10.3390/ijms20061305

Sawicka, B., Ziarati, P., Krochmal-Marczak, B., & Skiba, D. (2019). Nutraceuticals in food and pharmacy. A review. *Agronomy Science, 74*(4), 1–13.

Sen, T., & Samanta, S. K. (2015). Medicinal plants, human health and biodiversity: a broad review. *Advanced Biochemical Engineering and Biotechnology, 147*, 59–110. https://doi.org/10.1007/10_2014_273

Shah, M., Murad, W., Mubin, S., Ullah, O., Rehman, N. U., & Rahman, M. H. (2022). Multiple health benefits of curcumin and its therapeutic potential. *Environmental Science and Pollution Research, 29*(29), 43732–43744.

Shang, L., Qin, J., Chen, L. B., Liu, B. X., Jacques, M., & Wang, H. (2009). Effects of sodium ferulate on human osteoarthritic chondrocytes and osteoarthritis in rats. *Clinical and Experimental Pharmacology and Physiology, 36*(9), 912–918. https://doi.org/10.1111/j.1440-1681.2009.05171.x

Sharifi-Rad, M., Pezzani, R., Redaelli, M., Zorzan, M., Imran, M., Ahmed Khalil, A., … Sharifi-Rad, J. (2020). Preclinical pharmacological activities of epigallocatechin-3-gallate in signaling pathways: an update on cancer. *Molecules, 25*(3). https://doi.org/10.3390/molecules25030467

Sharma, D., Chaubey, P., & Suvarna, V. (2021). Role of natural products in alleviation of rheumatoid arthritis—a review. *Journal of Food Biochemistry, 45*(4), e13673.

Sharma, S., Chopra, K., Kulkarni, S. K., & Agrewala, J. N. (2007). Resveratrol and curcumin suppress immune response through CD28/CTLA-4 and CD80 co-stimulatory pathway. *Clinical and Experimental Immunology, 147*(1), 155–163. https://doi.org/10.1111/j.1365-2249.2006.03257.x

Sharma, V., & Sarkar, I. N. (2013). Bioinformatics opportunities for identification and study of medicinal plants. *Briefings in Bioinformatics, 14*(2), 238–250. https://doi.org/10.1093/bib/bbs021

Sheng, S., Wang, X., Liu, X., Hu, X., Shao, Y., Wang, G., … Chen, X. (2022). The role of resveratrol on rheumatoid arthritis: from bench to bedside. *Frontiers in Pharmacology, 13*, 829677.

Singh, R., Ahmed, S., Islam, N., Goldberg, V. M., & Haqqi, T. M. (2002). Epigallocatechin-3-gallate inhibits interleukin-1beta-induced expression of nitric oxide synthase and production of nitric oxide in human chondrocytes: suppression of nuclear factor kappaB activation by degradation of the inhibitor of nuclear factor kappaB. *Arthritis & Rheumatism, 46*(8), 2079–2086. https://doi.org/10.1002/art.10443

Singh, R., Akhtar, N., & Haqqi, T. M. (2010). Green tea polyphenol epigallocatechin-3-gallate: inflammation and arthritis. [corrected]. *Life Sciences, 86*(25–26), 907–918. https://doi.org/10.1016/j.lfs.2010.04.013

Wang, J., Zhang, Q., Jin, S., He, D., Zhao, S., & Liu, S. (2008). Genistein modulate immune responses in collagen-induced rheumatoid arthritis model. *Maturitas, 59*(4), 405–412. https://doi.org/10.1016/j.maturitas.2008.04.003

Wang, S., Li, H., Zhang, M., Yue, L. T., Wang, C. C., Zhang, P., … Duan, R. S. (2016). Curcumin ameliorates experimental autoimmune myasthenia gravis by diverse immune cells. *Neuroscience Letters, 626*, 25–34. https://doi.org/10.1016/j.neulet.2016.05.020

Wei, D., Le, X., Zheng, L., Wang, L., Frey, J. A., Gao, A. C., … Xie, K. (2003). Stat3 activation regulates the expression of vascular endothelial growth factor and human pancreatic cancer angiogenesis and metastasis. *Oncogene, 22*(3), 319–329. https://doi.org/10.1038/sj.onc.1206122

Xian, M.-L., Geng, N.-N., Deng, L., Kuang, B., Fan, M.-T., & Guo, F.-J. (2023). Kaempferol exerts therapeutic effect on rheumatoid arthritis by regulating cell senescence: a study based on network pharmacology and in vitro experiments. *Journal of Army Medical University, 45*(12), 1281–1291.

Yang, G., Chang, C.-C., Yang, Y., Yuan, L., Xu, L., Ho, C.-T., & Li, S. (2018). Resveratrol alleviates rheumatoid arthritis via reducing ROS and inflammation, inhibiting MAPK signaling pathways, and suppressing angiogenesis. *Journal of Agricultural and Food Chemistry, 66*(49), 12953–12960. https://doi.org/10.1021/acs.jafc.8b05047

Yoon, C.-H., Chung, S.-J., Lee, S.-W., Park, Y.-B., Lee, S.-K., & Park, M.-C. (2013). Gallic acid, a natural polyphenolic acid, induces apoptosis and inhibits proinflammatory gene expressions in rheumatoid arthritis fibroblast-like synoviocytes. *Joint Bone Spine, 80*(3), 274–279. https://doi.org/10.1016/j.jbspin.2012.08.010

Yoon, H. Y., Lee, E. G., Lee, H., Cho, I. J., Choi, Y. J., Sung, M. S., … Yoo, W. H. (2013). Kaempferol inhibits IL-1β-induced proliferation of rheumatoid arthritis synovial fibroblasts and the production of COX-2, PGE2 and MMPs. *International Journal of Molecular Medicine, 32*(4), 971–977. https://doi.org/10.3892/ijmm.2013.1468

Zhang, Y., Dong, J., He, P., Li, W., Zhang, Q., Li, N., & Sun, T. (2012). Genistein inhibit cytokines or growth factor-induced proliferation and transformation phenotype in fibroblast-like synoviocytes of rheumatoid arthritis. *Inflammation, 35*(1), 377–387. https://doi.org/10.1007/s10753-011-9365-x

Zhu, L., Zhang, Z., Xia, N., Zhang, W., Wei, Y., Huang, J., … Yang, L. (2020). Anti-arthritic activity of ferulic acid in complete Freund's adjuvant (CFA)-induced arthritis in rats: JAK2 inhibition. *Inflammopharmacology, 28*(2), 463–473. https://doi.org/10.1007/s10787-019-00642-0

13 Adverse Effects of Phytochemicals Used to Treat Rheumatoid Arthritis

Shakila Sabir, Muhammad Ejaz-ul-Haq, Ammara Saleem, Syed Haroon Khalid, and Muhammad Furqan Akhtar

WHAT WILL YOU LEARN FROM THIS CHAPTER?

This chapter provides you with the knowledge regarding the toxic effects of those phytochemicals that have been extensively explored for the treatment of rheumatoid arthritis. This chapter also highlighted the doses at which these phytochemicals produce adverse effects. We have also discussed the studies regarding the formulation development in order to ameliorate the adverse effects and improve efficacy.

LIST OF ABBREVIATIONS

CYPs	Cytochrome P450 enzymes
DNA	Deoxyribonucleic acid
GSH	Glutathione
LD50	Lethal dose 50
NOAEL	No observable adverse effect level
QQ	o-Quinone/quinonmethide
ROS	Reactive oxygen species

13.1 INTRODUCTION

Plant-based therapy has been used since ancient times, as it is considered safe owing to its natural origin (Niaz et al., 2020). However, certain phytochemicals even obtained from edible plants may cause carcinogenesis, hepatotoxicity, or nephrotoxicity, particularly at higher doses or prolonged therapy (Süntar & Yakıncı, 2020). Despite these concerns, phytonutrients are preferred because they are easily accessible, cost-effective, and cause comparatively low side effects at therapeutic doses. Herbal products are gaining much attention due to various phytonutrients having multiple beneficial health effects. Phytonutrients may be taken alone or in combination. Phytochemicals may be used as an adjuvant therapy along with allopathy for the treatment of multiple ailments like cardiovascular disorders, cancer, rheumatoid arthritis, or diabetes (Fernando et al., 2019). The big challenge associated with the use of phytonutrients is their bioavailability. Bioavailability of phytonutrients depends on age or gender. Nowadays, farmers massively treat the plants with herbicides or insecticides for the pest control and proper growth of plants. But, on the other hand, these pesticides have a great impact on the secondary metabolites. These pesticides are a great threat to human health (Saleem et al., 2023).

DOI: 10.1201/9781003535201-16

13.2 ADVERSE EFFECTS OF PHYTONUTRIENTS

Although there is a wide range of phytonutrients, curcumin, quercetin, apigenin, ferulic acid, gallic acid, genistein, kaempferol, and resveratrol are commonly used for the treatment of rheumatoid arthritis. A brief literature about the adverse effects of these abovementioned phytonutrients is given below.

13.2.1 Curcumin

The required therapeutic concentration of curcumin in plasma and tissues could not be achieved after oral administration due to its considerable metabolism in liver and intestine. It indicated that the therapeutic potential of oral curcumin administration is limited and poses a great hurdle in clinical drug development (Ireson et al., 2002). Literature reported that curcumin worked as an iron chelator during in vivo studies. Curcumin induced iron deficiency anemia in animal models. A study also suggested that curcumin intake by individuals with suboptimal iron levels severely affected systemic iron load (Jiao et al., 2009). The 2 α,β-unsaturated ketone groups in the curcumin structure are found to be involved in mediating the toxic effects of curcumin. Through Michael addition reactions, these ketone groups reacted with the thiol group of the cysteine residue found in proteins. This reaction of curcumin was responsible for the generation of reactive oxygen species (ROS) due to irreversible modification of antioxidant enzyme (thioredoxin reductase) (Fang, Lu, Holmgren, 2005; Sandur et al., 2007), induction of topoisomerase II-mediated deoxyribonucleic acid (DNA) damage, and inactivation of tumor suppressor proteins p53 (Martín-Cordero, López-Lázaro, Gálvez, Ayuso, 2003; Sandur et al., 2007).

A clinical study has reported mild toxicity at moderate doses of curcumin. Even curcumin, at a high dose of 12 g per day, did not produce serious adverse reactions against placebo. However, mild side effects were observed including diarrhea, nausea, and increased bowl movements. Such side effects were resolved within 1–3 days without any intervention. Curcumin may induce biliary colic in patients suffering from gallstones. Studies also reported that curcumin was safe to use during pregnancy (Asher & Spelman, 2013).

Alicia and his colleagues have conducted a systematic review in order to investigate the safety of curcumin nanoparticles. Twenty-six studies were included in this review. A study reported that curcumin nanoparticles are safe to use in comparison to free curcumin for the treatment of breast cancer. The study also suggested the oral route for the best results of these phytochemicals. All of the included studies reported little or no adverse effects of curcumin nanoparticles in terms of hematological and biochemical markers (Ombredane et al., 2021). Urbina et al. reported that curcumin at a dose of 50 μM induced DNA strand breakage. This adverse effect was further exaggerated when curcumin (at a dose of 50 mM) was co-administered with different concentrations of copper (Urbina-Cano et al., 2006).

Cao et al. reported that curcumin induced DNA damage in a dose-dependent manner. At low doses, curcumin performed as an antioxidant in human hepatoma G2 cells, while curcumin increased the oxidative stress and caused DNA damage at high doses. It reflected the dual role of curcumin. At high doses, curcumin caused mitochondrial DNA damage and subsequently cell death occurred (Cao et al., 2006). During a phase I clinical trial, curcumin at a dose of 3.6 g/day was administered to the 15 patients with advanced colorectal cancer for 4 months. Curcumin caused diarrhea and increased serum lactate dehydrogenase and alkaline phosphatase levels in few patients (Cheng et al., 2001; Sharma et al., 2004). Verschoyle et al. reported that curcumin did not have a mutagenic effect as tested by the *Salmonella typhimurium* test (Verschoyle, Steward, & Gescher, 2007).

When experimental mice received curcumin by an intraperitoneal route at a dose of 25–200 mg/kg, there was a dose-dependent increase in sister chromatid exchange observed in bone marrow. Although, chromosome aberration was not reported. Chronic administration of diet containing curcumin at a dose of 100–1,000 ppm for 6 months did not induce chromosome aberration.

TABLE 13.1
Toxicity of Phytochemicals Used to Treat Rheumatoid Arthritis

Phytochemical	Species/Cell Line	Dose/Route	Toxic Effect	Reference
Curcumin	Balb-C mouse lymphocytes	50 μM	DNA damage	Urbina-Cano et al. (2006)
	Human hepatoma G2 cells	2.5–40 μg/mL	Mitochondrial and nuclear DNA damage	Cao et al. (2006)
	Human	500–12,000 mg/day/ orally	Diarrhea, ↑ Serum lactate dehydrogenase ↑ alkaline phosphatase level	
	RKO cells	0–60 mM	Disrupts the conformation of p53 proteins	Moos et al. (2004)
	Mice	0.2, 0.5, and 2.0% of diet w/w	↓ Hemoglobin ↓ Hematocrit ↓ Serum iron	Jiao et al. (2009)
Resveratrol	Human	300–1000mg/oral	↑ sVCAM-1 ↑ tPAI-1	Mankowski et al. (2020)
	GRX cells	50 μM	Dose dependently increased ROS level	Martins et al. (2014)
	Rat	300-1000 mg/kg	Renal toxicity	Crowell et al. (2004)

sVCAM-1: soluble vascular cell adhesion molecule-1; tPAI-1: total plasminogen activator inhibitor-1; ROS: reactive oxygen species.

Furthermore, administration of diet containing curcumin at a dose of 500–1,000 ppm for 9 months caused chromosomal aberrations considerably (Giri, Das, Talukder, & Sharma, 1990). Curcumin at a dose of 15 μM or above caused DNA damage analyzed by the in vitro Comet assay in lymphocytes and gastric mucosal cells (Blasiak, Trzeciak, & Kowalik, 1999). Another study reported that curcumin at a dose of 40–60 μM hindered the activity of NADPH: quinone oxidoreductase in human cultured colon carcinoma and myeloid leukemia cells. Furthermore, curcumin also caused the degradation of the tumor suppressor gene called p53; subsequently, there was inhibition of apoptosis mediated by p53 (Tsvetkov et al., 2005). In line with the previous study, Moos et al. also reported degradation of p53 in colon cancer cells by curcumin treatment at a dose of 50 μM (Moos, Edes, Mullally, & Fitzpatrick, 2004).

Loo et al. reported the toxicity of curcumin nanoparticles in NR8383 depending upon the surface charges. Positively charged nanoparticles showed maximum toxicity followed by negatively charged and neutral curcumin-loaded nanoparticles. Positively charged nanoparticles induced the toxicity by mitochondrial and lysosomal destabilization (Loo, Siew, Young, Traini, & Lee, 2022). It is crucial to remember that any tactic that increases the curcumin's tissue level will also increase its toxicity in addition to its effectiveness. Literature has shown that curcumin has an antioxidant effect at low concentrations while it increases the oxidative stress at high concentration (Cao et al., 2006) (Table 13.1).

13.2.2 Quercetin

Literature has shown that quercetin is a comparatively safer compound. Recently, Cunningham and his colleagues reported that quercetin is an equally safe compound for both males and females at a dose of 50 mg/kg for 14 weeks (Cunningham et al., 2022). The results of the aforementioned study were also in line with the study conducted by Riuz. Quercetin did not cause mortality or any hazardous effects at higher doses (3,000 mg/kg/day) administered for 28 days in experimental mice (Ruiz, Estela, Asensi, Mañes, & Picó, 2006). Pamukcu et al. reported that quercetin showed carcinogenicity in the intestine, kidneys, and urinary bladder in experimental rats after administration of higher doses (1,000 and 40,000 ppm) for 58 and 104 weeks (Pamukcu, Yalçiner, Hatcher, & Bryan, 1980).

The International Agency for Research on Cancer has been assessing the carcinogenic risk of various compounds. In 1999, the Agency reported that quercetin did not cause carcinogenicity in human beings. Therefore, quercetin supplements are available commercially in various advanced western countries. Various beneficial effects of quercetin have been verified by clinical trials. However, quercetin caused genotoxicity in Salmonella but its safety in human beings has been approved (Okamoto, 2005). During another study, the genotoxic effect of quercetin was assessed by a bone marrow micronucleus assay and an unscheduled DNA synthesis assay in rats. Quercetin was given at a dose of 2,000 mg/kg orally for 24–48 hours. Quercetin did not increase the number of micronucleated polychromatic erythrocytes. Similarly, quercetin did not cause unscheduled DNA synthesis in liver cells of rats. Therefore, this study suggested that quercetin is not a genotoxic agent (Utesch et al., 2008).

Recently, Tiwari et al. conducted an acute toxicity study on polyherbal preparation containing quercetin, curcumin, and rutin. The study reported no abnormality in general behavior or any serious adverse effects in experimental animals. There were no considerable changes in body weight, hematological, and biochemical parameters. Zero mortality rate has been reported. Histopathological investigations showed no alterations at the given doses. The study concluded that polyherbal preparation of aforementioned phytochemicals did not cause any toxic effect in albino mice (Tiwari et al., 2020). o-quinone/quinonmethide (QQ) was produced through oxidation of quercetin. QQ has been considered a highly reactive compound. QQ reacted immediately with the surrounding molecules at the site of production. The primary target for QQ reaction was GSH. QQ reacted with protein-thiols in order to produce protein–quercetin adducts at a low concentration of GSH. QQ reacted slowly with H_2O in vivo. QQ regeneration by either ascorbate or DT-diaphorase was negligible in comparison to thiol-adduct production. Although GSQ was not a stable adduct and dissociated in GSH and QQ, the dissociated QQ reacted with GSH again in the presence of a higher concentration of GSH. If there was a low concentration of GSH, dissociated QQ reacted with other protein thiol groups. Generally, protein-QQ adducts were more stable than GSQ. These stable adducts accumulated in the body and caused toxicity (Boots, Balk, Bast, & Haenen, 2005).

13.2.3 Apigenin

Aside from in vitro models, highly employable models in the accessible pool of alternative modalities are computer-aided forecasting algorithms or prediction tools. With computerized methods, a compound's toxicological characteristics can be quickly and effectively evaluated. It was investigated by in silico prediction techniques that apigenin has drug-like properties without being carcinogenic or mutagenic (Divya Rajaselvi et al., 2023). Aglycone form of apigenin undergoes metabolism (phase I) through CYP enzymes. The special structural characteristics of this compound (including A ring hydroxylation and C2–C3 double bond) made it a potent inhibitor of CYP1A1, CYP1A2, and CYP1B1 (Kimura, Ito, Ohnishi, & Hatano, 2010). Another study also confirmed that apigenin was a potent inhibitor of CYP1A2 and CYP2C9 (Kale, Gawande, & Kotwal, 2008; Si et al., 2009). Co-administration of apigenin and a few other drugs resulted in either class C-D drug–drug interaction. These combinations are completely contraindicated according to the healthcare professional resource named Lexicomp (DeRango-Adem & Blay, 2021).

During interactions involving prescription drugs that are the substrate of CYP enzymes impacted by apigenin, this may be particularly concerning since they may cause toxicity or supratherapeutic effects of the prescribed drugs at higher doses of apigenin (Kleinschmidt & Delaney, 2015). It is crucial to remember that prodrugs that are processed by CYPs inhibited by apigenin have a risk of the active medication having subtherapeutic effects. Similarly, drugs that are processed for elimination by CYPs that are inhibited by apigenin would have higher therapeutic effects with higher toxicity. Additionally, it was also reported that apigenin at a dose of 300 mg/kg for 68 days did not cause toxicity (Ai et al., 2017; Tong et al., 2019).

13.2.4 Ferulic Acid

Choi et al. reported that ferulic acid was a safer compound. Even at a higher dose of 300 µg/mL, ferulic acid exhibited no effect on cell count as well as the viability of erythrocytes, leukocytes, and platelets. Ferulic acid exhibited mild toxicity in NIH-3T3 and 3T3-L1 cells at a dose of 500 µg/mL (Choi, Park, Kim, Lee, Kim, 2018). Peng et al. reported that ferulic acid caused further renal damage when administered for the 28-week treatment of chronic kidney disease (Peng et al., 2012).

13.2.5 Gallic Acid

A study has reported that gallic acid did not produce considerable toxicity in experimental animals when a 900 mg/kg/day dose of quercetin was administered for 28 days. No significant alterations were observed in histopathological, morphological, and behavioral parameters. The study also suggested that LD50 of gallic acid was more than 2,000 mg/kg/day in mice (Variya, Bakrania, Madan, & Patel, 2019). Nicho and his colleagues performed subchronic toxicity in rats and administered the diet containing 5% gallic acid for 13 weeks. Toxic effects after the ingestion of gallic acid-containing diet include reduced hemoglobin, hematocrit, red blood cells, and centrilobular hepatocyte hypertrophy (Niho et al., 2001).

During a subacute toxicity study, gallic acid did not produce toxicity at a dose of 1,000 mg/kg orally. In another acute oral toxicity study, gallic acid was administered to the experimental animals at a very high dose of 5,000 mg/kg orally and still there were no signs of lethal toxicity. The study suggested that gallic acid was safe up to a dosage of 5,000 mg/kg body weight orally. Furthermore, the subacute investigation demonstrated no cumulative toxicity, as seen by the nonsignificant changes in the hematological parameters examined. The highest dose tested had a NOAEL (no observable adverse effect level) of 5,000 mg/kg body weight (Zanwar, Badole, Shende, Hegde, & Bodhankar, 2014). In line with the previous studies, Variya et al. also reported that the LD50 of gallic acid was greater than 2,000 mg/kg in albino mice. Hematological investigations exhibited no changes in the transaminase level. Gross necropsy also exhibited no significant alterations. During histopathological investigations, it was observed that there were slightly fatty changes in bone marrow without bone marrow suppression. Gallic acid administration at a dose of 900 mg/kg for 28 days exhibited nonsignificant alterations in behavioral and morphological parameters. Histopathological investigations also suggested the safety of gallic acid in experimental animals.

13.2.6 Genistein

McClain et al. conducted a micronucleus test in order to evaluate the mutagenicity of genistein. Genistein was administered to rats at doses up to 2,000 mg/kg. There was no increase in micronuclei in treated rats. During the lymphoma assay, there was an increased number of small colonies, showing that genistein acted as a clastogen (McClain et al., 2006a). There are bundle of studies showing that genistein acts as a phytoestrogen, and most of its effects are because of its estrogenic activity. Thus, in most of the cases, genistein-induced hormonal alterations are not categorized under adverse effects of this compound. During a study, rats were treated with genistein at a dose of 50 mg/kg/day, nonsignificant hormonal-based effects were observed; thus, the genistein dose of 50 mg/kg/day was suggested as NOAEL. Genistein at a dose of 500 mg/kg produced mild hepatic effects such as bile duct proliferation and slightly enhanced glutamyl transferase level (McClain et al., 2006b).

The National Toxicology Program provided a report on the toxicity of genistein based on a 2-year toxicology and carcinogenesis study. When Sprague-Dawley rats were treated with 5, 100, and 500 ppm of genistein, there were no signs of carcinogenicity. Although, there were equivocal evidences of carcinogenicity of genistein on the basis of slightly enhanced incidence of pituitary gland neoplasm, adenocarcinoma, or mammary gland adenoma. It has also been demonstrated that genistein exposure accelerated the onset of abnormal estrous cycle in female Sprague-Dawley rats. The occurrence of common hormonally associated spontaneous neoplasms and the effects on the estrous cycle in female rats were consistent with an estrogenic activity of genistein ("Toxicology and carcinogenesis studies of genistein (Cas No. 446–72-0) in Sprague-Dawley rats (feed study)," 2008). Another study suggested that at a human consumption dose level, i.e., 0.4–4.1 mg/kg, maternal exposure to genistein could not have a deleterious impact on the reproductive organs of F1 offspring (Kang, Che, & Lee, 2002).

McClain and his colleagues recommended that genistein did not exhibit teratogenic activity even at a very high dose of up to 1,000 mg/kg/day orally in a pilot study of embryo-fetal toxicity or at a dose of up to 500 mg/kg/day by food admix in the prenatal developmental study. Although, these doses caused maternal toxicity as well as fetal toxicity. During a whole embryo culture screening assay, genistein exhibited teratogenic potential in vitro at high dosages; however, these results did not correspond with the in vivo observations. The NOAEL for maternal toxicity and detrimental impact on embryonic development was determined to be 100 mg/kg/day when given orally via food admixture based on the conclusive prenatal development study (McClain, Wolz, Davidovich, Edwards, & Bausch, 2007).

13.2.7 Kaempferol

Lee et al. conducted a toxicity study on kaempferol-rich goji berry leaf extract. It was found that the fetal dose of this extract was greater than 2,000 mg/kg based on the single-dose toxicity study for two weeks. Repeated daily doses of 0–1,000 mg/kg of extract for 14 days did not cause any signs of toxicity or gross pathological manifestations. Extract did not exhibit genotoxic activity in bacterial reverse mutation up to 5,000 µg/plate. Furthermore, chromosomal aberration was not observed regardless of metabolic activation with the use of CHO-K1 cells. For carcinogenic activity, chromosomal aberration was not observed at a dose of 2,000 mg/kg of extract by using the in vivo bone marrow micronucleus test (Lee, Lee, Ryu, Shim, 2023). Devi et al. reported that kaempferol has genotoxic and mutagenic activities. It was commonly recognized that kaempferol's in vitro pro-oxidant activity was responsible for its genotoxic effect. Although, there is a lack of literature on in vivo data supporting the genotoxic and carcinogenic effects of kaempferol, despite the large number of in vitro studies on these effects (Devi et al., 2015).

Akiyama et al. recently reported a randomized, double-blind, and placebo-controlled study. The kaempferol group was given a daily capsule containing 50 mg of kaempferol for four weeks, which was around five times more than the expected daily consumption of kaempferol found in food. General characteristics of participants, hematological and blood biochemical parameters, urine analysis, and adverse events were evaluated for general toxicity. When comparing kaempferol group's blood and urine values, anthropometric measurements, blood pressure, and other data to those in the placebo group, no clinical differences were seen. Additionally, there were no adverse effects of kaempferol. The study recommended that it is safe for a healthy person to take 50 mg of kaempferol for four weeks (Akiyama, Mizokami, Ito, & Ikeda, 2023). Neiring et al. suggested that food supplements with high-dose kaempferol should not be recommended as kaempferol at a high dose may cause DNA damage and apoptosis (Niering et al., 2005).

13.2.8 Resveratrol

It is well known that resveratrol has anticancer properties. However, low-dose resveratrol (5 μM) enhanced cancerous cells proliferation. Likewise, resveratrol at a dose of 5 μM did not alter the viability and functional role of INS-1 pancreatic cells of rats, while resveratrol at a dose of 50 μM increased the rate of apoptosis (Lin, Shen, Ni, 2011). Resveratrol exhibited antitumor activity in GRX cells at a dose of 0.1 and 1 μM (Souza et al., 2008). Resveratrol, at a dose of ≥50 μM, triggered a time- and dose-dependent increase in ROS level with subsequent cell death (Martins et al., 2014). Munoz et al. suggested that higher doses of resveratrol (2–5 g/day) may cause mild nausea, diarrhea, anal pruritus, and hypersensitivity reactions (Muñoz & Bustamante, 2015).

During a phase I randomized, double-blind study conducted by Howells et al., resveratrol at a dose of 5 g daily for 14 days was given to patients suffering from colorectal cancer and hepatic metastasis. Resveratrol was well tolerated but there were few side effects observed in these patients including nephrotoxicity and gut problems (Howells et al., 2011). Similar adverse effects were also reported in a randomized, placebo-controlled clinical trial conducted by Poulsen et al. (2013). Another study reported that a 450 mg per day dose of resveratrol was safe to use by a 60-kg person (Moon, Kohn, De Ferrari, & Kaykas, 2004). Resveratrol inhibited cytochrome P450 iso-enzymes including CYP2C9. CYP3A4 and CYP2D6, while activated CYP1A2 at a dose of 1,000 mg/day. These findings clearly suggested that resveratrol interacts with other multiple drugs (Detampel, Beck, Krähenbühl, Huwyler, 2012).

Recently, Mankowski and his fellows conducted a pilot study. This study reported that high-dose resveratrol (1,000 mg/day) elevated the biomarkers associated with cardiovascular diseases including, soluble E-selectin 1, oxidized low-density lipoprotein, soluble intracellular adhesion molecule-1, total plasminogen activator inhibitor, and soluble vascular cell adhesion in overweight old-age adults. But resveratrol at low doses did not produce any effect on these biomarkers (Mankowski et al., 2020). The alteration in aforementioned biomarkers could be because of resveratrol-induced oxidative stress at higher doses, which leads to proteolysis and DNA damage (Fox et al., 2012). During a study, resveratrol was administered to experimental animals at a wide range of dosages. Resveratrol at a dose level up to 300 mg/kg/day did not produce any adverse effect, while resveratrol at a dose of 1,000 or 3,000 mg/kg per day produced renal toxicity (Crowell, Korytko, Morrissey, Booth, & Levine, 2004).

Various types of drug carriers have been explored and are currently being used in order to improve the poor bioavailability and stability of resveratrol, which subsequently means decreased requirement for consumption of high resveratrol doses and lower side effects. These preparations include nanoparticles, emulsions, and liposomes (Intagliata, Modica, Santagati, & Montenegro, 2019; Wan, Zhang, Quan, & Wei, 2018). Solid lipid nanoparticle is a novel approach that can easily incorporate drugs with lipophilic nature and enhance their bioavailability, safety, water solubility, safety, biocompatibility, bio-distribution, and stability (Müller, Radtke, & Wissing, 2002). A study has reported that incorporating resveratrol in poly-lactic-co-glycolic acid nanoparticles enhanced its oral bioavailability up to 335.7% as compared to resveratrol alone (Müller et al., 2002). Nanoparticle preparation of resveratrol increased its efficacy and therapeutic potential in various cancer types. Aforementioned resveratrol preparation decreased the size of gliomas, ovarian, and colorectal cancer (Santos et al., 2019; Thipe et al., 2019) (Table 13.1).

13.3 FUTURE SUGGESTIONS

Phytonutrients have a promising role in the treatment of RA; on the other hand, evidence regarding their toxicity is very limited. Studies have shown that the prolonged use or higher doses of phytonutrients produced toxic effects. Low bioavailability is a major challenge regarding the therapeutic use of these agents. Although multiple approaches have been adopted to increase the bioavailability of phytonutrients without increasing their doses, it is imperative that studies should be conducted for the systematic evaluation of the safety profile. Until more robust data is available regarding the safety, phytonutrients should be taken with caution.

REFERENCES

Ai, X. Y., Qin, Y., Liu, H. J., Cui, Z. H., Li, M., Yang, J. H., … Yang, C. (2017). Apigenin inhibits colonic inflammation and tumorigenesis by suppressing STAT3-NF-κB signaling. *Oncotarget, 8*(59), 100216–100226. https://doi.org/10.18632/oncotarget.22145

Akiyama, M., Mizokami, T., Ito, H., & Ikeda, Y. (2023). A randomized, placebo-controlled trial evaluating the safety of excessive administration of kaempferol aglycone. *Food Science & Nutrition, 11*(9), 5427–5437. https://doi.org/10.1002/fsn3.3499

Asher, G. N., & Spelman, K. (2013). Clinical utility of curcumin extract. *Alternative Therapies in Health and Medicine, 19*(2), 20–22.

Blasiak, J., Trzeciak, A., & Kowalik, J. (1999). Curcumin damages DNA in human gastric mucosa cells and lymphocytes. *Journal of Environmental Pathology, Toxicology, and Oncology, 18*(4), 271–276.

Boots, A. W., Balk, J. M., Bast, A., & Haenen, G. R. (2005). The reversibility of the glutathionyl-quercetin adduct spreads oxidized quercetin-induced toxicity. *Biochemical and Biophysical Research Communications, 338*(2), 923–929. https://doi.org/10.1016/j.bbrc.2005.10.031

Cao, J., Jia, L., Zhou, H. M., Liu, Y., & Zhong, L. F. (2006). Mitochondrial and nuclear DNA damage induced by curcumin in human hepatoma G2 cells. *Toxicological Science, 91*(2), 476–483. https://doi.org/10.1093/toxsci/kfj153

Cheng, A. L., Hsu, C. H., Lin, J. K., Hsu, M. M., Ho, Y. F., Shen, T. S., … Hsieh, C. Y. (2001). Phase I clinical trial of curcumin, a chemopreventive agent, in patients with high-risk or pre-malignant lesions. *Anticancer Research, 21*(4b), 2895–2900.

Choi, J. H., Park, J. K., Kim, K. M., Lee, H. J., & Kim, S. (2018). In vitro and in vivo antithrombotic and cytotoxicity effects of ferulic acid. *Journal of Biochemical and Molecular Toxicology, 32*(1). https://doi.org/10.1002/jbt.22004

Crowell, J. A., Korytko, P. J., Morrissey, R. L., Booth, T. D., & Levine, B. S. (2004). Resveratrol-associated renal toxicity. *Toxicological Science, 82*(2), 614–619. https://doi.org/10.1093/toxsci/kfh263

Cunningham, P., Patton, E., VanderVeen, B. N., Unger, C., Aladhami, A., Enos, R. T., … Velázquez, K. T. (2022). Sub-chronic oral toxicity screening of quercetin in mice. *BMC Complementary Medicine and Therapies, 22*(1), 279. https://doi.org/10.1186/s12906-022-03758-z

DeRango-Adem, E. F., & Blay, J. (2021). Does oral apigenin have real potential for a therapeutic effect in the context of human gastrointestinal and other cancers? *Frontiers in Pharmacology, 12*, 681477. https://doi.org/10.3389/fphar.2021.681477

Detampel, P., Beck, M., Krähenbühl, S., & Huwyler, J. (2012). Drug interaction potential of resveratrol. *Drug Metabolism Review, 44*(3), 253–265. https://doi.org/10.3109/03602532.2012.700715

Devi, K. P., Malar, D. S., Nabavi, S. F., Sureda, A., Xiao, J., Nabavi, S. M., & Daglia, M. (2015). Kaempferol and inflammation: from chemistry to medicine. *Pharmacological Research, 99*, 1–10. https://doi.org/10.1016/j.phrs.2015.05.002

Divya Rajaselvi, N., Jida, M. D., Nair, D. B., Sujith, S., Beegum, N., & Nisha, A. R. (2023). Toxicity prediction and analysis of flavonoid apigenin as a histone deacetylase inhibitor: an in-silico approach. *In Silico Pharmacology, 11*(1), 34. https://doi.org/10.1007/s40203-023-00170-4

Fang, J., Lu, J., & Holmgren, A. (2005). Thioredoxin reductase is irreversibly modified by curcumin: a novel molecular mechanism for its anticancer activity. *Journal of Biological Chemistry, 280*(26), 25284–25290. https://doi.org/10.1074/jbc.M414645200

Fernando, W., Rupasinghe, H. P. V., & Hoskin, D. W. (2019). Dietary phytochemicals with anti-oxidant and pro-oxidant activities: a double-edged sword in relation to adjuvant chemotherapy and radiotherapy? *Cancer Letters, 452*, 168–177. https://doi.org/10.1016/j.canlet.2019.03.022

Fox, J. T., Sakamuru, S., Huang, R., Teneva, N., Simmons, S. O., Xia, M., … Myung, K. (2012). High-throughput genotoxicity assay identifies antioxidants as inducers of DNA damage response and cell death. *Proceedings of the National Academy of Sciences, 109*(14), 5423–5428.

Giri, A. K., Das, S. K., Talukder, G., & Sharma, A. (1990). Sister chromatid exchange and chromosome aberrations induced by curcumin and tartrazine on mammalian cells in vivo. *Cytobios, 62*(249), 111–117.

Howells, L. M., Berry, D. P., Elliott, P. J., Jacobson, E. W., Hoffmann, E., Hegarty, B., … Gescher, A. J. (2011). Phase I randomized, double-blind pilot study of micronized resveratrol (SRT501) in patients with hepatic metastases--safety, pharmacokinetics, and pharmacodynamics. *Cancer Prevention Research, 4*(9), 1419–1425. https://doi.org/10.1158/1940-6207.capr-11-0148

Intagliata, S., Modica, M. N., Santagati, L. M., & Montenegro, L. (2019). Strategies to improve resveratrol systemic and topical bioavailability: an update. *Antioxidants, 8*(8), 244.

Ireson, C. R., Jones, D. J., Orr, S., Coughtrie, M. W., Boocock, D. J., Williams, M. L., … Gescher, A. J. (2002). Metabolism of the cancer chemopreventive agent curcumin in human and rat intestine. *Cancer Epidemiology and Biomarkers Prevention, 11*(1), 105–111.

Jiao, Y., Wilkinson, J. T., Di, X., Wang, W., Hatcher, H., Kock, N. D., … Torti, S. V. (2009). Curcumin, a cancer chemopreventive and chemotherapeutic agent, is a biologically active iron chelator. *Blood, 113*(2), 462–469. https://doi.org/10.1182/blood-2008-05-155952

Kale, A., Gawande, S., & Kotwal, S. (2008). Cancer phytotherapeutics: role for flavonoids at the cellular level. *Phytotherapy Research, 22*(5), 567–577. https://doi.org/10.1002/ptr.2283

Kang, K. S., Che, J. H., & Lee, Y. S. (2002). Lack of adverse effects in the F1 offspring maternally exposed to genistein at human intake dose level. *Food and Chemical Toxicology, 40*(1), 43–51. https://doi.org/10.1016/S0278-6915(01)00091-6

Kimura, Y., Ito, H., Ohnishi, R., & Hatano, T. (2010). Inhibitory effects of polyphenols on human cytochrome P450 3A4 and 2C9 activity. *Food and Chemical Toxicology, 48*(1), 429–435. https://doi.org/10.1016/j.fct.2009.10.041

Kleinschmidt, K., & Delaney, K. (2015). Biochemical and metabolic principles. In L. S. Nelson, M. A. Howland, N. A. Lewin, S. W. Smith, L. R. Goldfrank, & R. S. Hoffman (Eds.), *Goldfrank's Toxicologic Emergencies* (10th Edition, pp. 155–171). McGraw-Hill.

Lee, H. J., Lee, S., Ryu, H. Y., & Shim, S.-M. (2023). Safety evaluation of kaempferol glycosides-rich standardized roasted goji berry leaf extract. *Regulatory Toxicology and Pharmacology, 140*, 105382. https://doi.org/10.1016/j.yrtph.2023.105382

Lin, C. Y., Shen, Y. T., & Ni, C. C. (2011). *Effect of Various Concentrations of Resveratrol on Pancreatic Beta-Cell.* Wiley Online Library.

Loo, C.-Y., Siew, E. L., Young, P. M., Traini, D., & Lee, W.-H. (2022). Toxicity of curcumin nanoparticles towards alveolar macrophage: effects of surface charges. *Food and Chemical Toxicology, 163*, 112976. https://doi.org/10.1016/j.fct.2022.112976

Mankowski, R., You, L., Buford, T., Leeuwenburgh, C., Manini, T., Schneider, S., … Anton, S. (2020). Higher dose of resveratrol elevated cardiovascular disease risk biomarker levels in overweight older adults–a pilot study. *Experimental Gerontology, 131*, 110821.

Martín-Cordero, C., López-Lázaro, M., Gálvez, M., & Ayuso, M. J. (2003). Curcumin as a DNA topoisomerase II poison. *Journal of Enzyme Inhibition and Medicinal Chemistry, 18*(6), 505–509. https://doi.org/10.1080/14756360310001613085

Martins, L. A., Coelho, B. P., Behr, G., Pettenuzzo, L. F., Souza, I. C., Moreira, J. C., … Guma, F. C. (2014). Resveratrol induces pro-oxidant effects and time-dependent resistance to cytotoxicity in activated hepatic stellate cells. *Cell Biochemistry Biophysics, 68*(2), 247–257. https://doi.org/10.1007/s12013-013-9703-8

McClain, R. M., Wolz, E., Davidovich, A., & Bausch, J. (2006a). Genetic toxicity studies with genistein. *Food and Chemical Toxicology, 44*(1), 42–55. https://doi.org/10.1016/j.fct.2005.06.004

McClain, R. M., Wolz, E., Davidovich, A., Pfannkuch, F., Edwards, J. A., & Bausch, J. (2006b). Acute, subchronic and chronic safety studies with genistein in rats. *Food and Chemical Toxicology, 44*(1), 56–80. https://doi.org/10.1016/j.fct.2005.05.021

McClain, R. M., Wolz, E., Davidovich, A., Edwards, J., & Bausch, J. (2007). Reproductive safety studies with genistein in rats. *Food and Chemical Toxicology, 45*(8), 1319–1332. https://doi.org/10.1016/j.fct.2007.01.009

Moon, R. T., Kohn, A. D., De Ferrari, G. V., & Kaykas, A. (2004). WNT and beta-catenin signalling: diseases and therapies. *Nature Reviews Genetics, 5*(9), 691–701. https://doi.org/10.1038/nrg1427

Moos, P. J., Edes, K., Mullally, J. E., & Fitzpatrick, F. A. (2004). Curcumin impairs tumor suppressor p53 function in colon cancer cells. *Carcinogenesis, 25*(9), 1611–1617. https://doi.org/10.1093/carcin/bgh163

Müller, R. H., Radtke, M., & Wissing, S. A. (2002). Solid lipid nanoparticles (SLN) and nanostructured lipid carriers (NLC) in cosmetic and dermatological preparations. *Advanced Drug Delivery Reviews, 54*, S131–S155.

Muñoz, O., & Bustamante, S. (2015). Pharmacological properties of resveratrol. A pre-clinical and clinical review. *Biochemical Pharmacology, 4*(184), 1–12.

Niaz, K., Shah, M. A., Khan, F., Saleem, U., Vargas, C., & Panichayupakaranant, P. (2020). Bioavailability and safety of phytonutrients. In S. M. Nabavi, I. Suntar, & H. Khan (Eds.), *Phytonutrients in Food* (pp. 117–136). Elsevier.

Niering, P., Michels, G., Wätjen, W., Ohler, S., Steffan, B., Chovolou, Y., ... Kahl, R. (2005). Protective and detrimental effects of kaempferol in rat H4IIE cells: implication of oxidative stress and apoptosis. *Toxicology and Applied Pharmacology, 209*(2), 114–122. https://doi.org/10.1016/j.taap.2005.04.004

Niho, N., Shibutani, M., Tamura, T., Toyoda, K., Uneyama, C., Takahashi, N., & Hirose, M. (2001). Subchronic toxicity study of gallic acid by oral administration in F344 rats. *Food and Chemical Toxicology, 39*(11), 1063–1070. https://doi.org/10.1016/s0278-6915(01)00054-0

Okamoto, T. (2005). Safety of quercetin for clinical application. *International Journal of Molecular Medicine, 16*(2), 275–278.

Ombredane, A. S., Silva, V. R., Andrade, L. R., Pinheiro, W. O., Simonelly, M., Oliveira, J. V., ... Garcia, M. P. (2021). In vivo efficacy and toxicity of curcumin nanoparticles in breast cancer treatment: a systematic review. *Frontiers in Oncology, 11*, 612903.

Pamukcu, A. M., Yalçiner, Ş., Hatcher, J. F., & Bryan, G. T. (1980). Quercetin, a rat intestinal and bladder carcinogen present in bracken fern (Pteridium aquilinum). *Cancer Research, 40*(10), 3468–3472.

Peng, C. C., Hsieh, C. L., Wang, H. E., Chung, J. Y., Chen, K. C., & Peng, R. Y. (2012). Ferulic acid is nephrodamaging while gallic acid is renal protective in long term treatment of chronic kidney disease. *Clinical Nutrition, 31*(3), 405–414. https://doi.org/10.1016/j.clnu.2011.11.003

Poulsen, M. M., Vestergaard, P. F., Clasen, B. F., Radko, Y., Christensen, L. P., Stødkilde-Jørgensen, H., ... Jørgensen, J. O. (2013). High-dose resveratrol supplementation in obese men: an investigator-initiated, randomized, placebo-controlled clinical trial of substrate metabolism, insulin sensitivity, and body composition. *Diabetes, 62*(4), 1186–1195. https://doi.org/10.2337/db12-0975

Ruiz, M. J., Estela, J. M., Asensi, M. Á., Mañes, J., Picó. (2006). Short-term oral toxicity of quercetin and pterostibene in Swiss mice. *Toxicology Letters*, *164*, 275–276.

Saleem, A., Akhtar, M. F., Shah, M. A., Saleem, M., & Siddique, R. (2023). Phytonutrients: adverse drug reactions. In S. M. Nabavi, I. Suntar, & H. Khan (Eds.), *Phytonutrients and Neurological Disorders* (pp. 353–368). Elsevier.

Sandur, S. K., Ichikawa, H., Pandey, M. K., Kunnumakkara, A. B., Sung, B., Sethi, G., & Aggarwal, B. B. (2007). Role of pro-oxidants and antioxidants in the anti-inflammatory and apoptotic effects of curcumin (diferuloylmethane). *Free Radical and Biology Medicine, 43*(4), 568–580. https://doi.org/10.1016/j.freeradbiomed.2007.05.009

Santos, A. C., Pereira, I., Magalhães, M., Pereira-Silva, M., Caldas, M., Ferreira, L., ... Veiga, F. (2019). Targeting cancer via resveratrol-loaded nanoparticles administration: focusing on in vivo evidence. *The AAPS Journal, 21*, 1–16.

Sharma, R. A., Euden, S. A., Platton, S. L., Cooke, D. N., Shafayat, A., Hewitt, H. R., ... Steward, W. P. (2004). Phase I clinical trial of oral curcumin: biomarkers of systemic activity and compliance. *Clinical Cancer Research, 10*(20), 6847–6854. https://doi.org/10.1158/1078-0432.ccr-04-0744

Si, D., Wang, Y., Zhou, Y. H., Guo, Y., Wang, J., Zhou, H., ... Fawcett, J. P. (2009). Mechanism of CYP2C9 inhibition by flavones and flavonols. *Drug Metabolism and Disposition, 37*(3), 629–634. https://doi.org/10.1124/dmd.108.023416

Souza, I. C., Martins, L. A., Coelho, B. P., Grivicich, I., Guaragna, R. M., Gottfried, C., ... Guma, F. C. (2008). Resveratrol inhibits cell growth by inducing cell cycle arrest in activated hepatic stellate cells. *Molecular and Cellular Biochemistry, 315*(1–2), 1–7. https://doi.org/10.1007/s11010-008-9781-x

Süntar, I., & Yakıncı, Ö. F. (2020). Potential risks of phytonutrients associated with high-dose or long-term use. In S. M. Nabavi, I. Suntar, & H. Khan (Eds.), *Phytonutrients in Food* (pp. 137–155). Elsevier.

Thipe, V. C., Panjtan Amiri, K., Bloebaum, P., Raphael Karikachery, A., Khoobchandani, M., Katti, K. K., ... Katti, K. V. (2019). Development of resveratrol-conjugated gold nanoparticles: interrelationship of increased resveratrol corona on anti-tumor efficacy against breast, pancreatic and prostate cancers. *International Journal of Nanomedicine, 14*, 4413–4428. https://doi.org/10.2147/ijn.s204443

Tiwari, R., Siddiqui, M. H., Mahmood, T., Farooqui, A., Bagga, P., Ahsan, F., & Shamim, A. (2020). An exploratory analysis on the toxicity & safety profile of Polyherbal combination of curcumin, quercetin and rutin. *Clinical Phytoscience, 6*(1), 82. https://doi.org/10.1186/s40816-020-00228-2

Tong, J., Shen, Y., Zhang, Z., Hu, Y., Zhang, X., & Han, L. (2019). Apigenin inhibits epithelial-mesenchymal transition of human colon cancer cells through NF-κB/Snail signaling pathway. *Bioscience Reports, 39*(5). https://doi.org/10.1042/bsr20190452

Toxicology and carcinogenesis studies of genistein (Cas No. 446–72-0) in Sprague-Dawley rats (feed study). (2008). *Natl Toxicol Program Tech Rep Ser*(545), 1–240.

Tsvetkov, P., Asher, G., Reiss, V., Shaul, Y., Sachs, L., & Lotem, J. (2005). Inhibition of NAD(P)H:quinone oxidoreductase 1 activity and induction of p53 degradation by the natural phenolic compound curcumin. *Proceedings of the National Academy of Sciences of the United States of America, 102*(15), 5535–5540. https://doi.org/10.1073/pnas.0501828102

Urbina-Cano, P., Bobadilla-Morales, L., Ramírez-Herrera, M. A., Corona-Rivera, J. R., Mendoza-Magaña, M. L., Troyo-Sanromán, R., & Corona-Rivera, A. (2006). DNA damage in mouse lymphocytes exposed to curcumin and copper. *Journal of Applied Genetics, 47*(4), 377–382. doi:10.1007/bf03194648

Utesch, D., Feige, K., Dasenbrock, J., Broschard, T. H., Harwood, M., Danielewska-Nikiel, B., & Lines, T. C. (2008). Evaluation of the potential in vivo genotoxicity of quercetin. *Mutation Research/Genetic Toxicology and Environmental Mutagenesis, 654*(1), 38–44. https://doi.org/10.1016/j.mrgentox.2008.04.008

Variya, B. C., Bakrania, A. K., Madan, P., & Patel, S. S. (2019). Acute and 28-days repeated dose sub-acute toxicity study of gallic acid in albino mice. *Regulatory Toxicology and Pharmacology, 101*, 71–78. https://doi.org/10.1016/j.yrtph.2018.11.010

Verschoyle, R. D., Steward, W. P., & Gescher, A. J. (2007). Putative cancer chemopreventive agents of dietary origin-how safe are they? *Nutrition and Cancer, 59*(2), 152–162. https://doi.org/10.1080/01635580701458186

Wan, S., Zhang, L., Quan, Y., & Wei, K. (2018). Resveratrol-loaded PLGA nanoparticles: enhanced stability, solubility and bioactivity of resveratrol for non-alcoholic fatty liver disease therapy. *Royal Society Open Science, 5*(11), 181457. https://doi.org/10.1098/rsos.181457

Zanwar, A. A., Badole, S. L., Shende, P. S., Hegde, M. V., & Bodhankar, S. L. (2014). Chapter 80 – Role of gallic acid in cardiovascular disorders. In R. R. Watson, V. R. Preedy, & S. Zibadi (Eds.), *Polyphenols in Human Health and Disease* (pp. 1045–1047). Academic Press.

Section D

Novel Dosage Forms of Phytochemicals

14 Novel Pharmaceutical Dosage Forms of Phytochemicals for the Treatment of Rheumatoid Arthritis

Samia Gul Niazi, Mahdia Arshad, Chanda Javed, Shahida Shaeen, Anil Ahsan, Muhammad Furqan Akhtar, and Ammara Saleem

WHAT YOU WILL LEARN?

In this chapter, the introduction to various targeted dosage forms will be provided that are used in numerous research studies to cope with the problems of drug adverse reaction, solubility, bioavailability, and target site delivery and enhance therapeutic effectiveness. In this chapter, the details of various dosage forms along with their pros and cons have been provided.

LIST OF ABBREVIATIONS

COX-2	Cyclooxygenase-2
DMARDs	Disease-modifying antirheumatic drugs
IL-1	Interleukin-1
LOX	Lipoxygenase
MMP	Matrix metalloproteinase
MAPK	Mitogen-activated protein kinase
MTX	Methotrexate
NF-κB	Nuclear factor kappa beta
NSAIDs	Nonsteroidal anti-inflammatory drugs
ORMS	Oral modified release systems
PCLPEG	Polycaprolactone–polyethylene glycol–polycaprolactone
PLGA	Polylactic-co-glycolic acid
PGE2	Prostaglandin E2
ROS	Reactive oxygen species
RA	Rheumatoid arthritis
TNF	Tumor necrosis factor

14.1 INTRODUCTION

Rheumatoid arthritis (RA) is an autoimmune disease in which a person's own immune system acts against the lining of joints all over the body. It is also referred to as a systemic inflammatory disease mainly affecting the diarthrodial joints. This inflammatory disease leads to cartilage and bone damage. It is a chronic systemic articular disease with extra-articular manifestations

DOI: 10.1201/9781003535201-18

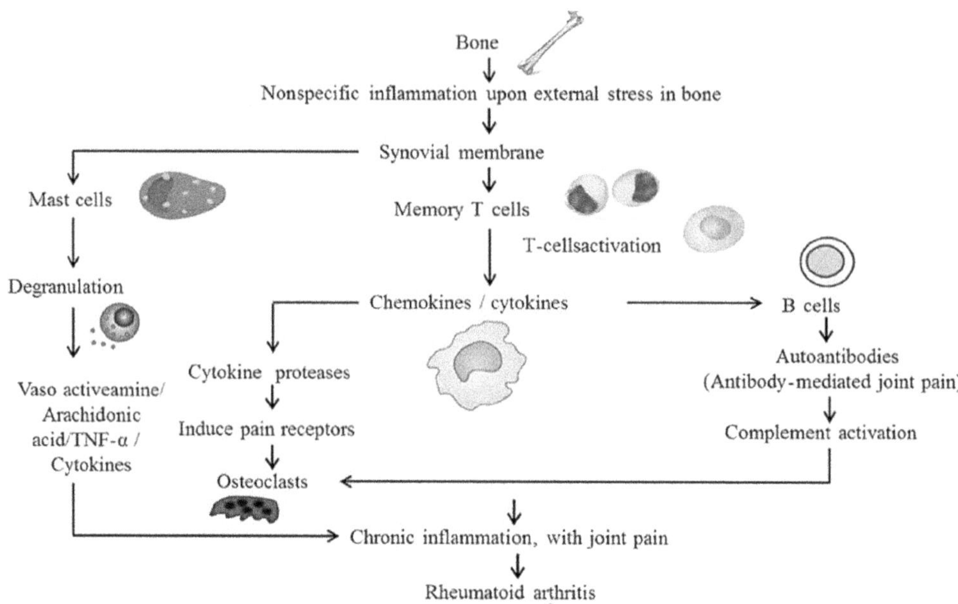

FIGURE 14.1 Pathophysiology of rheumatoid arthritis.

(Venetsanopoulou, 2022). Uncontrolled inflammation due to RA damages cartilage that acts as a shock absorber in the joints eventually causing erosion of bones and may also result in the fusion of joints as a body's protective effort against constant irritation. This disease has periods of increased activity such as flare alternating with periods of remission in which the swelling and pain decrease. This disease affects 0.5%–1% adult population worldwide. Pathophysiology of RA is shown in Figure 14.1.

14.2 PLANTS FOR THE TREATMENT OF RA

There are many phytochemicals that are commonly used for the treatment of RA. Some of these plants are listed below:

14.2.1 Calendula Oil

Calendula officinalis is a plant of the Asteraceae family. Calendula contains many phytochemicals such as triterpene glycosides like calendula glycoside A, triterpene alcohols such as taraxterol and faradiol, flavonoids like quercetin, volatile oils, and polyphenols (Arora D, 2013). Calendula oil is used in many topical formulations to study its beneficial effects against RA. Pain is one of the most common and intense symptoms of RA. Diclofenac gel is one of the most commonly used and most effective gels for the pain noticed in RA (Efe et al., 2014). The combination of calendula extract, diclofenac sodium, and methyl salicylate in the form of a gel showed improved results. Calendula extract proved to increase the percutaneous absorption of diclofenac sodium (Bodhankar, Kavade, & Patil, 2012). Calendula extract has anti-inflammatory and antioxidant activity that helps with the management of RA. Topical 70% ethanolic extract of calendula flower at a dose of 1.2mg/ear reduced croton oil-induced edema by 20% in mice (Ashwlayan, Kumar, Verma, Garg, & Gupta, 2018).

14.2.2 Aloe Vera Gel

Aloe barbadensis is a plant of the Aloaceae family. Phytochemicals extracted from the aloe plant contain chromones like aloesin, aloeresins, phenol pyrones like aloenin, anthrones like aloin, aloe emodin, flavonoids like quercetin, sterols, polyphenols, and volatile oils (Nalimu F, 2021). Research has been done to show the topical application of aloe vera gel and its application in the case of RA. A combination of calendula extract, aloe vera gel, and glycine max has recently been studied and has shown beneficial topical effects. The formulated gel contains many flavonoids, and the result of its application has shown the same effect as that of diclofenac gel. It causes a decrease in the inflammatory mediators and cytokines that result in its anti-inflammatory and antioxidant activity.

14.2.3 Agar Tree

Aquilaria agallocha is a plant of the family Thymelaeaceae. The major compound it contains includes volatile oils, valerianol, chromones, and sesquiterpenes (Lin, 2016). Its paste is applied on the joint to reduce pain and inflammation in RA (Chandrasekar, 2017). Both the *Aquilaria agallocha* oil from Heart Wood and the ethanolic extract of its leaves (EEAA) were investigated utilizing Freund's adjuvant-induced arthritic rat model. The increase in paw volume in the arthritic rats was decreased by EEAA, with the greatest inhibition occurring on day 21 at 21.20% and 25.34%, respectively (Habibur Rahman, 2016).

14.2.4 Ferula asafetida

It is a plant of Apiaceae family. The genus Ferula contains large amounts of ferunitinin, a phytoestrogen that has strong osteoprotective properties. When taken daily for two months, ferunitin significantly reduces the risk of osteoporosis (Palumbo C, 2009). The active ingredients of *F. persica*, including its crude extract, persica sulfide, and umbelliprenin, significantly inhibited the endopeptidases MMP-2 and 9, which are implicated in inflammatory arthritis, and control the degradation of an extracellular matrix (Shahverdi AR, 2006).

14.3 HERBAL FORMULATIONS IN CLINICAL USE

Various plant-based herbal formulations have been clinically tested and are found to be beneficial against RA. Some of these formulations that are mostly used are given in Table 14.1.

14.3.1 Herbal Dosage Forms from Different Systems of Medication

14.3.1.1 Ayurvedic Medicinal Systems

Ayurveda originates from two words: ayur meaning life and veda meaning science or study. Its origin is more than 5000 years old. It is a traditional Indian medicinal system and is used heavily in many Asian as well as Western countries due to its minimal or no side effects. It mainly focuses on mind, body, and spirit or three doshas that are the three energy sources for humans and their equilibrium. In the Ayurveda system, RA disease is named as Amavata (Chopra A, 2010). Amavata is defined as a defect of vata doshas or the energy systems and accumulation of Ama or toxic, pro-inflammatory waste products in the joints, resulting in inflammation and pain. Ayurvedic medicine not only tries to reduce the symptoms of RA but also tries to bring the body back to the normal state. Ayurvedic medicines contain medicinal plants or plant extracts. Some of the ayurvedic medicines used for the treatment of RA are given in Table 14.2.

TABLE 14.1
Herbal Formulations That are Clinically Used for Rheumatoid Arthritis

Preparation	Botanical Source	Major Phytoconstituents	Mechanism of Action	References
Amyla joint care capsule	*Withania somnifera* (Ashwagandha)	Withanolides, withaferin A.	Repress IL-6, IL-1, COX-2, TNF-α Inhibit NF-kB.	(Balkrishna A, 2019)
BCM-95	*Curcuma longa*	β-sitosterol, Campesterol	Repress mediators involved in inflammation Inhibits NF-kB and LOX.	(Haroyan A, 2018) (R., 2013)
Cesamet	*Cannabis sativum*	Cannabidiol	Decrease reactive oxygen intermediate Reduces TNF-α, INF-γ, and collagen II-specific antibodies.	(Bruni N, 2018) (Brunetti P, 2020) (Mohiuddin MM, 2020)
Divya Amavatri Ras	*Triphala*	Flavanoid, tannins, sterols	Inhibits LOX1, COX2, and IL-1β. Reduces the action of IL-6, TNF-α, and NF-ₖB	(Balkrishna A, 2019)
Guggulutiktam Kwatham	*Aconitum heterophyllum*	Aconitine, glycosides, terpenoids	Inhibits TLR-4 MYD88 Hinders NF-ₖB and MAPK activation.	(Paramanick D, 2017) (Aswathy IS, 2019)
Aflapin	*Benksia serrata*	Monoterpenoids, β-boswellic acid inhibition	Inhibition of LOX-5, COX-2, and NF-ₖB.	(Sumantran VN, 2011) (Suva MA, 2018)
Kapiva Artho Sure Juice	Cassia fistula	Glycosides	Reduction in levels of inflammatory mediators.	(Farooq H, 2018)

IL-1: interleukin-1; TNF: tumor necrosis factor; LOX: lipoxygenase, NF-κB: nuclear factor kappa beta; COX-2: cyclooxygenase-2; PGE2: prostaglandin E2, MAPK: mitogen-activated protein kinase; ROS: reactive oxygen species.

14.3.1.1.1 Churna or Powdered Dosage Form

Churna is the most widely used dosage form in the Ayurvedic system. It is a powdered formulation that is made by mixing two or more finely divided powders of natural origin. It is mostly taken orally and is usually taken with oils or honey. They can be bitter in taste, and their taste masking is quite difficult.

14.3.1.1.1.1 Ajmodadi Churna Ajmodadi churna or choorna is a polyherbal ayurvedic medicine. It is used as a carminative, anti-spasmodic, wormifuge that helps in painful conditions such as stiffness and also helps in restoring the digestive system (Sriwastava, 2010). It contains powders of *Carum roxburghianum*, *Piper longum*, *Cedrus deodara*, *Plumbago zeylanica*, *Piper nigrum*, *Ricinus communis*, *Embelia ribes*, *Foeniculum vulgare*, *Terminalia chebula*, *Zingiber officinale*, *Argyria speciosa*, and rock salt (Aswatha et al., 2010). It was reported that treatment with ajmodadi choorna improved conditions like body ache, pain, swelling and stiffness of joints, and indigestion (Mishra PK, 2013). It is used as an anti-inflammatory medicine and its aqueous extract reduces paw edema in rats (S. N. Aswatha Ram HN, Makhija I, Shreedhara CS., 2012a). The extracts of ajmodadi churna reduce paw swelling, which shows its inhibiting effect in the prostaglandin-mediated inflammatory pathway (Aswatha et al., 2012b). It also helps in healing tissue injury.

TABLE 14.2
Ayurvedic Medicines Used for the Treatment of Rheumatoid Arthritis

Name of Medicine	Type of Formulation	Ingredients of Formulation
Ajmodadi Churna	Churna or powdered form	*Carum roxburghianum, Piper longum, Cedrus deodara, Plumbago zeylanica, Piper nigrum, Ricinus communis, Embelia ribes, Foeniculum vulgare, Terminalia chebula, Zingiber officinale, Argyria speciosa*, and rock salt.
Amrutadyam Churna	Churna or powdered form	*Tinospora cordifolia, Zingiber officinale, Tribulus terrestris, Sphaeranthus indicus*, and *Crataeva nurvala.*
Alambusadi Churna	Churna or powdered form	*Mimosa pudica, Tribulus terrestris, Emblica officinalis, Terminalia chebula, Terminalia bellirica, Zingiber officinale, Tinospora cardifolia,* and *Operculina turpethum.*
Bhallaatakaadi Churna	Churna or powdered form	*Semecarpus anacardium, Sesamum indicum, Terminalia chebula,* and jiggery.
Chitrakadi Churna	Churna or powdered form	*Plumbago zeylanicum, Zingiber officinale, Asafoetida, Piper longum, Piper chaba, Trachyspermum roxburghianum, Piper nigrum, swarjikakshara, yavakshara* (alkaline preparation from barley), rock salt, sauvarchal salt (Black salt), vida salt (ammonium chloride), common salt, romaka salt (lake salt), and *Citrus medica* or *Punica granatum*
Hingwadi Churna	Churna or powdered form	*Asafoetida, Garcinia morella, Piper nigrum, Plumbago zeylanicum, Piper longum, Coriandrun sativum, Nigella sativa, Cuminum cyminum, Acorus calamus, Piper chaba, Cyclea peltate, Hedychium spicatum, Garcinia indica,* rock salt, sochal salt, bidal salt, *Zingiber officinale, Hordeum vulgare, Punica granatum, Terminalia chebula, Inula racemose, Sphaeranthus indicus, ginger juice, and lemon juice*
Panchasama Churna	Churna or powdered form	*Zingiber officinale, Terminalia chebula, Piper longum, Operculina terpethum* and black salt or *Sauvarchala*
Panchakola Churna	Churna or powdered form	*Piper longum, Piper chaba, Plumbago zeylanica,* and *Zingiber officinale*
Pathyadi Churna	Churna or powdered form	*Terminalia chebula, Zingiber officinale,* and *Anethum sowa*
Pippali Churna	Churna or powdered form	*Piper longum*
Punarnavadi Churna	Churna or powdered form	*Boerhavia diffusa, Tinospora cordifoila, Zingiber officinale, Anethum sowa, Argyreia speciosa, Hedychium spicatum,* and *Sphaeranthus indicus*
Triphala	Churna or powdered form	*Embelica officinale, Terminalia belerica,* and *Terminalia chebula*
Trikatu Churna	Churna or powdered form	*Zingiber officinale, Piper longum*, and *Piper nigrum*
Vaishnawar Churna	Churna or powdered form	Rock salt, *Trachyspermum ammi, Trachyspermum rocburghianum, Zingiber officinale,* and *Terminalia chebula*
Alambushadi Ghana Vati	Tablet	*Mimosa pudica, Tribulus terrestris, Terminalia chebula, Terminalia bellirica, Phyllanthus emblica, Zingiber officinale, Tinospora cordifolia,* and *Operculina turpethum.*
Amavata Pramathini Vati	Tablet	Potassium nitrate, *Aristolochia indica,* purified sulfur, calcined iron, and calcined mica
Patoladi Yoga Tablet	Tablet	*Trichochosanthes dioica, Tinospora cordifolia, Zingiber officinale,* and *Picrorhiza kurrooa*

(Continued)

TABLE 14.2 (*Continued*)
Ayurvedic Medicines Used for the Treatment of Rheumatoid Arthritis

Name of Medicine	Type of Formulation	Ingredients of Formulation
Shatapatrayadi Churna Tablet	Tablet	*Tinospora cordifolia, Terminalia chebula, Eclipta alba, Shatapatri, Azadirachta indica, Andrographis paniculate, Trichochosanthes dioica, Glycyrrhiza glabra, Emblica officinalis, Barberis aristate, Santalum album, Vitis vinifera, Phyllanthus urinaria, Cedrus deodara,* and *Curcuma longa.*
Rasona Pinda	Pill	*Anethum sowa, Allium sativum, Saussurea lappa, Curcuma longa, Sesamum indicum, Ferula asafoetida, Trikatu churna,* salt, *Piper longum, Plumbago zeylanica, Trachyspermum roxburghianum,* and *Coriandrum sativum*
Rasonadi Kwath	Decoction	*Vitex negundo, Zingiber officinale,* and *Allium sativum.*
Shunthi Gokshura Kwatha	Decoction	*Zingiber officinale* and *Tribulus terrestris*
Rasna Saptak Kwath	Decoction	*Boerhavia diffusa, Pluchea lanceolata, Ricinus communis, Tribulus terrestris, Cedrus deodara, Zingiber officinale, Cassia fistula,* and *Tinospora cordifolia*
Shigru Twak Kwath	Decoction	*Moringa oleifera*
Mahaushadhi Kwath	Decoction	*Tinospora cordifolia* and *Zingiber officinale*
Amrita Gharita	Medicated Ghee	Tinospora *cardifolia and Zingiber officinale*
Eranda Taila	Medicated oil	Castor oil
Saindavadi Taila	Medicated oil	Sodium chloride, *Cedrus deodara, Acorus calamus, Zingiber officinale, Myrica esculenta, Anethum sowa, Cyperus rotundus, Piper retrofractum, Polygonatum verticillatum, Polygonatum cirrhifolium, Croton tinglium, Operculina turpethum, Barringtonia acutangular, Pavonia odorata, Plumbago zeylanica, Clerodendrum serratum,Curcuma zedoaria, Embelia ribes, Glycyrrhiza glabra, Vitex negundo, Aconitum heterophyllum, Ricinus communis, Cissampleos pareira, Indigofera tinctoria, Baliospermum montanum, Piper nigrum, Trachyspermum ammi, Piper longum, Sausurea lappa, Pluchea lanceolata,* and *Brassica campestris*
Shiva guggul	Oleo gum resin	*Terminalia chebula, Terminalia belerica, Emblica officinalis, Ricinus communis,* purified sulfur, *Commiphora mukul, Alpinia galanga, Embelia ribes, Piper nigrum, Piper longum, Baliospermum monatum, Nardostachys jatamansi, Zingiber officinale,* and *Cedrus deodara*
Simhanaada Guggul	Oleo gum resin	*Terminalia chebula, Terminalia belerica, Emblica officinalis, Ricinus communis,* purified sulfur, and *Commiphora mukul*

14.3.1.1.1.2 Amrutadyam Churna Amrutadyam churna is a well-known ayurvedic powder formulation that is made up of *Tinospora cordifolia, Zingiber officinale, Tribulus terrestris, Sphaeranthus indicus,* and *Crataeva nurvala.* The experiment showed that after the treatment of RA patients with amrutadyam churna symptoms like anorexia and vomiting were completely relieved and also showed a moderate reduction in pain and a slight reduction in tenderness of muscles (Patel HD, 2020).

14.3.1.1.1.3 Alambusadi Churna This powdered formulation has many types of useful phytochemicals such as reducing sugars, tannins, phenolic compounds, saponins, glycosides, and gum (Saroj Kumar Debnath). It is mainly used in chronic inflammatory diseases. It is composed of

Mimosa pudica, Tribulus terrestris, Emblica officinalis, Terminalia chebula, Terminalia bellirica, Zingiber officinale, Tinospora cardifolia, and *Operculina turpethum* (Kamble S, 2015). According to the research, significant changes in pain, swelling, grip power, walking time, and stiffness of joints were seen after the treatment of RA with alambusadi churna (Kumar S, 2019).

14.3.1.1.1.4 Bhallaatakaadi Churna It contains Bhallataka (*Semecarpus anacardium*), Tila (*Sesamum indicum*), Haritaki (*Terminalia chebula*), and jiggery. It is used mainly with oils such as castor oil and olive oil. It was used with castor oil to treat 60 patients with RA. All the patients showed improvement in conditions, improved RA factor, and sedimentation rats (Khot, 2015). Its pharmacological evaluation showed its antioxidant, anti-inflammatory, and anti-nociceptive effects (Parihar, 2021).

14.3.1.1.1.5 Chitrakadi Churna Chitrakadi choorna is said to be katu (pungent) and tikkta rasa (bitter). Hence, it is used in digestion, as a blood purifier, as an anti-oxidant, and in the normal functioning of body organs (Baheti, 2022). It consists of *Plumbago zeylanicum, Zingiber officinale, Asafoetida, Piper longum, Piper chaba, Trachyspermum roxburghianum, Piper nigrum, swarjikakshara, yavakshara* (alkaline preparation from barley), rock salt, sauvarchal salt (Black salt), vida salt (Ammonium chloride), common salt, romaka salt (lake salt), and *Citrus medica* or *Punica granatum*. It contains β-sitosterols that inhibit the secretion of IL-6 and TNF-a, both of these factors are implicated in RA. It helps in decreasing joint pain and tenderness. Its analgesic and anti-inflammatory effects are highly helpful in the case of RA. It also acts as an immunosuppressant and also contains ascorbic acid, which acts as an antioxidant (Rajput NO, 2016).

14.3.1.1.1.6 Hingwadi Churna In Ayurveda, it is mainly used in the treatment of indigestion, malnutrition syndrome, bloating, abdominal pain, etc. (Hebbar). It is composed of various plants such as *Asafoetida, Garcinia morella, Piper nigrum, Piper chaba, Cyclea peltate, Hedychium spicatum, Garcinia indica,* rock salt, sochal salt, bidal salt, *Zingiber officinale, Hordeum vulgare, Punica granatum, Terminalia chebula, Inula racemose, Sphaeranthus indicus,* ginger juice, and lemon juice. The treatment of arthritic patients with hingwadi choorna exhibited a decrease in the intensity of pain and changes in functional parameters such as foot pressure. It also decreases the blood sedimentation rate (Negi et al., 2017).

14.3.1.1.1.7 Panchasama Churna Panchasama churna is an ayurvedic formulation that is mainly used for piles, abdominal pain, bloating, and appetite loss (Chauhan, 2019). It is made up of five components *Zingiber officinale, Terminalia chebula, Piper longum, Operculina terpethum,* and black salt or *Sauvarchala*. Various controlled clinical trials were conducted in order to assess its effects in the case of RA. It was noticed that its use causes a decrease in body pain, anorexia, thirst, laziness, and fever.

14.3.1.1.1.8 Panchakola Churna The drugs present in panchakola churna help improve the digestive system as well as reduce the inflammation of joints. It was also used against gouty arthritis (Mohan A, 2021). It contains *Piper longum, Piper chaba, Plumbago zeylanica*, and *Zingiber officinale*. It is used as a supportive therapy in many conditions of arthritis and rheumatism (Upadhyay S, 2012).

14.3.1.1.1.9 Pathyadi Churna This ayurvedic medicine has several uses and is rich in medicinal properties like anti-oxidant, astringent, carminative, expectorant, and laxative. It is composed of *Terminalia chebula, Zingiber officinale*, and *Anethum sowa*. It was effective for relieving symptoms of RA such as pain in joints, swelling, morning stiffness, tenderness, fatigue, grip strength, and foot pressure ("Churna, Health Benefits Of Pathyadi," 2022).

14.3.1.1.1.10 Pippali Churna It is a famous ayurvedic medicine used for weight loss, diabetes, indigestion, and tuberculosis. It consists of *Piper longum*. In a study, 40 RA patients were taken randomly and were divided into two groups. The first group received pippali churna with honey and the second group received *Abhayadi leha*. Both formulations were found to be effective against RA. *Abhayadi leha* was found to be more effective in terms of relieving conditions of RA.

14.3.1.1.1.11 Punarnavadi Churna Punarnavadi churna contains various plants and is effective in improving conditions of arthritic patients (Thote AS, 2015).

14.3.1.1.1.12 Triphala It is an ayurvedic dosage form made up of three plants *Emblica officinalis, Terminalia belerica*, and *Terminalia chebula*. It causes immune suppression during inflammation and hence is used to treat autoimmune diseases. It decreases the serum inflammatory mediators level in cases of RA (Rohit, 2014).

14.3.1.1.1.13 Trikatu Churna Trikatu churna is an ayurvedic medicine used mostly for general health problems like fever, cold, and cough. It shows antibacterial and analgesic activity (Kalaiselvan S, 2015), antihelminthic activity (Reddy, 2019), and hepatoprotective activity (Reddy, 2009). It is made up of *Zingiber officinale, Piper longum*, and *Piper nigrum*. Its effect against inflammation was studied and then compared with that of the standard indomethacin. It decreased the inflammatory mediators and circulating immune complexes in the rats (S. S. Kumar, & Mishra, S. H., 2004).

14.3.1.1.1.14 Vaishnawar Churna Vaishnawar churna has a history of use in the case of stomach disorders such as constipation (Benni JM, 2011). It is prepared from rock salt, *Trachyspermum ammi, Trachyspermum rocburghianum, Zingiber officinale*, and *Terminalia chebula (Parwe, 2021)*. Its clinical therapeutic assessment was done for RA, and it was concluded that 62% cases were completely recovered, 26% were highly improved, and 12% were slightly improved (Mengi SA, 1995).

14.3.1.1.2 Vatika and Gutika

Vatika is similar to tablets. It is a solid dosage form of flat circular shape. Gutika is a circular dosage form often compared with pills. Some of the vatika used for RA are given below.

14.3.1.1.2.1 Alambushadi Ghana Vati It is an ayurvedic medicine that reduces the swelling in the joints of RA patients (Soman I, 2004). A randomized, open clinical study was done on 30 patients of RA who were divided into two groups. One of the groups was treated with Alambushadi Ghana Vati, and the second was treated with Vaitarana basti (an enema). The first group showed an improvement in signs and symptoms of RA (Gowri PM, 2007).

14.3.1.1.2.2 Amavata Pramathini Vati This is an ayurvedic medicine that shows very significant effects in the case of RA (Sharma A, 2010). It contains potassium nitrate, *Aristolochia indica*, purified sulfur, calcined iron, and calcined mica.

14.3.1.1.2.3 Patoladi Yoga Tablet This medicine is known to be useful in many conditions and is mainly used in the case of acid gastritis (Paschapur MS, 2009). It is composed of *Trichochosanthes dioica, Tinospora cordifolia, Zingiber officinale*, and *Picrorhiza kurrooa*. It is being used and its effects against RA are being noticed.

14.3.1.1.2.4 Shatapatrayadi Churna Tablet It is used in the treatment of tiredness, stress, fatigue, as an antiaging treatment, and as a natural aphrodisiac (J. Kumar, Dave, A. R., & Vyas, M. G.). A comparative study of shatapatryadi churna tablet and patoladi yoga tablet was done. The results indicated that Shatapatrayadi Churna tablet was more effective in alleviating the signs of RA (Hebbar).

14.3.1.1.2.5 Rasona Pinda It is extremely effective in reducing pain, swelling, tenderness, and stiffness, which are the main symptoms of RA (Gupta S, 2013). It is prepared from *Anethum sowa, Allium sativum, Saussurea lappa, Curcuma longa, Sesamum indicum, Ferula asafoetida, Trikatu churna,* salt, *Piper longum, Plumbago zeylanica, Trachyspermum roxburghianum*, and *Coriandrum sativum.* Its effect was studied in comparison with etoricoxib. Two groups of patients were taken randomly, the first group received Rosana panda while the second group was treated with etoricoxib. The first group showed a decrease in RA factor after three months of therapy. A decrease in joint pain, tenderness, and swelling was seen along with an improvement in grip strength and foot pressure (Singh JP, 2010).

14.3.1.1.3 Kvaatha or Decoctions

Kvaatha or kwath or kashaya or kashayam is a decoction that is prepared by boiling plants, herbs, or drugs in water. Various decoctions used in RA are as follows:

14.3.1.1.3.1 Rasonadi Kwath It is a widely used decoction for the treatment of RA and osteoarthritis and in the case of postviral fever, body muscle and joint pain, and shoulder dislocation pain (Marwat SK). Rasonadi Kwath contains *Vitex negundo, Zingiber officinale*, and *Allium sativum.* It contains many effective phytochemicals such as adenosine, ajoene, alpha tocopherol, caffeic acid, luteolin, isorhamnetin, and citral. These phytochemicals have anti-inflammatory, antiarthritic, analgesic, and immunomodulatory properties (Hebbar). In a study, the treatment of RA patients was done by giving 30ml of Rasonadi kwath with lukewarm water twice a day for a month. It was seen that 40% patients showed major improvement, 50% patients showed minor improvement, and 10% patients showed no improvements but no patient was completely cured (S., 2006).

14.3.1.1.3.2 Shunthi Gokshura Kwatha This formulation is mainly used for inflammation and edema. Several studies showed its effects against RA. It contains *Zingiber officinale* and *Tribulus terrestris.* In the study, patients were divided into three groups of 15 patients. The first group was treated with Shunthi Gokshura Kwatha, the second group was given aceclofenac, and the third group was the combination group where patients were treated with both Shunthi Gokshura Kwatha and aceclofenac. Shunthi Gokshura Kwatha was more effective in improving digestion, anorexia, and stiffness. Aceclofenac showed more improvement than group one but the patients of group three showed significant relief from *most signs of RA* (Gaikwad, 2017; Prakash, 2010).

14.3.1.1.3.3 Rasna Saptak Kwath It is a liquid ayurvedic medicine that is used mostly for back pain associated with injury or other conditions (Jeyadevi R). It contains various plants. Due to *Tribulus terrestris,* it has a surface charge neutralizing effect; due to other ingredients, it acts as an antiarthritic agent. It also has anti-inflammatory properties due to *Zingiber officinale (Habber).*

14.3.1.1.3.4 Shigru Twak Kwath It is prepared from *Moringa oleifera.* A group of 20 patients was treated with Shigru Twak Kwath. After the treatment, it was seen that Shigru Twak Kwath reduces the symptoms of RA and also corrects the pathogenesis of RA (2007).

14.3.1.1.3.5 Mahaushadhi Kwath It has only two ingredients: *Tinospora cordifolia* and *Zingiber officinale.* It was given to 30 patients for 60 days to evaluate its effects. Improvement was seen in different parameters of RA.

14.3.1.1.4 Medicated Ghee and Medicated Oils

Ghee in Ayurveda is called as gharitas and oils are referred to as taila or thailam. Some of these formulations are as follows:

14.3.1.1.4.1 Amrita Gharita It is composed of *Tinospora cardifolia* and *Zingiber officinale.* According to the clinical investigation, Amrita Gharita causes significant relief from the signs and symptoms of RA (Waskar R, 2019).

14.3.1.1.4.2 Eranda Taila Eranda taila or castor oil is widely used to prevent inflammation. Its effects were seen along with *Zingiber officinale*, which showed its effects in improving signs and symptoms of RA (Lekurwale PS, 2010).

14.3.1.1.4.3 Saindavadi Taila It is a polyherbal formulation and is also composed of sodium chloride. After treatment with this oil, RA patients showed a significant decrease in many parameters of RA (Mishra G, 2017).

14.3.1.1.5 Guggul

It is an oleo gum resin obtained from *Commiphora mukul.* Some of the gugguls used for RA are as follows:

14.3.1.1.5.1 Shiva Guggul It is composed of *Terminalia chebula, Terminalia belerica, Emblica officinalis, Ricinus communis,* purified sulfur, *Commiphora mukul, Alpinia galanga, Embelia ribes, Piper nigrum, Piper longum, Baliospermum monatum, Nardostachys jatamansi, Zingiber officinale*, and *Cedrus deodara* and is very effective in the case of RA.

14.3.1.1.5.2 Simhanaada Guggul It is made up of *Terminalia chebula, Terminalia belerica, Emblica officinalis, Ricinus communis,* purified sulfur, and *Commiphora mukul.* Its effect was compared with Shiva guggul in a study. The results showed that both formulations improve grip strength and walking time of the patient (Negi et al., 2017).

14.3.1.2 Unani Medical System

According to the Unani system, Waja ul Mafaasil or RA is caused by four natural humors of the body such as blood, phlegm, yellow bile, and black bile. Balancing out the humors reduces the complications and risks of RA. Some of the formulations used in the case of RA are as follows:

14.3.1.2.1 Majoon-e-Kundur

It is a compound Unani formulation that contains *Quercus incana, Zingiber officinale, Cyperus rotundus, Piper nigrum, Saussurea lappa*, and *Boswellia serrata.* Its beneficial effects were seen after oral administration in the case of osteoarthritis (Pandey S, 2012). A study conducted a comparative clinical trial with knee osteoarthritis patients that concluded that Majoon-e-kundur was more effective and had better tolerability than valdicoxib (Kimmatkar N, 2003). It inhibits the synthesis of leukotrienes in the case of RA that are the main cause of inflammation.

14.3.1.2.2 Majoon Suranjan

It is one of the most popular polypharmaceutical formulations of the Unani system that is used for RA (Thawani V, 2007). It was administered at 880 mg/kg to arthritic rats for 20 days via the oral route. Methotrexate, on the other hand, was used as a reference or standard drug. Reduction in serum autoantibodies and C reactive protein levels was seen. Significant inhibition of foot swelling in rats was also observed (Khan, 2018).

14.3.1.2.3 Habb-e-Asgand

It is a potent analgesic unani formulation (Y. S. Subramaneyaan M, Ahmed RS, Arora VK, Tripathi AK, Banerjee BD., 2013). It is well known for its anti-inflammatory and is used in the case of hepatotoxicity and arthritis. It is prepared from various plants. Arthritis-induced rats were treated by Habb-e-Asgand at the dose of 115 mg/kg once by oral route for 20 days and the effects were noticed and compared with methotrexate. Habb-e-Asgand caused a decrease in C-reactive protein levels and in serum autoantibodies more than methotrexate (Suhail, 2017).

14.3.1.2.4 UNIM-301

It is a polyherbal formulation well known for its anti-inflammatory properties (Y. S. Subramaneyaan M, Arora VK, Tripathi AK, Banerjee BD, Ahmed RS., 2013). It is composed of *Withania somnifera, Alpinia galanga, Operculina turpenthum, Colchicum luteum*, and *Zingiber officinale*. It caused dose-dependent reduction in paw edema and in the concentration of proinflammatory mediators when given to arthritic animals (S. Singh, et al., 2015). It reduces paw edema and lysosomal enzyme activity in the animal model of CFA-induced arthritis (Singh JP, 2010).

14.3.1.2.5 Qurs-e-Mafasil

It is a tablet dosage form made up of powders of *Curcuma longa, Delphinium denudatum, Colchicum luteum, Terminalia chebula*, and *Commiphora mukul*. Its use causes an improvement in morning stiffness, joint swelling, and pain (Rasool M, 2000).

14.3.1.3 Siddha System of Medicine

In the Siddha medicinal system, RA is termed as Uthiravathasuronitham. It is characterized by loss of appetite, pain, and swelling of joints. Some of siddha herbal formulations are given below:

14.3.1.3.1 Amukkara Choornam

It is a well-known formulation that is used in many cases such as *Candida albicans* (Yasmeen S, 2012) and ulcer (Rajamohamed, 2019). It is prepared from *Zingiber officinale, Withania somnifera, Piperlongum, Cinnamomum verum, Piper nigrum, Elettaria cardamomum, Syzygium aromaticum*, and sugar. It's *in vitro* use along with diclofenac sodium shows highly anti-inflammatory action (Patra, 2014).

14.3.1.3.2 Kalpaamruthaa

This formulation has beneficial antioxidant, anti-inflammatory, and antiapoptotic effects in many cardiovascular diseases (Amala, 2018). It is made of only two plants: *Semecarpus anacardium* and *Emblica officinalis*. Its efficacy was studied in Freund's adjuvant-induced arthritic rats. One group of rats was treated with *Semecarpus anacardium* and the second group was treated with Kalpaamruthaa for 14 days. Kalpaamruthaa group showed a decrease in activities of lysosomal enzymes and the levels of glucosaminoglycans (Latha R, 2015).

14.3.1.3.3 Mudakuvatha Legium

It is a polyherbal formulation. Many components of Mudakuvatha Legium have anti-inflammatory and antiarthritic activities (Mythilypriya R, 2008).

14.3.1.3.4 Karpoora Chindaani Mathirai

This formulation is given internally as well as externally for the treatment of RA (Singh A, 2009). It is prepared from minerals and herbs like calomel, *Myristica fragrans, Croton tiglium*, gum acacia, and juice of *Aloe vera*.

14.3.1.3.5 Mannennaikalavai Thylam

It is used as a tonic for the nervous system and musculoskeletal system and when taken internally it acts as an anti-inflammatory agent and supports bone, muscles, and joints (Meena, 2018). It is prepared from camphor, wax, neem oil, and peacock feather. After treatment with Mannennaikalavai thylam, swelling was reduced and RA factors and C-reactive protein levels were decreased in patients with RA (D. J. Singh, 2019).

14.3.1.3.6 *Vayu Keelagallagam*

It is composed of many herbs that are proven to be antiarthritic and anti-inflammatory (Meena R, 2015).

14.3.1.4 Chinese Medicinal System

In the Chinese medicinal system, RA falls in the category of Bi symptoms, which means a condition of obstruction. In Chinese medicinal systems like the other systems, this disease is treated by balancing out all the systems of the body. Some important Chinese formulations for the treatment of RA are as follows:

14.3.1.4.1 *Bi Qi Capsule*

It is a polyherbal formulation containing *Strychnos nux-vomica, Pheretima aspergillum, Codonopsis pilosula, Poria cocos, Atractylodis macrocephalae, Glycyrrhiza uralensis, Ligusticum wallichii, Salvia miltiorrhiza, Panax notoginseng*, and *Achyranthes bidentata.* Its main component is *Strychnos nux-vomica* that regulates the immune system. This capsule is effective in improving C-reactive protein levels and tender joint count (Gohila M. Nalini sofia H, 2019).

14.3.1.4.2 *Gui-Zhi-Shao-Yao-Zhi-Mu Decoction*

It is prepared from *Cinnamomum cassia, Cynanchum otophyllum, Anemarrhena asphodeloides, Glycyrrhiza uralensis, Ephedra sinica, Zingiberofficinale, Atractylodes macrocephala, Saposhnikovia divaricata*, and *Aconitum carmichaelii.* After clinical experimentation, it was seen that this drug is equal or may be better in effects than allopathic medicine (Chen XM, 2018).

14.3.1.4.3 *Han Shi Bi Pill*

It is a pill composed of *Aconiti lateralis, Cinnamom ramulus, Atractylodis macrocephalae, Chaenomelis fructus,* and *Asari radix.* It is used in the long-term treatment of RA as it has fewer side effects (Du TX, 2002; Man Y, 2010).

14.3.1.4.4 *Huo Luo Xiao Ling Dan*

It is made up of *Boswellia carteri, Commiphora myrrha, Angelica sinensis,* and *Salvia miltiorrhiza.* Its experimental studies showed its anti-arthritic activity as it reduces paw swelling, TNF-α level, and interleukin (IL)-6 levels (Zhang RX, 2009).

14.3.1.4.5 *Tong Bi Capsule*

It is prepared by *Gentiana macrophylla, CyperiRhizoma, Notopterygii, Angelicae pubescentis, Spatholobi caulis, Trachelospermi caulis, Taxilliherba Cyathulae*, and *Angelica sinensis.* It is used along with methotrexate for more effective treatment of RA (He, X.C., et al., 2014).

14.3.2 Other Novel Dosage Forms (Drug Carrier Systems) for Rheumatoid Arthritis

For ten years, significant progress has been made in the treatment of RA patients through new drug delivery methods, and second-line medications have been demonstrated to reduce activated T-cell proliferations. These are widely used in medicine delivery as it immediately reaches the afflicted area and control the release of the drugs over an extended length of time with lesser adverse effect in contrast to conventional dosage forms such as tablets and capsules. Novel dosage forms encapsulating phytochemicals/synthetic drugs for RA treatment are discussed here.

14.3.2.1 Nanoparticles

Polymeric nanoparticles are used as drug delivery vehicles for the treatment of RA. Researchers used polylactic-co-glycolic acid (PLGA) nanoparticles to control the pace of drug release from encapsulation and prolong the circulation time. It has been demonstrated that PLGA betamethasone

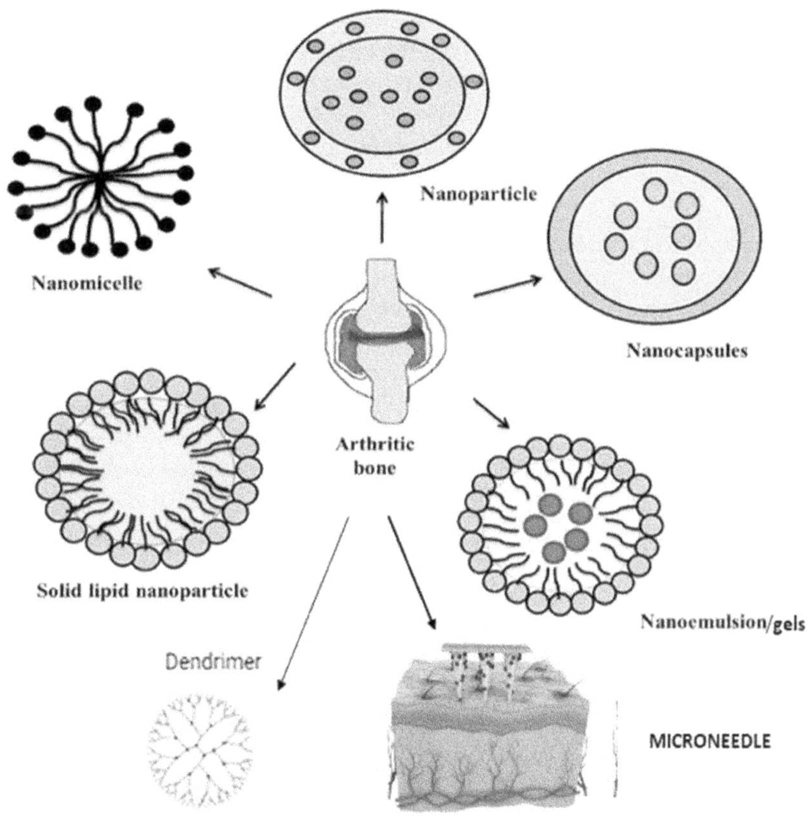

FIGURE 14.2 Novel dosage forms for the treatment of rheumatoid arthritis.

systems are substantially more effective than free glucocorticoids at lowering the inflammatory state when administered intravenously to arthritic rats and mice. Nagei et al. used solid ketoprofen nanoparticles with an average particle size of 83 nm to make a nanogel ointment using a bead mill process and observed its anti-inflammatory effects in Adjuvant-induced arthritis (AIA) rats. To treat RA, the FDA used both metallic and gold nanoparticles. However, it mostly favors iron-based nanoparticles for passive RA targeting. However, gold nanoparticles used to treat RA show nonspecific toxicity due to the adverse side effects (Higaki, 2005). Different forms of nanoparticles used in inflammation are given in Figure 14.2 and discussed below.

Two advantages of using a nanoparticle drug delivery technique are decreased dose frequency and enhanced drug solubility. It exhibits tissue distribution, altered pharmacokinetics, prolonged drug release, and a reduction in negative side effects. Higher prices, more difficult production, less flexibility in dose modification, highly complicated technology, the need for manufacturing knowledge, and the need to maintain dosage form stability are some of its disadvantages.

14.3.2.2 Transportation Challenges for Nanomedicines

Notwithstanding the advantages that nanoparticles offer, some of the challenges that drug delivery systems must overcome are as follows:

The untargeted nanoparticles might be detected by the mononuclear phagocyte system, which is present in bone marrow, blood, and organs including the liver, spleen, and lung (Andón F. T., 2017).

I. Because the particles are hydrophobic, there can be more blood components adhering to the nanoparticles shell.

II. The nanoparticles must travel a long distance to reach the target site for *in vivo* injection.
III. The enhanced permeability and retention effect may be impeded by atypical cell assembly in drug targeting, which would lead to less efficient drug internalization.

Numerous biological barriers, such as endosomal, nuclear, and cell membranes, severely impair drug delivery.

1. When biological fluid comes into contact with nanoparticles, molecules are gathered on their surface to create a "protein corona," which might affect how well the particles function as a therapy in the future. The biomolecular overlay of a nanoparticle determines its identity and influences its perception and interaction with cells, facilitating diffusion.
2. Respiratory endothelial system-derived activated macrophages assist the circulation in removing waste products and aging blood cells from the RES organs.
3. The nanoparticles experience a range of intricate fluid forces as they pass through the curvatures of both healthy and diseased blood vessels (M. J. Gomez-Garcia, 2018).
4. The extracellular matrix, which contains more collagen and provides tissues with their rigidity and structural integrity, also serves as a barrier to prevent nanoparticles from penetration.
5. Endocytosis occurs when the particles must stick to the cell membrane after extravasation from the blood arteries to the infection site.

Following their localization in the late endosomes and nucleus, the nanotherapeutic agents either return to the plasma membrane or escape lysosomal degradation. This implies that the nanoparticles face a number of difficulties when injected or applied topically; however, most of the transport issues can be resolved if a nanomedicine is created that targets a molecular component and is applied where inflammation is present.

14.3.2.2.1 Chitosan-Based Nanoparticle Modifications

Chitosan is a naturally occurring polysaccharide that is harmless and compostable and is comprised of amino chemical moieties and is widely used in biomedicine because of its biocompatibility and activity. Research on the amphiphilic qualities of chitosan by-products has gained attention recently. A thiolated glycol-chitosan polymer, which easily assembles in inflammatory joints and produces prolonged cycle physical characteristics in vivo, was used in one study to contain RNA that targets the Notch pathway (M. J. Kim, 2015).

14.3.2.2.2 Biomimetic Nanocarriers

According to recent research, PEG polymer nanoparticles can boost blood clearance by triggering a strong immunological reaction (M. Mohamed, 2019). Consequently, a biomimetic drug delivery system utilizing nanoparticles that are natural was created. To decrease their harmful and off-target effects, lengthen circulation, improve aggregation at the target location, circumvent immune defense, and trigger a strong immunological response, the technique disguises the nanoparticles as autologous materials (K. Jin, 2018).

14.3.2.2.3 Actively Targeting Macrophages Using Nanoparticles

Active drug delivery describes special interactions between the therapeutic medications and the nanocarrier and the target cells, often via a specific ligand–receptor interaction (F. Canal, 2010). When these components are adjacent to one another, interactions between the ligand and the receptor are possible (N. Gaspar, 2019). Surface modification allows the drug-transporting carrier system to reach the target area without the RES physiologically absorbing it. The modification method involves applying a biological bonding agent and albumin protein, a specific cell, or an antibody to the surface. The lytic and digesting enzymes present in the lysosomal organelles, the extracellular matrix, or the cell surface can all release the medication from its carrier. This ensures the drug will not adhere to the colloidal carrier moiety when it is released into the cytosol.

14.3.2.2.4 Liposomes Actively Target Macrophages in the Synovium

Liposomes are spherical nanocarriers composed of a phospholipid bilayer membrane on the exterior and a polar center (A. Akbarzadeh, 2013). The chemical and physical properties of liposomes make them an efficient delivery mechanism since therapeutic agents can be enclosed in the polar cavity or between the bilipid outer layers. Liposomes have long been used to treat RA. PEG is a hydrophilic polymer that can decrease the RES's capacity to detect and absorb liposomes, hence extending the body's circulation of these particles (G. Bozzuto and A. Molinari, 2015). Stealth polymeric liposomes can be used as nanocarriers to boost dexamethasone's therapeutic efficacy in RA. By preferentially accumulating in inflammatory joints, liposomes can inhibit excessive amounts of attributes that promote inflammation in RA joints, hence reducing inflammation and joint enlargement in arthritic animals. This illustrates polymer stealth liposomes' promise as a cutting-edge drug delivery technique for various medicinal uses (M. Zheng, 2021).

14.3.2.2.5 Synovial Macrophages Are Actively Targeted by Polymer-Based Nanocarriers

Due to their biocompatibility and ability to target specific areas, polymer nanocarriers are frequently used for medication delivery in RA. Therapeutic medications can be applied to the surface or enclosed in polymer nanoparticles to reach the RA joints. The most popular method for altering the surface of nanoparticles is PEG (E. Kabia, 2018). This technique can make nanodrugs more stable, less immunogenic, and more resistant to the body's elimination. By delivering a small dose of dexamethasone, polycaprolactone–polyethylene glycol–polycaprolactone (PCLPEG) micelles aim to treat arthritis. The micelles showed a lengthy lifetime in the systemic circulation and were extensively dispersed at the location of inflamed joints. Micelle-based dexamethasone delivery may successfully lower bone loss, inflammatory factor expression, and blood and joint edema. PCLPEG micelles significantly reduced adverse effects on body mass, T-cell and B-cell counts, and blood glucose levels in comparison to bare dexamethasone. Furthermore, they have a considerably diminished capacity to activate the complement system (Q. Wang J. J., 2016).

14.3.2.2.6 Microenvironment-Specific Nanoparticles

Priming agents, which alter the host environment in ways that promote drug accumulation unique to the target location, are a less popular method of enhancing drug delivery.

Inhibiting the complement system may hinder the mononuclear phagocyte system from identifying and eliminating nanoparticles; minor hypothermia increases vascular permeability; inhibiting fibroblasts decreases the production of extracellular matrix elements; target cell apoptosis increases the interstitial space; improving the target distribution of nanoparticles; and suppressing resident macrophages in the liver and spleen reduces the uptake of nanoparticles in healthy organs. These strategies can be used to achieve drug delivery to prevent drug uptake by healthy cells (C. I. Colino, 2020) (Nasra, 2022). A range of stimuli-sensitive nanoparticles release the therapeutic medications at the site of inflammatory joints in response to the disease-modified microenvironment.

14.3.2.2.7 Nanoparticles That Respond to Stimuli and a Dual Delivery System

Target-specific delivery techniques are stimulated by several stimulants, such as temperature, pH, and oxidation-reduction potential. In certain cases, two stimulants are concurrently present close to the target environment. When temperature and pH promote drug transfer, smart drug transport strategies might be helpful (C. Stefanadis, 2001). Numerous bifurcated polymers may be paired with binary stimulation, which has been developed with logical applications in a range of illnesses (Y. Pang, 2011).

14.3.2.2.8 Dendrimers

Dendrimers, often known as cascade molecules, are multipurpose, regularly branching macromolecules. Its unique branching structure and globular shape create a lot of surface groups that can be tailored to act as a template for drug administration, increasing drug loading. In the production of folic acid and methotrexate-conjugate poly (amid amine) dendrimers for the exact targeted activation of macrophages, the therapeutic effects of dendrimers are studied using Collegen-induced arthritis (CIA) model in mice. Methotrexate, when combined with amino groups of dendrimers exposed to blood circulation before being injected into the inflammatory areas, decreased the body weight of CIA mice due to its high cationic charge density (Khandare, 2006). The use of dendrimers has several advantages, such as improved drug loading, controlled release, and greater drug delivery to the targeted location; on the other hand, because dendrimers include an amino functional group, they are very hazardous.

Three main sites are included in the dendrimer design for drug entrapment via a range of mechanisms: (i) hydrogen bonding-mediated branching points; (ii) empty space-mediated molecule entrapment; and (iii) charge–charge interactions-mediated outer surface groups. Drug and dendrimer structures affect drug entrapment. Understanding the chemistry underlying the dendrimer design is crucial. Commercially accessible are PAMAM (polyamidoamine) dendrimers with amine, hydroxyl, carboxylate, and pyrrolidinone surfaces, along with the corresponding generations. This is a recommended method for determining a medicinal molecule's suitable dendrimer.

14.3.2.2.9 Lipogelosome

According to Turkey et al., lipogelosomes exhibit anti-inflammatory qualities. When administered intra-articularly, the lipogelosome containing diclofenac sodium shows an anti-inflammatory effect, in contrast to the topical version of the drug that is sold commercially. Furthermore, lipogelosome formulations employed to treat inflammatory changes in the synovium joints showed considerably better outcomes than those of the contralateral joints (control) in histopathology and biodistribution tests (Thakur S., 2018). It works wonderfully for treating joint issues in general and enhances retention time. However, stability issues and drug loss are among lipogelosome's disadvantages. When used intra-articularly to treat RA, the diclofenac sodium lipogelosome formulation demonstrated increased activity.

14.3.2.2.10 Diclofenac Sodium Lipogelosome

Drug delivery was one of the many uses for DFNa-loaded lipogelosome formulations. Production, characterization, *in vitro* release, and scintigraphic imaging tests were conducted according to the previously described protocols. Region-of-interest analysis was performed after arthritic knee joints were manually segmented in order to determine the formulation retention in scintigraphic imaging tests. The investigation demonstrated a general improvement in diclofenac sodium's anti-inflammatory properties, as well as an increase in the drug's bioavailability and efficacy (Tu¨rker S, 2005).

14.3.2.2.11 Emulgels

Stability, in vitro penetration, pH, and skin irritation test are the parameters used in the production of emulgels. Vandana et al. used the Carrageenan-induced approach to cause hind paw edema in Wistar rats in order to manufacture an aloe vera gel of mesulide. The ability of the produced formulations to lower inflammation was assessed through both in vitro and in vivo studies. The authors discovered that mesulide emulgels had a 54% penetration rate at 30 minutes, which suggests better drug release without irritating the skin, compared to the commercially available gel of mesulide, which had a 44% penetration rate. Its higher drug-loading capacity of 86% compared to 70% for commercial formulations further qualifies it for a strong anti-inflammatory effect (Vandana K. R., 2014). Because they facilitate medication release at the intended site without irritating the skin, emulgels are effective. On the other hand, contact dermatitis and inadequate absorption of larger drug particles are the primary adverse effects. Emulgels, which are hybrid emulsion and gel formulations, provide special benefits such as enhanced bioavailability, extended release, and targeted drug delivery to inflamed joints. These benefits minimize systemic adverse effects and optimize therapeutic efficacy in treating the disorders. Oral medications and topical creams have disadvantages, including limited penetration, low effectiveness, and side effects (Sinha, 2024). Emulgels improve drug solubility and penetration into affected tissues, which raises the possibility of disease-modifying effects. By carefully reviewing recent research on emulgels' (micro- and nanoemulgels') potential as a therapeutic approach for the treatment of arthritis, this review article emphasizes the promising potential of emulgels in targeted therapy regimens. The study emphasizes the need for more research and clinical trial translation to ascertain the long-term safety, efficacy, and cost-effectiveness of emulgel-based arthritis therapies in comparison to conventional treatments, despite the fact that they have made tremendous progress (D'Souza, 2015). Emulgels provide a potential new treatment option for conditions associated with arthritis that might revolutionize the way arthritis is treated and significantly enhance patients' quality of life, especially as drug delivery continues to advance (Sinha, 2024).

14.3.2.2.12 Microemulsions

The microemulsion, which is given topically to the affected regions of inflammation, is a colloidal dispersion composed of a thermodynamically stable system. Its particles range in size from 10 to 100 nm and have remarkably high mean cumulative percentage values. When compared to the conventional medicine cream and solution, researchers argued that the Tenoxicam (TNX) microemulsion formulation showed a higher cumulative percent. The TNX formulation, which has good oral formulation results and is used to treat inflammation, was made using Tween 80, ethanol, and oleic acid. The resultant formulations are more effective when given topically as TNX in a range of inflammatory disorders (Goindi S., 2016). It exhibits greater cumulative percentage values and is more successful in reducing inflammation. Furthermore, a gelling agent is needed because of its low viscosity.

14.3.2.2.13 Micelles

Micelles are the medication delivery system used to treat RA due to the fact that micelle stability is an important consideration. To evaluate their solubility and stability, PEG phospholipids—which are used in pharmaceutical encapsulation—were used to create these. Activation of synovial macrophages is mostly caused by overexpression of the vasoactive intestinal peptide receptor. Micelles made from phospholipid preparation (phosphogliv) loaded with methotrexate show a greater ability to control inflammation when given to rats with arthritis (Aliabadi H. M., 2005). Similarly, compared to free camptothecin, camptothecin micelles showed less inflammation when administered to arthritic rats. Additionally, this method can aid in increasing the solubility of DMARDs. Other Novel drug delivery systems (NDDs), such as enhanced drug effectiveness, decreased toxicity, higher solubility of poorly soluble medicines, and superior physicochemical stability, provide more advantages than micelles. The main drawback, though, is dilution that causes disintegration.

14.3.2.3 Liposomes

A number of liposomal systems have been investigated to improve RA symptoms. These systems are administered intravenously and build up in the synovial tissue of RA patients. Cationic liposomes, immunoliposomes, superoxide dismutase liposomes, lactoferrin-loaded liposomes, clodronate-loaded liposomes, and stealth liposomes were among the liposome kinds that were produced and gathered. The rats with arthritis are given anti-inflammatory drugs that also help reduce bone resorption by encapsulating phosphatidylcholine and cholesterol-based liposomes with clodronate. The purpose of these liposome clodronates was to lower the quantity of synovial macrophages. Therefore, the resultant liposome lowers joint toxicity and inflammation in comparison to free medications (Hankey, A., 2001).

14.3.2.4 Significance of the Use of Liposomes in the Delivery of Antirheumatic Drugs

Treating arthritis with oral administration of antirheumatic drugs has proven to be a recurring problem for physicians since long-term oral use of these drugs is linked to substantial clinical outcomes. Long-term NSAID therapy for RA is associated with gastrodestructive side effects, such as ulcers and intra-abdominal bleeding. Oral or injectable steroid therapy is usually associated with immune system suppression that is irreversible. It is commonly known that DMARDs can impair immunological function when taken orally, intravenously, or intramuscularly (M. Gulati, 1998). To avoid the systemic effects of these drugs, especially when the illness only affects a few joints, they might be injected into the synovial capsule of the affected joint (N. Gerwin, 2006). However, the rapid transfer of drugs from the synovial cavity into the circulation defeats the purpose of intra-articular drug delivery. Liposomes have been demonstrated to be the most effective delivery systems for maintaining the drug in the synovial cavity due to their size and chemical composition. Liposomes, which resemble multilamellar vesicles, can get beyond the clearance of medications administered intrasynovially because of their size (J. T. Dingle, 1978). By promoting the drug's absorption by the target synovial cells and reducing exposure to nontarget sites, this gets rid of the undesirable side effects.

14.3.2.5 Liposomal Carrier System for Rheumatoid Arthritis

These days, the liposome method is widely used to improve RA treatment efficacy. Intravenous administration of liposomes causes them to accumulate in the synovial tissue of RA patients. Because of their anti-inflammatory qualities, cholesterol and phosphatidylcholine liposomes encapsulated with clodronate have been demonstrated to decrease bone resorption when administered to arthritic rats (Williams A. S., 1994). Both toxicity and joint inflammation are reduced in the resultant liposomes. One example is the use of stealth liposomes to increase the therapeutic effectiveness of glucocorticoids, which are frequently administered intravenously because of their nonspecific organ toxicity. PEG-modified liposomes and stealth liposomes that circulate with superoxide dismutase, however, show promising drug release (Corvo M. L., 1999). Novel dosage forms for RA are given in Table 14.3.

TALE 14.3
Novel Pharmaceutical Dosage Form for the Treatment of Rheumatoid Arthritis

Novel Dosage Form	Type	Drug Applied	Route
Nanoparticle-Based Drug Delivery Systems	a. Polymeric NP b. Lipid-Based NP c. Metal (gold & silver) NP	MTX, Prednisolone Tacrolimus	Oral, IV, IA
Liposomes and Niosomes	a. Encapsulated vesicles target inflamed synovial tissues	Biologics, Glucocorticoids Methotrexate, Etanercept	IV, IA
Topical (Transdermal Drug Delivery Systems)	a. Nanoemulsions and gels b. Transdermal Patches (Microneedles, iontophoresis)	NSAIDs, Tacrolimus, Methotrexate	Topical
Intra-Articular Drug Delivery	a. Injectable Hydrogels b. Micro/nanospheres c. Implants	Steroids, Biologics	Injectable
Biologicals	a. Monoclonal Antibodies b. Gene therapy vectors (adenoviral, lentiviral, lipid nanoparticles)	TNF blockers Cytokine modulation	SC, IV
Oral-Modified Release Systems (ORMS)	a. Floating tablets b. Colon targeted systems	MTX, SSZ	Oral

(Continued)

TALE 14.3 (*Continued*)
Novel Pharmaceutical Dosage Form for the Treatment of Rheumatoid Arthritis

Novel Dosage Form	Type	Drug Applied	Route
Implantable Devices	a. Osmotic pumps b. Biodegradable implants	Methotrexate, TNF-α inhibitors	IA, SC
mRNA and RNAi-Based Therapies	a. Targets proinflammatory cytokines b. Experimental, coding for anti-inflammatory proteins	miRNA, Biologics	SC, IV
Exosome-Based Delivery Systems	a. Mesenchymal Stem Cells b. Dendrimers	Exosome	IV, IA, Local
Smart Drug Delivery Systems	a. Stimuli-responsive systems b. Magnetic and ultrasound-responsive systems	Multiple (NSAIDs, DMARDs) Methotrexate, dexamethasone	IA, SC, Oral IA (via guided targeting)
3D-Bioprinted Scaffolds (with MSCs)	a. MSCs+biomaterials		Surgical implantation

14.3.2.6 Microneedles

Microneedles represent a novel pharmaceutical dosage form for RA, offering minimally invasive, pain-free transdermal drug delivery. They enable targeted delivery of anti-inflammatory drugs or biologics directly into the skin, bypassing the gastrointestinal tract and first-pass metabolism. This approach enhances drug bioavailability, reduces systemic side effects, and may improve patient compliance, especially for chronic RA management.

REFERENCES

Aho, K., & Heliövaara, M. J. A. o. m. (2004). Risk factors for rheumatoid arthritis. *Annals of Medicine, 36*(4), 242–251.

Akbarzadeh, A., Rezaei-Sadabady, R., Davaran, S., Joo, S. W., Zarghami, N., Hanifehpour, Y., … Nejati-Koshki, K. (2013). Liposome: classification, preparation, and applications. *Nanoscale Research Letters*, *8*, 1–9.

Aliabadi, H. M., Brocks, D. R., & Lavasanifar, A. (2005). Polymeric micelles for the solubilization and delivery of cyclosporine A: pharmacokinetics and biodistribution. *Biomaterials*, *26*(35), 7251–7259.

Amala, M., Vinodini, R., Hazel, A. A., Sundaram, M. M., & Banumathi, V. (2018). Anti-inflammatory property of Amukkara Choornam – a polyherbal preparation: an in-vitro study. *World Journal of Pharmaceutical Research, 8*(1), 804–808.

Andón, F. T., Digifico, E., Maeda, A., Erreni, M., Mantovani, A., Alonso, M. J., & Allavena, P. (2017). Targeting tumor associated macrophages: the new challenge for nanomedicine. In *Seminars in Immunology* (Vol. 34, pp. 103–113). Academic Press.

Arora, D., Rani, A., & Sharma, A. J. P. r. (2013). A review on phytochemistry and ethnopharmacological aspects of genus Calendula. *Pharmacognosy Reviews*, *7*, 179–187.

Ashwlayan, V., Kumar, A., Verma, M., Garg, V., & Gupta, S. (2018). Therapeutic potential of *Calendula officinalis. Pharmacy & Pharmacology International Journal*, *6*(2), 149–155.

Aswathy, I. S., Krishnan, S., Peter, J., Sabu, V., & Helen, A. (2019). Scientific validation of Antiarthritic effect of Kashayams–a polyherbal formulation in collagen induced arthritic rats. *Journal of Ayurveda and Integrated Medicine*, 12(1), 20–27.

Bader, R. A. J. R. A. T. R., & InTech, S. (2012). The development of targeted drug delivery systems for rheumatoid arthritis treatment. *Rheumatoid Arthritis–Treatment*, *18*, 111–32.

Baheti, A. M., Pawar, A. T., & Upaganlawar, A. (2022). Herbal formulations useful in the treatment of rheumatoid arthritis. In R. K. Gautam, L. Deb, & K. Dua (Eds.), *Natural Products for the Management of Arthritic Disorders* (pp. 164–192). Bentham Science Publisher.

Balkrishna, A., Sakat, S. S., Joshi, K., Paudel, S., Joshi, D., Joshi, K., . . . Varshney, A. J. F. i. p. (2019a). Antiinflammatory and antiarthritic efficacies of an Indian traditional herbo-mineral medicine 'Divya Amvatari Ras' in Collagen Antibody-Induced Arthritis (CAIA) mouse model through modulation of IL-6/IL-1β/TNF-α/NFκB signaling. *Frontiers in Pharmacology, 10*, 659.

Balkrishna, A., Sakat, S. S., Joshi, K., Paudel, S., Joshi, D., Joshi, K., . . . Varshney, A. J. S. r. (2019). Herbo-mineral formulation 'Ashwashila' attenuates rheumatoid arthritis symptoms in collagen-antibody-induced arthritis (CAIA) mice model. *Science Reports, 9*(1), 8025.

Balkrishna, A., Sakat, S. S., Joshi, K., Paudel, S., Joshi, D., Joshi, K., . . . Varshney, A. J. S. r. (2005). Socioeconomic status and the risk of developing rheumatoid arthritis: results from the Swedish EIRA study. *Annals of Rheumatic Disease*, *11*, 1588–1594.

Benita, S., Friedman, D., & Weinstock, M. (1986). Pharmacological evaluation of an injectable prolonged release emulsion of physostigmine in rabbits. *Journal of Pharmacy and Pharmacology*, *38*(9), 653–658.

Benni, J. M., Jayanthi, M., & Suresha, R. J. I. j. o. p. (2011). Evaluation of the anti-inflammatory activity of *Aegle marmelos* (Bilwa) root. *Indian Journal of Pharmacology, 43*(4), 393–397.

Bodhankar, M., Kavade, V., & Patil, A. (2012). Consequence of calendula oil on the in-vitro percutaneous absorption of diclofenac sodium. *Journal of Drug Delivery and Therapeutics*, *12*, 118–120.

Bozzuto, G., & Molinari, A. (2015). Liposomes as nanomedical devices. *International Journal of Nanomedicine*, *12*, 975–999.

Brunetti, P., Pichini, S., Pacifici, R., Busardò, F. P., & Del Rio, A. J. M. (2020). Herbal preparations of medical cannabis: avademecum for prescribing doctors. *Medicina (Kaunas), 56*(5), 237.

Bruni, N., Della Pepa, C., Oliaro-Bosso, S., Pessione, E., Gastaldi, D., & Dosio, F. J. M. (2018). Cannabinoid delivery systems for pain and inflammation treatment. *Molecules, 23*(10), 2478.

Canal, F., Vicent, M. J., Pasut, G., & Schiavon, O. (2010). Relevance of folic acid/polymer ratio in targeted PEG–epirubicin conjugates. *Journal of Controlled Release*, *146*(3), 388–399.

Chandrasekar, R., & Chandrasekar, S. (2017). Natural herbal treatment for rheumatoid arthritis-a review. *International Journal of Pharmaceutical Sciences and Research, 8*(2), 368.

Chauhan, M. (2019). Panchasama Churna. Retrieved from Planet Ayurveda: https://www.planetayurveda.com/library/panchasama-churna/

Chen, X.-M., Wu, J.-Q., Huang, Q.-C., Zhang, J.-Y., Pen, J.-H., and Huang, Z.-S. (2018). Systematic review and meta-analysis of the efficacy and safety of Biqi capsule in RA patients. *Experimental and Therapeutic Medicine, 15*(6), 5221–5230.

Choi, S. T., & Lee, K. H. (2018). Correction: clinical management of seronegative and seropositive rheumatoid arthritis: a comparative study. *PLoS One,13*(6), 1–13.

Chopra, A., Saluja, M., Tillu, G. J. J. o. A., & medicine, i. (2010). Ayurveda-modern medicine interface: a critical appraisal of studies of Ayurvedic medicines to treat osteoarthritis and rheumatoid arthritis. *Journal of Ayurveda and Integrated Medicine, 1*(3), 190–198.

Churna, Health Benefits of Pathyadi. (2022, May 20). Retrieved from the vaidyar: https://thevaidyar.com/blog/health-benefits-of-pathyadi-churna

Colino, C. I., Lanao, J. M., & Gutierrez-Millan, C. (2020). Targeting of hepatic macrophages by therapeutic nanoparticles. *Frontiers in Immunology*, *11*, 218.

Corvo, M. L., Boerman, O. C., Oyen, W. J., Van Bloois, L., Cruz, M. E. M., Crommelin, D. J., & Storm, G. (1999). Intravenous administration of superoxide dismutase entrapped in long circulating liposomes: II. In vivo fate in a rat model of adjuvant arthritis. *Biochimica et Biophysica Acta* (BBA)-Biomembranes, *1419*(2), 325–334.

Cross, M., Smith, E., Hoy, D., Carmona, L., Wolfe, F., Vos, T., ... & March, L. (2014). The global burden of rheumatoid arthritis: estimates from the global burden of disease 2010 study. *Annals of Rheumatic Disease*, *22*, 1316–1322.

D'Souza, W. Z., & Gude, M. R. (2015). Formulation, design, development and evaluation of emulgel for topical delivery of meloxicam in the treatment of rheumatoid arthritis. *IndoAmerican Journal of Pharmaceutical Research*, *5*(3), 1271–1279.

Daily, J. W., Zhang, T., Cao, S., & Park, S. (2017). Efficacy and safety of GuiZhi-ShaoYao-ZhiMu decoction for treating rheumatoid arthritis: a systematic review and meta-analysis of randomized clinical trials. *Journal of Alternative and Complementary Medicine*, *23*(10), 756–770.

Deol, P., & Khuller, G. K. (1997). Lung specific stealth liposomes: stability, biodistribution and toxicity of liposomal antitubercular drugs in mice. *Biochimica et Biophysica Acta (BBA)-General Subjects*, *1334*(2–3), 161–172.

Di Giuseppe, D., Alfredsson, L., Bottai, M., Askling, J., & Wolk, A. J. B. (2012). Long term alcohol intake and risk of RA in women: a population based cohort study. *BMJ, 345*, e4230.

Dingle, J. T., Gordon, J. L., Hazleman, B. L., Knight, C. G., Page Thomas, D. P., Phillips, N. C., ... Lowe, J. S. (1978). Novel treatment for joint inflammation. *Nature*, *271*(5643), 372–373.

Du, T. X., Ren, H. Y., & Li, G. L. (2002). A clinical observation of HanShiBi pill for rheumatism of type of simultaneous occurrence of cold dampness syndromes. *Journal of Traditional Chinese Orthopedics and Traumatology*, *14*(8), 5–7.

Duckworth, H. (2018). RA types: what distinguishes types of rheumatoid arthritis? Retrieved from RA: https://www.rheumatoidarthritis.org/ra/types/

Durgad, D. L. (2018). A *Comparative Clinical Study to Evaluate the Efficacy of Ajamodadi Churna and Punarnavadi Churna with Kati Basti in Gridhrasi WSR to Sciatica* (Doctoral dissertation). Rajiv Gandhi University of Health Sciences (India).

Efe, T., Sagnak, E., Roessler, P. P., Getgood, A., Patzer, T., Fuchs-Winkelmann, S., . . . Schofer, M. (2014). Penetration of topical diclofenac sodium 4% spray gel into the synovial tissue and synovial fluid of the knee: a randomised clinical trial. *Knee Surgery, Sports Traumatology, Arthroscopy*, *22*(2), 345–350.

Farooq, H., Piracha, M. I., & Alam, S. S. J. P. S. (2018). Prophylactic antiarthritic effect of Cassia fistula in murine RA model. *Proceedings SZPGMI*, *32*, 35–39.

Frisell, T., Holmqvist, M., Källberg, H., Klareskog, L., Alfredsson, L., and Askling, J. J. A. (2013). Familial risks and heritability of rheumatoid arthritis: role of rheumatoid factor/anti-citrullinated protein antibody status, number and type of affected relatives, sex, and age. *Arthritis & Rheumatism*, *22*, 2773–2782.

Gaikwad, B. D. (2017). A clinical evaluation of the effect of Rasonadi Kwatha in the management of Amavata. *Journal of Ayurveda and Integrated Medical Sciences*, *2*(01), 35–41.

Gaspar, N., Zambito, G., Löwik, C. M., & Mezzanotte, L. (2019). Active nano-targeting of macrophages. *Current Pharmaceutical Design*, *25*(17), 1951–1961.

Gerwin, N., Hops, C., & Lucke, A. (2006). Intraarticular drug delivery in osteoarthritis. *Advanced Drug Delivery Reviews*, *58*(2), 226–242.

Ghawi, H., Crowson, C. S., Rand-Weaver, J., Krusemark, E., Gabriel, S. E., & Juhn, Y. J. J. B. o. (2015). A novel measure of socioeconomic status using individual housing data to assess the association of SES with RA and its mortality: a population-based case-control study. *BMJ Open*, *5*(4), e006469.

Giancane, G., Consolaro, A., Lanni, S., Davì, S., Schiappapietra, B., and Ravelli, A. J. (2016). Juvenile idiopathic arthritis: diagnosis and treatment. *Rheumatology Therapeutics*, *3*(2), 187–207.

Gohila, M., Nalini Sofia, H., & M. S. (2019). An overview of a Siddha poly herbal formulation Vayu Keelaga Ilagam. *Siddha Papers*, *14*(2), 1–23.

Goindi, S., Narula, M., & Kalra, A. (2016). Microemulsion-based topical hydrogels of tenoxicam for treatment of arthritis. *Aaps Pharmscitech*, *17*, 597–606.

Gomez-Garcia, M. J., Doiron, A. L., Steele, R. R., Labouta, H. I., Vafadar, B., Shepherd, R. D., ... Rinker, K. D. (2018). Nanoparticle localization in blood vessels: dependence on fluid shear stress, flow disturbances, and flow-induced changes in endothelial physiology. *Nanoscale*, *10*(32), 15249–15261.

Gowri, P. M., Tiwari, A. K., Ali, A. Z., & Rao, J. M. (2007). Inhibition of α-glucosidase and amylase by bartogenic acid isolated from *Barringtonia racemosa* Roxb. seeds. *Phytotherapy Research*, *21*(8), 796–799.

Gulati, M., Grover, M., Singh, S., & Singh, M. (1998). Lipophilic drug derivatives in liposomes. *International Journal of Pharmaceutics*, *165*(2), 129–168.

Gupta, S. K., Gupta, A., Gupta, A., & Pakash, D. J. I. (2013). In-vitro antiarthritic activity of ethanolic extracts of *Callicarpa macrophyall* flower. *International Research Journal of Pharmacy*, *4*(3), 160–162.

Habber, J. V. (n.d.). *Rasna Saptak Kashayam Tablet, Kadha: Uses, Dose Content.* Retrieved from Ayurmed info: https://www.ayurmedinfo.com/2015/07/24/rasnasaptak-kashayam-tablet-kadha/

Habibur Rahman, M. C. (2016). Anti-arthritic activity of leaves and oil of *Aquilaria agallocha* Saudi J. *Life Science*, *33*, 34–43.

Hankey, A. (2001). Ayurvedic physiology and etiology: ayurvedo Amritanaam. The doshas and their functioning in terms of contemporary biology and physical chemistry. *Journal of Alternative Complementary Medicine*, *7*(5), 567–574.

Haroyan, A., Mukuchyan, V., Mkrtchyan, N., Minasyan, N., Gasparyan, S., and Sargsyan, A. (2018). Efficacy and safety of curcumin and its combination with boswellic acid in osteoarthritis: a comparative, randomized, double-blind, placebo-controlled study. *BMC Complementary and Alternative Medicine*, *18*(1), 7.

He, X. C., Hu, W. M., Yang, Y. H., & Xu, G. M. (2014). Clinical observation Tong-Bi-Granules (TBG) combined methotrexate (MTX) therapy in elderly RA. *China Health Care & Nutrition*, *6*, 3781.

Hebbar, J. V. (n.d.). Amrutadi churna - benefits, dose, side effects and ingredients. Retrieved from: https://www.ayurmedinfo.com/2012/05/13/amrutadi-churna-benefits-dose-side-effects-and-ingredients/

Hebbar, J. V. (n.d.). Hingvadi churna - benefits, dosage, ingredients, side effects. Retrieved from Ayurmed info: https://www.ayurmedinfo.com/2012/05/29/hingvadi-churna-benefits-dosage-ingredients-side-effects/

Hebbar, J. V. (n.d.). Rasnadi Kashayam (Kwath) - benefits, usage, indications and reference. Retrieved from Ayurmed info: https://www.ayurmedinfo.com/2012/02/15/rasnadi-kashayam-benefits-dose-side-effects-ingredients-and-reference/

Hebbar, J. V. (n.d.). Shatapatrayadi Churna benefits, dosage, ingredients, side effects. Retrieved from Ayurmed info: https://www.ayurmedinfo.com/2012/05/15/shatavaryadi-churna-benefits-dosage-ingredients-side-effects/

Higaki, M., Ishihara, T., Izumo, N., Takatsu, M., & Mizushima, Y. (2005). Treatment of experimental arthritis with poly (D, L-lactic/glycolic acid) nanoparticles encapsulating betamethasone sodium phosphate. *Annals of the Rheumatic Diseases*, *64*(8), 1132–1136.

Huh, J. E., Jung, I. T., Choi, J., Baek, Y. H., Lee, J. D., Park, D. S., & Choi, D. Y. (2013). The natural flavonoid galangin inhibits osteoclastic bone destruction and osteoclastogenesis by suppressing NF-kB in collagen-induced arthritis and bone marrow-derived macrophages. *European Journal of Pharmacology*, 698, 57–66.

Hunter, T. M., Boytsov, N. N., Zhang, X., Schroeder, K., Michaud, K., & Araujo, A. B. (2017). Prevalence of RA in the United States adult population in healthcare claims databases, 2004–2014. *Rheumatology International*, *23*, 1551–1557.

Jeyadevi, R., Sivasudha, T., Rameshkumar, A., & Dinesh Kumar, L.J.I.R. (2013). Anti-arthritic activity of the Indian leafy vegetable *Cardiospermum halicacabum* in Wistar rats and UPLC-QTOF-MS/MS identification of the putative active phenolic components. *Inflammation Research, 62*(1), 115–126.

Jin, K., Luo, Z., Zhang, B., & Pang, Z. (2018). Biomimetic nanoparticles for inflammation targeting. *Acta Pharmaceutica Sinica B*, *8*(1), 23–33.

Jutley, G. S., Latif, Z. P., Raza, K. J. B. P. (2017). Symptoms in individuals at risk of rheumatoid arthritis. *Best Practice & Research Clinical Rheumatology, 31*(1), 59–70.

Kabia, E., Mbau, R., Muraya, K. W., Morgan, R., Molyneux, S., & Barasa, E. (2018). How do gender and disability influence the ability of the poor to benefit from pro-poor health financing policies in Kenya? An intersectional analysis. *International Journal for Equity in Health*, *17*, 1–12.

Kalaiselvan, S., & Rasool, M. K. J. (2015). The anti-inflammatory effect of triphala in arthritic-induced rats. *Pharmaceutical Biology, 53*(1), 51–60.

Kale, V.M., & Namdeo, A.G. (2014). Antiarthritic effect of galangin isolated from rhizomes of Alpinia officinarum in complete Freund's adjuvant-induced arthritis in rats. *International Journal of Pharmacy and Pharmaceutical Sciences*, 34, 499–505.

Kamble, S., Mane, P., and Luxman, S K. (2015). A clinical evaluation of Alambushadi Churna (Y.R.) in Amavata. *International Journal of Current Research, 7*(5), 17525–17529.

Khan, I., Nisar, M., Ebad, F., Nadeem, S., Saeed, M., and Khan, H. (2009). Anti-inflammatory activities of Sieboldogenin from *Smilax china* Linn.: experimental and computational studies. *Journal of Ethnopharmacology, 121*(1), 175–177.

Khan, S., Dwivedi, C., Parmar, V., Srinivasan, K., & Shirwaikar, A. J. P. B. (2012). Methanol extract of dried exudate of *Commiphora mukul* prevents bone resorption in ovariectomized rats. *Pharmaceutical Biology, 50*(10), 1330–1336.

Khan, A. A., Bashir, F., Akhtar, J., Anjum, N., Alam, S., and Naushin, S. J. J. o. D. D. (2018). Majoon Suranjan: a potent Unani formulation for arthritis. *Journal of Drug Delivery and Therapeutics, 8*(6), 351–355.

Khan, M. A. (2018). Chronic toxicity evaluation of Majoon-e-Kundur: a polyherbal formulation. *Journal of Drug Research in Ayurvedic Sciences, 3.2,* 119–127.

Khandare, J. J., Jayant, S., Singh, A., Chandna, P., Wang, Y., Vorsa, N., & Minko, T. (2006). Dendrimer versus linear conjugate: influence of polymeric architecture on the delivery and anticancer effect of paclitaxel. *Bioconjugate Chemistry*, *17*(6), 1464–1472.

Khot, V. S. (2015). Comparative clinical study of Bhallatakadi Churna and Eranda Sneha in the management of amvata with special reference to rheumatoid arthritis. *International Journal of Research Ayurveda Pharmacy*, *6*(1), 35–38.

Kim, B. H., Yoon, J. H., Yang, J. I., Myung, S. J., Lee, J. H., and Jung, E. U. (2013). Guggulsterone attenuates activation and survival of hepatic stellate cell by inhibiting nuclear factor kappa B activation and inducing apoptosis. *Journal of Gastroenterology and Hepatology, 28*(12), 1859–1868.

Kim, M. J., Park, J. S., Lee, S. J., Jang, J., Park, J. S., Back, S. H., … Kim, K. (2015). Notch1 targeting siRNA delivery nanoparticles for rheumatoid arthritis therapy. *Journal of Controlled Release*, *216*, 140–148.

Kimmatkar, N., Thawani, V., Hingorani, L., & Khiyani, R. J. P. (2003). Efficacy and tolerability of *Boswellia serrata* extract in treatment of osteoarthritis of knee – a randomized double blind placebo controlled trial. *Phytomedicine, 10*(1), 3–7.

Kizhakkedath, R. (2013). Clinical evaluation of a formulation containing Curcuma longa and Boswellia serrata extracts in the management of knee osteoarthritis. *Molecular Medicine Reports*, 8(5), 1542–1528.
Kumar, S., & Singh, O. J. A. (2019). Clinical assessment of Alambushadi Churna and Dwipanchmuladhya Tail Vasti in the management of amavata vis-a-vis rheumatoid arthritis. *Ayushdhara, 6*(4), 2286–2304.
Kumar, J. D. (2011). A comparative clinical study of Shatapatrayadi churna tablet and Patoladi yoga in the management of Amlapitta. *AYU (An International Quarterly Journal of Research in Ayurveda), 32*(3), 361–364.
Kumar, S. S. (2004). Hepatoprotective activity of the Trikatu Churna-an Ayurvedic formulation. *Indian Journal of Pharmaceutical Sciences, 66*(3), 365–367.
Küpeli, E., & Yesilada, E. J. J. (2007). Flavonoids with anti-inflammatory and antinociceptive activity from *Cistus laurifolius* L. leaves through bioassay-guided procedures. *Journal of Ethnopharmacology, 112*(3), 524–530.
Latha, R., Shanthi, P., & Sachdanandam, P. J. P. M. (2015). Protective role of Kalpaamruthaa in type II diabetes mellitus-induced cardiovascular disease through the modulation of protease-activated receptor-1. *Pharmacognos Magazine, 11*(Suppl 1), S209–S216.
Lee, M. J., Lee, M. H., & Shim, C. K. (1995). Inverse targeting of drugs to reticuloendothelial system-rich organs by lipid microemulsion emulsified with poloxamer 338. *International Journal of Pharmaceutics, 113*(2), 175–187.
Lekurwale, P., Pandey, K., & Yadaiah, P. J. A. (2010). Management of Amavata with 'Amrita Ghrita': a clinical study. *Ayu, 31*(4), 430–435.
Lin, C. L. (2016). Chemical constituents of the wood of *Aquilaria agallocha. Chemistry of Natural Compounds, 52,* 316–317.
Lü, Y., Chen, D., Deng, J., & Tian, L. J. Z. (2003). Effect of Smilax china on adjunctive arthritis mouse. *Zhong Yao Cai, 26*(5), 344–346.
Man, Y., & Wang, W. J. W. H. D. (2010). The effectiveness of Hanshibi pill in 89 RA patients. *World Health Digest, 7*(7), 266–267.
Marwat, S. K., & U. K. (2011). Medico-ethnobotanical studies of edible wild fruit plants species from the flora of north western Pakistan (DI Khan district). *Journal of Medicinal Plants Research, 5*(16), 3679–3686.
Mechoulam, R. E. A. (2005). Cannabinoids in models of chronic inflammatory conditions. *Phytochemistry Reviews, 4*, 11–18.
Meena, R. R. R. (2015). Treatment of Uthiravatha Suronitham (rheumatoid arthritis) with a Siddha compound formulation – a case study. *IOSR Journal of Pharmacy, 5*(9), 50–52.
Meena, R. S. (2018). A dose related ADR of a siddha drug Karpoora chindhamani mathirai (KCM). *Research Journal of Pharmacy and Technology, 11*(5), 1877–1880.
Mengi, S., & Deshpande, S. J. I. J. P. (1995). Evaluation of ocular anti-inflammatory activity of *Butea frondosa. Indian Journal of Pharmacology, 27*(2), 116.
Mishra, G., & Pandya, D. H. J. J. (2017). Efficacy of Erand Sneha (castor oil) in the management of Amavata insamastage. *Journal of Research in Traditional Medicine, 3*(2), 28–35.
Mishra, P. K., & R. N. (2013). A clinical study on Ajmodadichurna and Erandataila in the management of Amavata (rheumatoid arthritis). *International Research Journal of Pharmacy, 4*(6), 211–213.
Mohamed, M., Abu Lila, A. S., Shimizu, T., Alaaeldin, E., Hussein, A., Sarhan, H. A., ... Ishida, T. (2019). PEGylated liposomes: immunological responses. *Science and Technology of Advanced Materials, 20*(1), 710–724.
Mohan, A., Mohanan, A., & Nair, V. P. K. (2021). Panchakola Siddha Yavagu (Medicated Rice Gruel) in management of Vatarakta (Gouty Arthritis): a review. *Biology and Life Sciences Forum, 6*(1), 22.
Mohiuddin, M. M., Mizubuti, G. B., Haroutounian, S., Smith, S. M., Rice, A. S., Campbell, F., . . . Gilron, I. J. T. C. (2020). Adherence to Consolidated Standards of Reporting Trials (CONSORT) guidelines for reporting safety outcomes in trials of medical cannabis and cannabis-based medicines for chronic noncancer pain: a systematic review. *Clinical Journal of Pain, 36*(4), 302–319.
Myasoedova, E., Crowson, C. S., Kremers, H. M., Therneau, T. M., Gabriel, S. E. J. A., & Rheumatism. (2010). Is the incidence of RA rising?: Results from Olmsted County, Minnesota, 1955–2007. *Arthritis & Rheumatism, 33*, 1576–1582.
Myasoedova, E., Davis, J., Matteson, E. L., & Crowson, C. S. J. A. (2020). Is the epidemiology of RA changing? Results from a population-based incidence study, 1985–2014. *Annals of the Rheumatic Diseases, 79*(4), 440–444.
Mythilypriya, R., & S. P. (2008). Efficacy of Siddha formulation Kalpaamruthaa in ameliorating joint destruction in RA in rats. *Chemico-Biological Interactions, 176*(2–3), 243–251.

Nair, V., Singh, S., & Gupta, Y. J. J. (2012). Investigation into the anti-inflammatory and antigranuloma activity of *Colchicum luteum* Baker in experimental models. *Inflammation, 35*(3), 881–888.

Nair, V., Kumar, R., Singh, S., & Gupta, Y. J. I. (2011). Evaluation of the disease modifying activity of *Colchicum luteum* Baker in experimental arthritis. *Journal of Ethnopharmacology, 133*(2), 303–307.

Nalimu, F., Oloro, J., Kahwa, I., & Ogwang, P. E. J. F. J. (2021). Review on the phytochemistry and toxicological profiles of Aloe vera and Aloe ferox. *Future Journal of Pharmaceutical Sciences, 145*, 1–11.

Nasra, S., Bhatia, D., & Kumar, A. (2022). Recent advances in nanoparticle-based drug delivery systems for rheumatoid arthritis treatment. *Nanoscale Advances, 4*(17), 3479–3494.

Negi, A., & P. P. (2017). A clinical study to evaluate the efficacy of Saindhavadi Taila Matra Vastiin the management of amavataw.s.r. rheumatoid arthritis. *Asian Journal of Pharmaceutical and Clinical Research, 10*(12), 435–439.

Negi, A., & P. P. (2017). A clinical study to evaluate the efficacy of Hingwadichurna and Rasnadashmula kwatha in amavata with special reference to rheumatoid arthritis. *International Journal of Research in Ayurveda and Pharmacy, 8*(3), 173–176.

Nikiphorou, E., Sjöwall, C., Hannonen, P., Rannio, T., & Sokka, T. J. B. (2016). Long-term outcomes of destructive seronegative (rheumatoid) arthritis – description of four clinical cases. *BMC Musculoskeletal Disorders, 17*(1), 246.

Okada, Y. E. (2014). Genetics of RA contributes to biology and drug discovery. *Nature, 506*(7488), 376–381.

Palumbo, C., & F. M. (2009). Influence of ferutinin on bone metabolism in ovariectomized rats. I: Role in preventing osteoporosis. *Journal of Bone and Mineral Metabolism, 27*(5), 538–545.

Pandey, S., & J. N. (2012). Clinical efficacy of Shiva Guggulu and Simhanada Guggulu in Amavata (Rheumatoid Arthritis). *Ayu, 33*(2), 247–254.

Pang, Y., Zeng, G., Tang, L., Zhang, Y., Liu, Y., Lei, X., … Xiong, Y. (2011). Preparation and application of stability enhanced magnetic nanoparticles for rapid removal of Cr (VI). *Chemical Engineering Journal, 175*, 222–227.

Paramanick, D., Panday, R., Shukla, S. S., & Sharma, V. J. J. (2017). Primary pharmacological and other important findings on the medicinal plant "*Aconitum heterophyllum*" (aruna). *Journal of Pharmacopuncture, 20*(2), 89–92.

Parihar, R. D. (2021). A comprehensive analysis of Bhallatakadi Churna in the management of amavata w.s.r to rheumatoid arthritis. *International Research Journal of Ayurveda and Yoga, 4*(2), 147–158.

Parwe, S. T. (2021). A critical review on formulations used in the management of Malavstambha (Constipation). *Journal of Pharmaceutical Research International, 33*(34A), 92–100.

Paschapur, M. S., Patil, M., Kumar, R., & Patil, S. R. J. J. (2009). Evaluation of anti-inflammatory activity of ethanolic extract of *Borassus flabellifer* L. male flowers (inflorescences) in experimental animals. *Journal of Medicinal Plants Research, 3*(2), 49–54.

Patel, H. D., & S. D. (2020). The effect of Amrutadyamchurna in amavataw.s.r. to rheumatoid arthritis: a case study. *World Journal of Pharmaceutical and Medical Research, 6*(5), 208–210.

Patra, K. C. (2014). Gastroprotective effect of standardized extract of Amukkara choornam on experimental gastric ulcer in rats. *Journal of Natural Medicines, 68,* 284–294.

Pikwer, M., & G. A. (2014). Association between testosterone levels and risk of future RA in men: a population-based case-control study. *Annals of the Rheumatic Diseases, 33*, 573–579.

Pradeepkiran, J. A. (2019). Insights of RA risk factors and associations. *Journal of Translational Autoimmunity*, 2(1), 1–11.

Prakash, S. (2010). A study of Shunthi-Gokshura Kwatha on Amavata. *Biomedical & Pharmacology Journal, 3*(1), 221.

Prashanth, G. S. (2010). A clinical comparative study of the management of chronic renal failure with Punarnavadi compound. *AYU (An International Quarterly Journal of Research in Ayurveda), 31*(2), 185–192.

Qadir, M. M. F., & B. A. (2018). Immunomodulatory and therapeutic role of *Cinnamomum verum* extracts in collagen-induced arthritic BALB/c mice. *Inflammopharmacology, 26*(1), 157–170.

Rajamohamed, B. S. (2019). Modulatory effects of Amukkara Choornam on *Candida albicans* biofilm: in vitro and in vivo study. *Molecular Biology Reports, 46,* 2961–2969.

Rajput, N. O., & S. K. (2016). A comparative study of Chitrakadi Churna with Rasnapancaka Kvatha in the management of amvatachikitsa siddhanta. *International Ayurvedic Medical Journal, 4*(2), 193–200.

Ram, H. A., Sriwastava, N. K., Makhija, I. K., & Shreedhara, C. S. (2012). Anti-inflammatory activity of Ajmodadi churna extract against acute inflammation in rats. *Journal of Ayurveda and Integrated Medicine, 3*(1), 33–37.

Rasool, M., Latha, L. M., Varalakshmi, P. J. P. (2000). Effect of *Withanias omnifera* on lysosomal acid hydrolases inadjuvant-induced arthritis in rats. *Pharmacy and Pharmacology Communications, 6*(4), 187–190.
Reddy, B. U. (2009). Anthelmintic activity of Trikatu churna and its ingredients. *Ethnobotanical Leaflets,* 4, 13.
Reddy, B. U. (2019). Antimicrobial and analgesic activities of Trikatu Churna and its ingredients. *Pharmacologyonline, 3,* 489–495.
Rohit, R. M. (2014). A clinical study on the effect of Punarnavadi churna, Singhanada guggulu in the management of Amavata (Rheumatoid Arthritis). *International Journal of Ayurveda and Pharma Research, 2,* 46–54.
Seror, R., & H. J.-R. (2019). Passive smoking in childhood increases the risk of developing rheumatoid arthritis. *Rheumatology (Oxford), 44,* 1154–1162.
Shahverdi, A. R., & S. F. (2006). Two matrix metalloproteinases inhibitors from Ferula persica var. persica. *Phytomedicine, 13*(9–10), 712–717.
Sharma, A., & B. S. (2010). Anti-inflammatory and analgesic activity of different fractions of *Boswellia serrata. International Journal of Phytomedicine, 2*(1), 1–11.
Sharma, H. D. (2021). Clinical study to evaluate the efficacy of Punarnavadi-Kanamula Yoga (hypothetical) in hypothyroidism. *International Research Journal of Ayurveda and Yoga, 4*(10), 41–49.
Silman, A. J., Pearson, J. E. J. A. (2002). Epidemiology and genetics of rheumatoid arthritis. *Arthritis Research, 4*(3), 265–272.
Singh, A., and Duggal, S. J. I. J. (2009). Piperine-review of advances in pharmacology. *International Journal of Pharmaceutical Sciences and Nanotechnology, 2*, 615–620.
Singh, J. P., Antiwal, M., Vaibhav, A., Tripathi, J., & Tiwari, S. J. A. (2010). Clinical efficacy of Rasona Pinda in the management of Amavata (rheumatoid arthritis). *Ayu, 31*(3), 280–286.
Singh, D. J. (2019). Mahanarayan Oil (Mahanarayana Thailam). Retrieved from Ayur times: https://www.ayurtimes.com/mahanarayan-oil-mahanarayana-thailam/
Singh, S. E. (2015). Anti-inflammatory and antiarthritic activity of UNIM-301 (a polyherbal unani formulation) in Wistar rats. *Pharmacognosy Research, 7*(2), 188.
Sinha, A., Garg, U., Nagaich, U., Chaudhary, A., Pandey, M., & Jain, N. (2024). Emulgels: a promising topical drug delivery system for arthritis management and care. *Pharmaceutical Development and Technology, 29*(1), 25–39.
Smolen, J. S., Aletaha, D., Koeller, M., and Weisman, M. H. (2007). New therapies for treatment of rheumatoid arthritis. *Lancet, 370*, 186174.
Smolen, J. S., & Steiner, G. J. N. (2004). Effect of leaves of *Butea frondosa* on stress, anxiety, and cognition in rats. *Pharmacology Biochemistry and Behaviour, 79*(1), 11–16.
Sriwastava, N. K., Shreedhara, C. S., & Ram, H. A. (2010). Standardization of Ajmodadi churna, a polyherbal formulation. *Pharmacognosy Research*, *2*(2), 98–101.
Stefanadis, C., Chrysochoou, C., Markou, D., Petraki, K., Panagiotakos, D. B., Fasoulakis, C., … Toutouzas, P. K. (2001). Increased temperature of malignant urinary bladder tumors in vivo: the application of a new method based on a catheter technique. *Journal of Clinical Oncology*, *19*(3), 676–681.
Subramaneyaan, M., & Y. S. (2013). Evaluation of therapeutic efficacy of Majoon Suranjan, a Unani formulation, in the treatment of rheumatoid arthritis: an experimental study. *Experimental Biology and Medicine, 238*(12), 1379–1387.
Subramaneyaan, M., & Y. S. (2013). Safety evaluation and therapeutic efficacy of Habb-e-Asgand, a commonly used antirheumatic polyherbal Unani formulation. *Alternative Therapies in Health and Medicine, 19*(5), 52–59.
Suhail, S. E. (2017). Habb-e-Suranjaan: a classical analgesic Unani formulation. *International Journal of Advances in Pharmacy Medicine and Bioallied Sciences, 24*, 1–6.
Sumantran, V. N., & J. A. (2011). Antiarthritic activity of a standardized, multiherbal, Ayurvedic formulation containing *Boswellia serrata*: in vitro studies on knee cartilage from osteoarthritis patients. *Phytotherapy Research, 25*(9), 1375–1380.
Suva, M. A. E. A. (2018). Aflapin®: a novel and selective 5-lipoxygenase inhibitor for arthritis management. *Indian Journal of Pain, 32*, 16.
Tang, X. X. (2018). Phytochemical profiles, and antimicrobial and antioxidant activities of greater galangal [*Alpinia galanga* (Linn.) Swartz.] flowers. *Food Chemistry, 255*, 300–308.
Thakur, S., Riyaz, B., Patil, A., Kaur, A., Kapoor, B., & Mishra, V. (2018). Novel drug delivery systems for NSAIDs in management of rheumatoid arthritis: an overview. *Biomedicine & Pharmacotherapy*, *106*, 1011–1023.
Thawani, V., & P. S. (2007). Open, randomized, controlled clinical trial of *Boswellia serrata* extract as compared to valdecoxib inosteoarthritis of knee. *Indian Journal of Pharmacology, 39*(1), 27–29.

The Chinese Pharmacopoeia. (2010). Chemical Industry Press.

Thote, A. S., & D. S. (2015). To evaluate the efficacy of Pathyadi churna in amavata. *International Ayurvedic Medical Journal, 3*(3), 740–745.

Tjendraputra, E., & T. V.-B. (2001). Effect of ginger constituents and synthetic analogues on cyclooxygenase-2 enzyme in intact cells. *Bioorganic Chemistry, 29*(3), 156–163.

Türker, S., Erdoğan, S., Özer, A. Y., Ergün, E. L., Tuncel, M., Bilgili, H., & Deveci, S. (2005). Scintigraphic imaging of radiolabelled drug delivery systems in rabbits with arthritis. *International Journal of Pharmaceutics, 296*(1–2), 34–43.

Upadhyay, S., & M. A. (2012). Toxicological evaluation of Panchakola Avaleha, an Ayurvedic classical formulation, in albino rats. *Ayu, 33*(2), 303–306.

Vandana, K. R., Yalavarthi, P., Sundaresan, C. R., N Sriramaneni, R., & C Vadlamudi, H. (2014). In-vitro assessment and pharmacodynamics of nimesulide incorporated Aloe vera transemulgel. *Current Drug Discovery Technologies, 11*(2), 162–167.

van Tuyl, L. H. (2016). Impact of symptoms on daily life in people at risk of rheumatoid arthritis. *Musculoskeletal Care 14.3,* 169–173.

Venetsanopoulou, A. I. (2022). Epidemiology of rheumatoid arthritis: genetic and environmental influences. *Expert Review of Clinical Immunology, 18*, 923–931.

Verma, S. K., & S. M. (2004). Protective effect of ginger, *Zingiber officinale* Rosc on experimental atherosclerosis in rabbits. *Indian Journal of Experimental Biology, 42*(7), 736–738.

Wang, Q., Jiang, J., Chen, W., Jiang, H., Zhang, Z., & Sun, X. (2016). Targeted delivery of low-dose dexamethasone using PCL–PEG micelles for effective treatment of rheumatoid arthritis. *Journal of Controlled Release, 230*, 64–72.

Waskar, R., and Sharma, J. (2019). A clinical study to evaluate the effect of Mahaushadhi Kwath in the management of Amavata Vis-À-Vis rheumatoid arthritis. *International Journal of Ayurveda and Pharmaceutical Chemistry, 10*(1), 85–97.

Whitbourne, K. (2024). What Type of RA Do You Have? Retrieved from https://www.webmd.com/rheumatoid-arthritis/rheumtoid-arthritis-types.

Williams, A. S., Camilleri, J. P., & Williams, B. D. (1994). Suppression of adjuvant-induced arthritis by liposomally conjugated methotrexate in the rat. *Rheumatology, 33*(6), 530–533.

Yasmeen, S., and Kumar, H. (2012). Randomized, double-blind study of the safety and effectiveness of a polyherbal drug in the treatment of active rheumatoid arthritis. *International Journal of Current Research Reviews, 4*(9), 129–142.

Zhang, R. X., Fan, A. Y., Zhou, A. N., Moudgil, K. D., Ma, Z. Z., Lee, D. Y. W., ... & Lao, L. (2009). Extract of the Chinese herbal formula Huo Luo Xiao Ling Dan inhibited adjuvant arthritis in rats. *Journal of Ethnopharmacology, 121*(3), 366–371.

Zheng, M., Jia, H., Wang, H., Liu, L., He, Z., Zhang, Z., … Gao, F. (2021). Application of nanomaterials in the treatment of rheumatoid arthritis. *RSC Advances, 11*(13), 7129–7137.

Zou, G. S. E. A. (2018). Lemon-Peel extract ameliorates RA by reducing xanthine oxidase and inflammatory cytokine levels. *Journal of Taiwan Institute of Chemical Engineering, 93*, 54–62.

15 Photodynamic Therapy Using Photoactive Phytochemicals

Maria Tahir, Ammara Saleem,
Qadeer Ahmad, Muhammad Furqan Akhtar,
Mirza Muhammad Faran Ashraf Baig,
and Muhammad Imran Khan

WHAT WILL YOU LEARN IN THIS CHAPTER?

This chapter will provide comprehensive knowledge about photodynamic therapy and its use in treating rheumatoid arthritis. It will also provide information about the antiarthritic mechanism of photoactive agents and phytocompounds that are used as photoactive agents.

LIST OF ABBREVIATIONS

AQ	Anthraquinone
BPD-M	Benzoporphyrin monoacid ring A
CDK4	Cyclin-dependent kinase 4
COX-2	Cyclooxygenase-2
DAG	Diacyl glycerol
ES	Excited state
FDA	Food and Drug Administration
IL-6	Interleukin-6
iNOS	Inducible nitric oxide synthase
MMP-9	Matrix metalloproteinase-9
NF-κB	Nuclear factor kappa-B
NO	Nitric oxide
PDT	Photodynamic therapy
PGE2	Prostaglandin E2
PS	Photosensitizer
PUVA	Longer wavelength ultraviolet light
RA	Rheumatoid arthritis
ROS	Reactive oxygen species
S°	Ground singlet state
Sl	Excited singlet state
TNF	Tumor necrosis factor
UV	Ultraviolet
VEGF	Vascular endothelial growth factor

15.1 INTRODUCTION

Rheumatoid arthritis (RA) is a chronic, immune-mediated inflammatory disease that affects 1%–3% of the global population. It occurs when the immune system attacks the synovial tissues in healthy joints, leading to joint inflammation and the degeneration of cartilage, tendons, and

DOI: 10.1201/9781003535201-19

ligaments (Radu & Bungau, 2023). Degeneration of these connective tissues leads to severe pain and immobility. Various environmental and genetic factors, such as socioeconomic status, diet, smoking, and obesity, play a role in the progression of the disease. The primary pharmacotherapeutic approaches in managing RA aim to alleviate pain, reduce inflammation, and prevent joint destruction (Chandrasekar & Chandrasekar, 2017). These include nonsteroidal anti-inflammatory drugs, glucocorticoids, and disease-modifying antirheumatic drugs. However, the clinical application of these medicinal substances is limited due to poor patient compliance, adverse effects, high costs, and the need for long-term usage. To overcome these challenges, there is a need for improvement in these conventional therapeutic approaches (Prasad et al., 2023).

Photodynamic therapy (PDT) is a Food and Drug Administration (FDA)-approved alternative therapeutic approach that delivers photo-activated medicinal substances to the target site without causing pain. In this therapy, photosensitizers of specific wavelengths are utilized (Correia et al., 2021). When light irradiates, these photosensitizers are activated to generate reactive oxygen species (ROS), which in turn inhibit inflammatory cells, causing inflammation and joint degeneration (Gallardo-Villagrán et al., 2019). The practice of using light in various ailments, including skin diseases, rheumatism, paralysis, muscle weakness, and tuberculosis, has been utilized since ancient times. This practice is also known as heliotherapy (Lee et al., 2020). In the field of medicine, phototherapy was pioneered by Danish physician Niels Finsen, the creator of phototherapy in dermatology. He was awarded the Nobel Prize in 1903 for his work on the ultraviolet (UV) treatment of lupus vulgaris using UV rays (Grzybowski & Pietrzak, 2012). Subsequently, Tappeiner and Jodlbauer discovered the role of oxygen in photosensitized reactions and coined the term "photodynamic action" (Algorri et al., 2021). In 1972, Ivan Diamond renamed the term PDT. When PDT is combined with standard therapeutic approaches, it gives positive therapeutic outcomes in RA by preventing the destruction of bones and cartilage. These positive approaches improve the quality of life in patients with RA (Gallardo-Villagrán et al., 2019).

15.2 PRINCIPLES OF PHOTODYNAMIC THERAPY

PDT primarily depends on three essential components: the photosensitizer, molecular oxygen, and light at a specific wavelength (dos Santos et al., 2019). These components are nonirritating to cells; however, when exposed to UV and visible light, they trigger a photochemical reaction that produces highly reactive singlet molecular oxygen, leading to cell toxicity and eventual death. Heliotherapy is a two-step procedure. In the first step, a nontoxic photosensitizer is introduced into the body either topically or intravenously, depending on the treated condition. The photosensitizer then accumulates at the targeted site. The next step involves the activation of the photosensitizer by the projection of light at a specific wavelength. The photosensitizer absorbs energy, causing its electrons to jump from the ground state to a higher energy level. This energy is then transferred to molecular oxygen to generate ROS, responsible for cellular destruction (Castano, Demidova, & Hamblin, 2004; Shi et al., 2019).

15.3 MAJOR COMPONENTS OF PHOTODYNAMIC THERAPY

15.3.1 Light Source

The light sources of PDT in RA range from 200 to 1,300 nm. The range of the near-infrared region (700–1300 nm) holds superior tissue penetration. This property makes them highly effective in promoting tissue regeneration and healing synovial tissues. In the clinical setting, various types of light sources, such as sunlight, X-rays, laser light, and light-emitting diodes, are employed (Zhang et al., 2018).

Ideally, a PDT light source must have the following properties:

1. The wavelength of the light source must match the absorption spectra of PS.
2. Enough penetration power to reach the desired tissue.
3. Optimal luminous power and stable output.
4. Cost-effective.

15.3.2 Photosensitizers

Photosensitizers are the second most crucial component of PDT. The therapeutic effectiveness of PDT primarily relies on the use of photosensitizers. These are the conjugated compounds having a strong light absorbance coefficient (a wavelength of 600–800 nm). These substances remain nonreactive and nontoxic in the absence of light. Furthermore, they exhibit high permeability and site-specific accumulation with low photo-bleaching property (Correia et al., 2021; Escudero et al., 2021). In RA, these substances are injected via intravenous or intra-articular injections. After administration, the light of a specific wavelength was projected to the affected area to activate these photosensitizers. Upon activation, ROS are generated to suppress the activity of various pro-inflammatory cytokines, including tumor necrosis factor alpha (TNF-α), interleukin-6 (IL-6), and IL-17, thereby alleviating symptoms of RA. In the case of infectious RA, PDT effectively eliminates microbial growth by interacting with the microbial DNA (Mi et al., 2023).

15.4 GENERATIONS OF PHOTOSENSITIZERS

Three generations of photosensitizers are used in PDT. The first generation includes hematoporphyrin and its derivatives, isolated in 1841 from hemoglobin. Porphyrin, the lyophilized form, treats various tumors and inflammatory disorders. Photofrin® (Porfimer sodium) is an FDA-approved photosensitizer utilized for treating various tumors. However, the use of Photoforin has been limited due to its toxic nature, extended half-life, low wavelength of light, poor penetration, high tissue retention, and high photosensitivity (Baskaran, Lee, & Yang, 2018; Tavakkoli Yaraki, Liu, & Tan, 2022). First-generation photosensitizers, Photosan-663, and Photofrin have shown cytotoxic effects on RA but have the above-mentioned limitations (Trauner et al., 1998). Benzoporphyrin monoacid ring A (BPD-MA) was the first-generation photosensitizer reported in various studies for RA. In a study, an intravenous injection of BPD-MA (0.5mg/kg) was administered to complete Freund's adjuvant-induced mice. After an incubation period, the red light (λ=690 nm) was used to irradiate the animals on various consecutive days (days 0, 10, 20, 30). The treatment results were compared to indomethacin, cyclosporin A, and 3 Gy sublethal whole-body irradiation. The effectiveness of the PDT treatment was comparable to conventionally used treatments without producing systemic toxicity (Ratkay et al., 1994).

The therapeutic effectiveness of second-generation photosensitizers is much better than that of first-generation photosensitizers. This may be due to their site-specific targeting, prolonged retention time, and widened absorption spectrum. The most predominant tetrapyrrole derivative, Protoporphyrin IX, is widely employed in PDT. They exhibit strong cytotoxic and immunosuppressive properties. 5-Aminolevulinic acid is another second-generation photosensitizer that also exhibits strong cytotoxic activity against tumor tissues.

The third-generation photosensitizer is conjugated with antibodies, ligands, receptors, or encapsulated in nanocarriers, thus increasing target selectivity and extending the wavelength absorption (Baskaran, Lee, & Yang, 2018; Montaseri, Kruger, & Abrahamse, 2021). In PDT, third-generation photosensitizers exhibit cytotoxic effects by replacing the calreticulin chaperone from the nucleus.

15.5 MECHANISM OF PHOTODYNAMIC REACTIONS

Photodynamic reactions can be categorized into two types based on the underlying chemical reactions. The initial step is identical in both mechanisms. When the photosensitizer absorbs light of a specific wavelength, it becomes excited and transitions from the ground singlet state (S°) to the excited singlet state (S^1). Some of the energy is emitted as fluorescence, while the remaining energy moves the photosensitizer from the excited singlet state (S^1) to the excited triplet state (Kwiatkowski et al., 2018).

15.6 TYPE I PHOTODYNAMIC REACTION

In type I photosensitized reactions, the photosensitizer in its excited triplet state interacts with a nearby substrate. The substrate transfers an electron to the photosensitizer, producing anionic and cationic free radicals based on redox potential. Oxygen then acts as a catalyst, and the photosensitizer donates an electron to oxygen to form a free radical such as superoxide anion radical (O2-). This initiates a series of reactions to generate more ROS, which are responsible for damaging inflammatory cells, contributing to the amelioration of RA.

15.7 TYPE II PHOTODYNAMIC REACTION

In type II photodynamic reactions, oxygen plays a crucial role. In the excited state (ES), the photosensitizer interacts with molecular oxygen if the energy level of the ES aligns with the threshold required to generate singlet oxygen. During this process, the photosensitizer transitions from the ES to the ground state, releasing energy to convert molecular oxygen into singlet oxygen (Schweitzer & Schmidt, 2003). This singlet oxygen is a highly reactive form of ROS (Hamblin, 2008) that interacts with various biological molecules, such as proteins, lipids, DNA, and RNA (Figure 15.1).

15.8 PHYTOCHEMICALS AS PHOTOSENSITIZERS

Phytochemicals derived from medicinal plants have been widely used in RA due to their potent anti-inflammatory, immunosuppressant, and antioxidant properties. These compounds are considered green pharmacotherapeutic approaches because of their lower toxicity and higher efficacy (Gandhi et al., 2022). Several natural medicinal compounds produce pharmacological activity upon irradiation with light at a specific wavelength. Photoactive phytochemicals have selective toxicity, biocompatible absorption spectra (400–700 nm), and a nontoxic nature in the dark. These characteristics make them valuable candidates as photosensitizers in PDT (Muniyandi et al., 2020). Various photoactive phytocompounds were isolated using advanced identification, characterization, and extraction techniques. However, the lack of in vivo and in vitro studies, phototoxicity due to high absorbance capacity, and regulatory approval pose a significant challenge in the pharmacotherapy of various diseases. The phototoxic potential of phytocompounds can be overcome by recent advances, including the use of a powerful controllable laser light source (Sulaiman et al., 2022). During photosynthesis in plants, light-sensitive molecules reach higher energy levels when they absorb photons (by chromophore) and become reactive, referred to as photosensitizers. One such molecule is chlorophyll. Naturally, chlorophyll is protected by the chloroplast, rendering it nonreactive, while other molecules are converted to functional PS. The molecule altered by a photosensitizer is known as a substrate, and if oxygen acts as a substrate, the process is referred to as photodynamic (Quinn, Kessell, & Weston, 2014).

15.9 FURANOCOUMARIN

Furanocoumarins belong to the Umbelliferae family of plants and encompass a range of linear and angular derivatives. Linear forms comprise psoralen derivatives, including 5-methoxypsoralen and 8-methoxypsoralen, wherein the furan ring is linked to the coumarin unit (Munier et al., 2014). Psoralens enhance the absorption of UV light from the skin. This allows UV radiation at 300–400 nm to effectively treat various inflammatory conditions such as psoriasis, dermatitis, and rheumatism (Panno et al., 2009). Photo-activated 5-methoxypsoralen produces a strong antioxidant

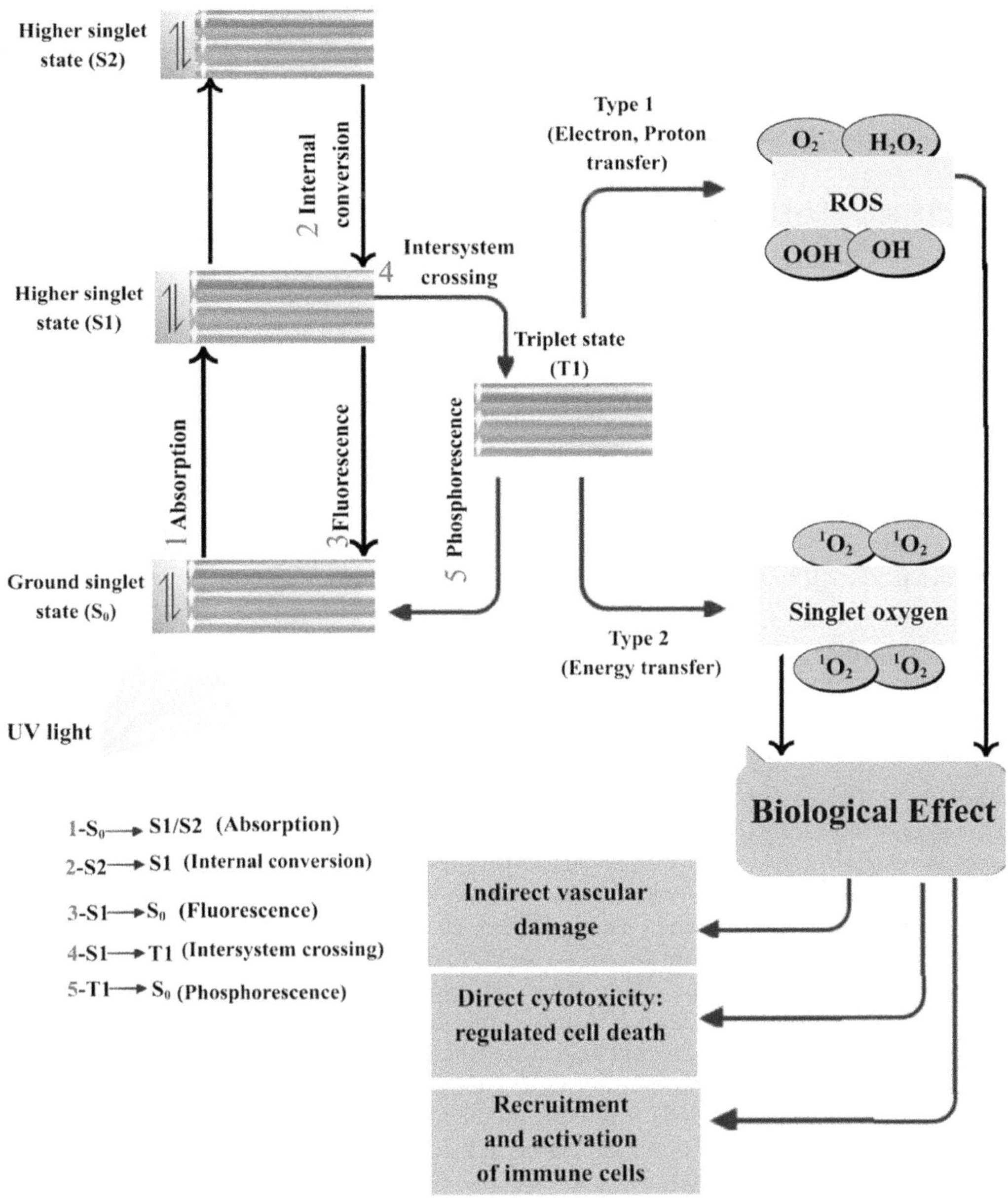

FIGURE 15.1 Schematic illustration shows the excitation and relaxation process of photosensitizers. This includes energy transition through absorption, fluorescence, phosphorescence, and internal conversion system. The triplet state photosensitizer interacts with oxygen and other molecules. This leads to the production of singlet molecular oxygen, hydrogen peroxide, superoxide anion radical, and hydroxyl radical. These ROS contribute to vascular damage, immune cell activation, and regulated cell death.

effect by inhibiting the JAK/STAT signaling pathway, thereby alleviating the symptoms of RA. The activity of various inflammatory mediators such as TNF-α, IL-1β, IL-6, prostaglandin E2 (PGE2), and nitric oxide (NO) was also suppressed, as evidenced in an in vitro study (Zhou et al., 2017). These compounds also exhibit a cytotoxic effect upon irradiation with UV light at a wavelength of

320–400nm (Bethea et al., 1999). They induce apoptotic cell death in T-cell lymphoma by interacting with DNA (Nagatani et al., 1990). Xanthotoxin is another FDA-approved photo-activated psoralen derivative that is used for treating T-cell lymphoma in a concentration-dependent manner. They are capable of absorbing UV light with a longer wavelength (PUVA), making them a valuable drug candidate in PDT (de Oliveira et al., 2016).

15.10 THIOPHENES

Thiophenes exhibit a broad range of pharmacological properties, such as anti-inflammatory, antimicrobial, antioxidant, antiarrythmic, and antimitotic properties (Shah & Verma, 2018). These biological properties are produced by type II photodynamic reactions upon irradiation with UV light at a wavelength of 314–350nm (Park & Lee, 2007). Activation of thiophenes by UV light produces ROS, which mediates apoptotic death of inflammatory cells (Bartekova et al., 2018).

Various photoactive analogs of thiophenes are used in PDT as documented in the literature. In the study conducted by Hu et al., thiophenes produce light-sensitive methoxy analogs by an isomerization process (Hu et al., 2013). In a lipopolysaccharide-induced inflammatory model, these analogs suppress the expression of extracellular signal-regulated kinase, p38, and NF-κB signaling pathway. Furthermore, the secretion of IL-6 and TNF-α was also downregulated (Bartekova et al., 2018).

15.11 ANTHRAQUINONES

Anthraquinones (AQ) collaborate with molecular oxygen via type II photodynamic reactions and act as a strong photosensitizer in PDT (Cogno et al., 2020). The biological properties of AQ result from the hydroxylation process. These compounds can absorb UV and visible light at a broad range of wavelengths (Gutiérrez et al., 1997; Pawłowska et al., 2003). Among them, emodin, aloe-emodin, hypericin, and rubiadin exhibit strong anti-inflammatory and antioxidant properties due to their broader wavelength. These characteristics make them strong candidates in PDT for treating various ailments, including RA (Nowak-Perlak, Ziółkowski, & Woźniak, 2023).

Emodin is derived from various plant families, such as *Aloe vera, Polygonum multiforum, Polygonum cuspidatum*, and *Giant knotweed.* In PDT, emodin produces strong anti-inflammatory and immunomodulating effects via both type I and type II photodynamic reactions (Muniyandi et al., 2020). Upon irradiation with UV and visible light, emodin suppresses the secretion of TNF-α, IL-6, IL-4, IL-10, IL-13, and IL-11. All these inflammatory mediators strongly contribute to the progression of RA. Furthermore, the production of PGE2 and cyclooxygenase-2 (COX-2) enzyme in synovial tissue was also inhibited by the photo-activated form of emodin (Zhu et al., 2013). They also downregulate the activation of JAK/STAT and NF-κB signaling pathways (Schett, 2007). All these physiological modifications positively contribute to the amelioration of RA symptoms.

Aloe-emodin is also used as a photosensitizer for alleviating inflammation in RA (Liu et al., 2018). Photoactivated aloe-emodin at a wavelength of 5–40μM inhibits the expression of inducible nitric oxide synthase (iNOS), mRNA, and PGE2 (Park, Kwon, & Sung, 2009). Due to their high light-detecting properties, phototoxicity occurred. This effect was tested in the K3 cell line when illuminated with blue light at a wavelength of 405nm (Wamer, Vath, & Falvey, 2003).

Rubiadin, another derivative of AQ phytocompounds, belongs to the Rubiaceae family of plants. They exhibit strong anti-inflammatory and antioxidant properties upon irradiation with UV and visible light (Nowak-Perlak, Ziółkowski, & Woźniak, 2023). The photoactivated form of rubiadin is reported to be a potential cytotoxic candidate by generating singlet oxygen (Comini et al., 2011). The research by Montoya et al. (2008) and Vittar et al. (2010) indicated that AQs demonstrate photosensitization effects in Balb/c mice and can inhibit leukocyte activity in a dose-dependent manner

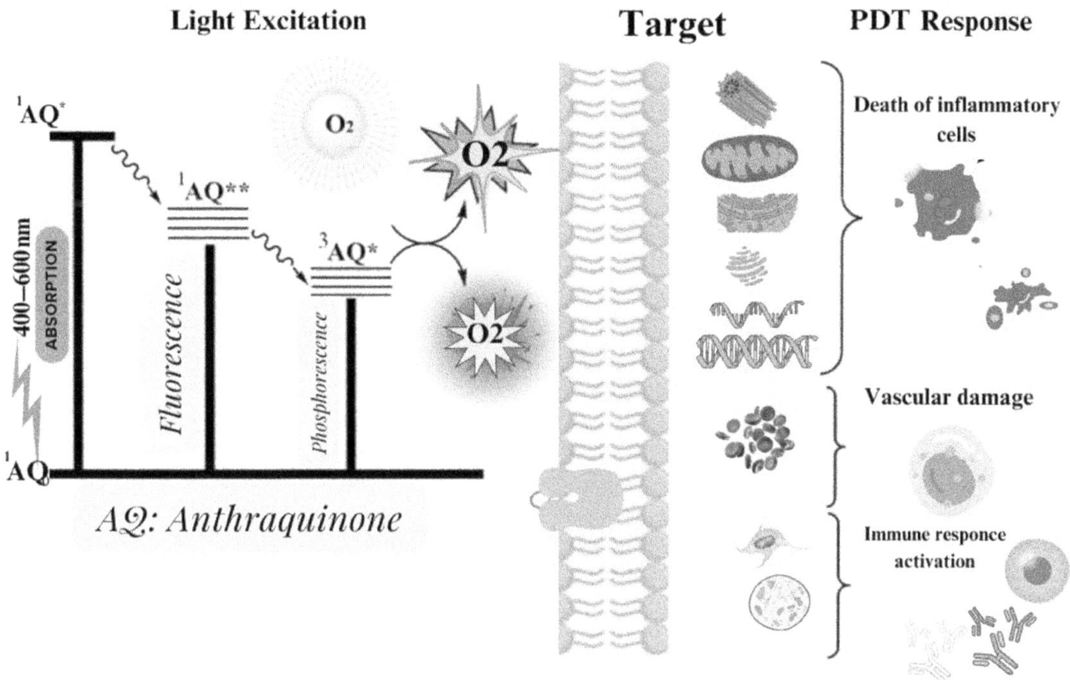

FIGURE 15.2 This is a schematic illustration of the mechanism of anthraquinone in photodynamic therapy. Anthraquinones absorb light in the range of 400–460 nm, transforming from their ground state (S°) to an excited singlet state (S^{1}). The excited singlet state undergoes a triplet state through fluorescence and intersystem crossing. This state interacts with the molecular oxygen to produce singlet oxygen and ROS, which are crucial for PDT. The generated singlet oxygen interacts with the biological targets, including tumor cells in cancer therapy, inflammatory cells in autoimmune diseases, and microorganisms for antimicrobial applications. This leads to the death of inflammatory cells, thereby reducing chronic inflammation in rheumatic patients. In cancer, the vascular damage results in the disruption of the abnormal tissue growth.

through apoptosis, necrosis, or autophagy. Their findings illustrate the photoactive properties of AQs in impeding the growth of cancer cells (Figure 15.2).

15.12 PERYLENEQUINONES

Perylenequinones are the natural medicinal compounds that attribute specific characteristics to PDT. They produce singlet molecular oxygen through type II photodynamic reactions (Jendželovská et al., 2016). Various photo-activated forms of perylenequinones possess potent anti-inflammatory, immunomodulatory, and cytotoxic properties. These properties make them valuable for treating a range of pathological conditions, including RA, skin diseases, and malignancies.

Hypericin (HYP), a naturally occurring polycyclic quinone, originates from *Hypericum perforatum* plants, commonly known as St. John's wort (Luksiene & de Witte, 2003). HYP has also been used for treating depression and viral infections in the past (Kashef, Karami, & Djavid, 2015). HYP absorbs light around 599 nm to produce the most potent photosensitizing effect (Nakajima & Kawashima, 2012). Upon irradiation with light, HYP inhibits the multiplication of MH7A cells, initiates apoptosis, and stimulates the regulation of ROS, thereby reducing the symptoms of RA as reported earlier. In PDT of RA, HYP modulates the NF-κB signaling pathway (Zhang et al., 2018). Numerous studies have shown that HYP-based PDT exhibits potent tumor-specific cytotoxicity with minimal adverse effects (Xu et al., 2015).

Hypocrellin, primarily synthesized from *Shiraia bambusicola*, acts as a strong PS in the PDT by producing singlet oxygen, superoxide, hydroxy radicals, and hydrogen peroxide. These free radicals can suppress the immune response, thereby decreasing the symptoms of RA (Table 15.1).

TABLE 15.1
List of Photoactivated Phytochemicals Used in Preclinical Studies of Photodynamic Therapy

Phytochemical	Source	In Vivo/ In Vitro	Dose	Results	References
Hypericin	*Hypericum perforatum*	In vivo In vivo In vivo	2.5 mg/kg 5 mg/kg 1 mg/kg 5 mg/kg 2 mg/kg 5 mg/kg	Inhibition of tumor cell growth No measurable tumor mass after 9–10 days Induce apoptosis, decrease vascular endothelial growth factors (VEGFs)	Zupko et al. (2001) Vandenbogaerde et al. (1996) Thong et al. (2006).
Blepharismin	*Blepharisma undulans*	In vitro	200ng/ml	Inhibit the production of superoxide anion radicals via diacyl glycerol (DAG)-induced protein kinase-c (PKC) and arachidonate-induced PKC reactions	Watanabe et al. (1995)
Elsinochrome A	*Elsinoe annonae*	In vitro In vitro	50.97ng/ml 8,16,32 μg/mL	Inhibit the proliferation of Hce-8693 cells in human colorectal cancer. Inhibit the growth of *Candia albicans* after irradiation	Li et al. (2015) Pan et al. (2022)
Cercosporin	*Cercospora* spp.	In vitro	4.0 mg	Generate superoxide radicals, singlet molecular oxygen to induce death of hepatocellular carcinoma cells	Ye et al. (2020)
Isocercosporin	*Cercospora* spp.	In vitro	10,20 ng/ml	Generate superoxide anion radicals, singlet molecular oxygen, hydroxyl radicals, and hydrogen peroxide, which leads to the death of inflammatory cells.	Diwu and Lown (1992)
Hypocrellin A	*Hypocrella bambusae*	In vitro In vitro	10^{-3} M 0.08 μmol/L	Significant reduction in the survival of two Gram-positive bacteria. Generate ROS that causes the death of inflammatory cells involved in the progression of RA. Mitochondrial disruption, release of cytochrome c, and various caspases	Su et al. (2011) Qi et al. (2019)

(*Continued*)

TABLE 15.1 (*Continued*)
List of Photoactivated Phytochemicals Used in Preclinical Studies of Photodynamic Therapy

Phytochemical	Source	In Vivo/ In Vitro	Dose	Results	References
Hypocrellin B	*Hypocrella bambusae*	In vitro In vivo In vitro	4.6 μg/ml 5mg/kg in the tail vein 2.5 μM for 5 h	Generate singlet molecular oxygen that induces mitochondrial damage Significant reduction in tumor volume Increase in early and late apoptotic rates in human ovarian cancer HO-8910 cells	Zhao et al. (2014) Jiang et al. (2014)
Phleichrome	*Cladosporium phlei*	In vitro In vitro	5-50μg/mL 5,10,25,50 μg/disc	Significant reduction in the viability of MCF-7 human adenocarcinoma cells. Inhibit the growth of all Gram-positive bacteria in a dose-dependent manner	So, Chun, and Kim (2018)

15.13 ALKALOIDS

Alkaloids belong to a diversified group of natural substances. Their chemical structure consists of a heterocyclic structure associated with a nitrogen atom in the ring (Benyhe, 1994). Several scientific investigations reveal numerous derivatives of alkaloids that are used in the PDT. Among them, berberin is considered the most widely used alkaloid in PDT due to its broad photo-pharmacological properties (Och, Podgórski, & Nowak, 2020). They absorb UV and visible light at a wavelength of 410 nm. Upon activation, the growth of cancerous cells in the brain was inhibited (Lopes et al., 2020). Photo-activated form of berberine inhibits mitochondrial oxidative phosphorylation, thereby ameliorating RA symptoms (Fan et al., 2018).

Various quinoline-based alkaloids also demonstrate photo-pharmacological properties by absorbing light in the 300–410 nm wavelength range. Among them, quinine exhibits a high absorption peak at 331 nm and generates singlet oxygen, producing ROS such as superoxide, hydroperoxyl, and peroxyl. The production of these ROS can suppress the immune response, thereby alleviating the symptoms of RA (López-Molina et al., 2021; Morten et al., 1999). Aporphine, belonging to the isoquinoline alkaloid family of phytocompounds, possesses strong free radical scavenging and photosensitizing properties. Therefore, it is a strong candidate for managing symptoms associated with RA (Ubeda et al., 1993).

β-Carbolines, another alkaloidal phytocompound, serve as a potent photosensitizer in the PDT of tumors and various inflammatory diseases, including arthritis. These compounds become active upon absorbing moderate light, generating singlet oxygen quantum. In an aqueous solution, β-carbolines undergo multiple protonation steps, leading to alterations in their photophysical and sensitizing properties (López-Molina et al., 2021). Harmine, obtained from the *Peganum harmala* family of plants, exerts robust light-detecting properties. When exposed to UV light, it induces a cytotoxic effect by photooxidizing the DNA of cancerous cells. Furthermore, the photo-activated form of harmine promotes lysosomal localization, thereby facilitating autophagy to prevent inflammation and maintain cellular health (Yañuk et al., 2018).

Norharmane, another analog of photoactivated carboline alkaloids, absorbs light at 50 nm. Upon UV light illumination, Norharmane generates ROS and oxidizes adenosine monophosphate nucleotide via type I PD reactions (Gonzalez et al., 2012). These ROS induce apoptosis of inflammatory cells within the joints, thereby reducing inflammation in RA (Table 15.2).

TABLE 15.2
List of Photoactivated Phytochemicals Used in Preclinical Studies of Photodynamic Therapy

Phytochemical	Source	In Vivo/ In Vitro	Dose	Results	References
Isomoschatoline	*Guatteria melosma*	In vitro	50 µL	Generate singlet molecular oxygen upon irradiation at 560 nm via type II PD reactions to reduce the viability of microorganisms	Andreazza et al. (2016)
Glycoborinine Arborin	*Glycosmis pentaphylla*	In vitro In vitro	1.8 kb DNA fragments 250 µg/ml and 1000 µg/ml	Upon irradiating with UVA, the compound shows moderate inhibition on Nde, Xba, Nco 1, and Bcl 1 restriction enzymes The strongest photoactivated enzymatic restriction inhibitor against EcoR1 and Sall, thereby potentiating antimicrobial effect in combination with ciprofloxacin and erythromycin	Chen et al. (2012) Taib et al. (2022)
Flavopereirine	*Rauvolfia nukuhivensis*	In vitro	15.33, 10.52, 10.76, 8.15, and 9.58 µM	Inhibit the MAP kinase signaling pathway to reduce inflammation, Downregulate the expression of p53 for the suppression of colorectal cancer	Li et al. (2019)
Palmatine	*Hydrastis canadensis*	In vitro In vitro	1, 2, 4, 8, and 16 µM 8.3 µmol/L, 4.1 µmol/L and 2.1 µmol/L	Induce apoptosis of tumor cells in colon adenocarcinoma HT-29 cells and breast cancer MCF-7 cells. Decrease the expression of cyclin B1 and CDK1 and produce ROS to suppress the immune response	Qi et al. (2019). Sun et al. (2021).
Harmine	*Pegnum harmala*	In vitro Raw 264.7 cells	10 mg/kg/day	Inhibit osteoclast formation and nuclear factor of activated T cells c1	Zhang, Li, and Yu (2020).

(Continued)

TABLE 15.2 (*Continued*)
List of Photoactivated Phytochemicals Used in Preclinical Studies of Photodynamic Therapy

Phytochemical	Source	In Vivo/ In Vitro	Dose	Results	References
Berberine Berbamine	*Hydrastis canadensis*	In vivo/in vitro In vitro In vivo	200 mg/ml to prepare a Ti3c2 scaffold 100μm	Upon irradiation with laser light, repair the mandible damage by the healing of bone Potent antibacterial effect against Gram-negative E. *coli* Inhibit the antioxidant enzymes to produce an antioxidant effect and disrupt the regulation of calcium ions	Tan et al. (2024) Wimmer et al. (2024) Zhu et al. (2024)

15.14 CURCUMIN

Curcumin belongs to the Zingiberaceae family of plants and is synthesized from *Curcuma longa*. It exhibits a wide range of biological activities, including antiproliferative (Karunagaran, Rashmi, & Kumar, 2005), antioxidant (Wright, 2002), antinociceptive (Zhu et al., 2014), anti-inflammatory (Chainani-Wu, 2003), and antiparasitic effects (Rai et al., 2020). Numerous scientific investigations have uncovered that curcumin acts as a strong photosensitizer and holds promise in the PDT for treating various conditions, including inflammatory rheumatism (Haukvik et al., 2010; Tønnesen et al., 1987). Upon irradiation with light at a wavelength of 300–500 nm, curcumin produces singlet oxygen species through type II PD reactions (Rego-Filho et al., 2014). Photoactivated form of curcumin performs its action by the inhibition of various inflammatory mediators, including COX-2, iNOS, NF-κB, 5-lipooxygenase, matrix metalloproteinases-9 (MMP-9), MMP-2, and IL-8 (Maheshwari et al., 2006; Yance Jr & Sagar, 2006). These inflammatory mediators play a pivotal role in the progression of inflammation and cartilage destruction in RA. Inhibition of these mediators is crucial for ameliorating the symptoms of RA (Zhang et al., 2022).

Furthermore, curcumin inhibits the proliferation of cancerous cells in various forms of cancer, including colon, breast, lungs, kidneys, and ovaries (Karunagaran, Rashmi, & Kumar, 2005). Upon irradiation with UV light, curcumin inhibits the manifestation of p53 genes, cyclin-dependent kinase-4 (CDK-4), while also producing ROS (Bernd, 2014).

15.15 PHENOLS

Phenols are a class of phytocompounds that exhibit various pharmacological activities, including antimicrobial (Vaquero, Alberto, & de Nadra, 2007), anti-inflammatory (Azuma et al., 1986), and antioxidant (Foti, 2007) effects. Various analogs of phenols have photosensitizing properties and, therefore, act as potential PS in the PDT of various pathogenic conditions, including RA. Among them, resveratrol, a polyphenol especially present in grapes, improves the quality of life by producing anti-inflammatory, antiangiogenic (Lançon, Frazzi, & Latruffe, 2016), and antimicrobial effects (Bostanghadiri et al., 2017). The photoactivated form of resveratrol produces singlet oxygen through type II PD reactions, which induce apoptosis of inflammatory cells within the joints. This reduces inflammation and alleviates RA symptoms (de Sá Coutinho et al., 2018; Zhou et al., 2018). Pyrogallol and terpinolene are reported to be used as PS, generating a strong ROS activity, thus acting as a cytotoxic agent against bacteria (Sarker & Ahn, 2022). Phytoalexin is an isoflavonoid phytochemical obtained from plants due to biotic and abiotic stresses. Upon irradiation with UV light, phytoalexin produces ROS, including OH, H_2O_2, and singlet molecular oxygen. These ROS contribute to the demise of inflammatory cells and the suppression of the immune response, thereby ameliorating the symptoms of RA by inhibiting the progression of inflammation (Bakker et al., 1983).

15.16 PROS AND CONS OF PDT

PDT is superior to conventionally available therapies in various aspects. First and foremost, it is noninvasive in nature and the patient suffers little to no pain or physical damage during its application, thus improving the patient's compliance. As in PDT, radiation time and position are controllable, and minimal systemic toxicity and minimal disturbance of physiological functions occur. Thus, PDT is a good treatment option for RA for improving the overall life quality of patients (Mi et al., 2023).

For successful PDT application, the presence of oxygen is very important to form the singlet oxygen molecule to destroy the cytotoxic cells. In case of arthritis, the joint already has a hypoxic environment, so PDT application is very suitable for RA treatment (Gunaydin, Gedik, & Ayan, 2021; Quiñonez-Flores, González-Chávez, & Pacheco-Tena, 2016). PDT has very less adverse effects as compared to available therapies. Unlike many therapies, fertility remains intact, and it is safe for pregnant patients (Chizenga, Chandran, & Abrahamse, 2019).

There are various shortcomings of PDT. As light is an important component in PDT therapy, it is required to excite the chromophore to generate a singlet oxygen molecule to produce a cytotoxic effect. If light doesn't penetrate the tissues, due to its wavelength or the optical properties of the tissues, it fails to produce a therapeutic effect. So, the choice of light source is important. The therapeutic efficiency of PDT is also limited by the choice of photosensitizers (water solubility, purity, dosimetry) (Gunaydin, Gedik, & Ayan, 2021). Other disadvantages for the patient include repeated application, complex scheduling, and post-therapy care. PDT may induce photosensitivity weeks after therapy (Chhatre et al., 2021).

15.17 FUTURE STRATEGIES

To date, various treatment regimens and alternative treatments such as synovectomy are available for RA, but they only provide symptomatic relief and have an invasive or destructive nature, high treatment cost, postsurgery care, and adverse effects that are associated with all treatment options, but PDT is a promising technique with fewer side effects. The side effects of PDT can be overcome by selecting the appropriate photosensitizers. More of the choice of light source, its wavelength, and penetration ability should be optimized by trials.

Other possibilities to enhance the treatment efficiency are to combine PDT with various other available therapies (photothermal therapy, chemotherapy, immunotherapy) to overcome the limitations of PDT and combine the good aspects of such therapies, and these combination therapies should undergo lab and clinical trials to find the best treatment with minimum side effects.

REFERENCES

Algorri, J. F., Ochoa, M., Roldan-Varona, P., Rodriguez-Cobo, L., & Lopez-Higuera, J. M. (2021). Photodynamic therapy: a compendium of latest reviews. *Cancers, 13*(17), 4447.

Andreazza, N. L., de Lourenco, C. C., Hernandez-Tasco, A. J., Pinheiro, M. L. B., Stefanello, M. É. A., Costa, E. V., & Salvador, M. J. (2016). Antimicrobial photodynamic effect of extracts and oxoaporphine alkaloid isomoschatoline from *Guatteria blepharophylla. Journal of Photochemistry and Photobiology B: Biology, 160*, 154–162.

Azuma, Y., Ozasa, N., Ueda, Y., & Takagi, N. (1986). Pharmacological studies on the anti-inflammatory action of phenolic compounds. *Journal of dental research, 65*(1), 53–56.

Bakker, J., Gommers, F., Smits, L., Fuchst, A., & Vries, F. d. (1983). Photo activation of isoflavonoid phytoalexins: involvement of free radicals. *Photochemistry and Photobiology, 38*(3), 323–329.

Bartekova, M., Radosinska, J., Jelemensky, M., & Dhalla, N. S. (2018). Role of cytokines and inflammation in heart function during health and disease. *Heart Failure Reviews, 23*, 733–758.

Baskaran, R., Lee, J., & Yang, S.-G. (2018). Clinical development of photodynamic agents and therapeutic applications. *Biomaterials Research, 22*(1), 25.

Benyhe, S. (1994). Morphine: new aspects in the study of an ancient compound. *Life Sciences, 55*(13), 969–979.

Bernd, A. (2014). Visible light and/or UVA offer a strong amplification of the anti-tumor effect of curcumin. *Phytochemistry Reviews, 13*, 183–189.

Bethea, D., Fullmer, B., Syed, S., Seltzer, G., Tiano, J., Rischko, C., Gillespie, L., Brown, D., & Gasparro, F. P. (1999). Psoralen photobiology and photochemotherapy: 50 years of science and medicine. *Journal of Dermatological Science, 19*(2), 78–88.

Bostanghadiri, N., Pormohammad, A., Chirani, A. S., Pouriran, R., Erfanimanesh, S., & Hashemi, A. (2017). Comprehensive review on the antimicrobial potency of the plant polyphenol Resveratrol. *Biomedicine & Pharmacotherapy, 95*, 1588–1595.

Castano, A. P., Demidova, T. N., & Hamblin, M. R. (2004). Mechanisms in photodynamic therapy: part one—photosensitizers, photochemistry and cellular localization. *Photodiagnosis and Photodynamic Therapy, 1*(4), 279–293.

Chainani-Wu, N. (2003). Safety and anti-inflammatory activity of curcumin: a component of turmeric (*Curcuma longa*). *The Journal of Alternative & Complementary Medicine, 9*(1), 161–168.

Chandrasekar, R., & Chandrasekar, S. (2017). Natural herbal treatment for rheumatoid arthritis-a review. *International Journal of Pharmaceutical Sciences and Research, 8*(2), 368.

Chen, Y., Yang, B., Xu, J., Zheng, T., Fan, H., & Yang, G.-Z. (2012). Photo-activated DNA binding and antimicrobial activities of alkaloids from *Glycosmis pentaphylla. Yao xue xue bao= Acta Pharmaceutica Sinica, 47*(12), 1646–1652.

Chhatre, S., Vachani, A., Allison, R. R., & Jayadevappa, R. (2021). Survival outcomes with photodynamic therapy, chemotherapy and radiation in patients with stage III or stage IV non-small cell lung cancer. *Cancers, 13*(4), 803.

Chizenga, E. P., Chandran, R., & Abrahamse, H. (2019). Photodynamic therapy of cervical cancer by eradication of cervical cancer cells and cervical cancer stem cells. *Oncotarget, 10*(43), 4380.

Cogno, I. S., Gilardi, P., Comini, L., Nunez-Montoya, S. C., Cabrera, J. L., & Rivarola, V. A. (2020). Natural photosensitizers in photodynamic therapy: in vitro activity against monolayers and spheroids of human colorectal adenocarcinoma SW480 cells. *Photodiagnosis and Photodynamic Therapy, 31*, 101852.

Comini, L. R., Fernandez, I. M., Vittar, N. R., Montoya, S. N., Cabrera, J. L., & Rivarola, V. (2011). Photodynamic activity of anthraquinones isolated from *Heterophyllaea pustulata* Hook f.(Rubiaceae) on MCF-7c3 breast cancer cells. *Phytomedicine, 18*(12), 1093–1095.

Correia, J. H., Rodrigues, J. A., Pimenta, S., Dong, T., & Yang, Z. (2021). Photodynamic therapy review: principles, photosensitizers, applications, and future directions. *Pharmaceutics, 13*(9), 1332.

de Oliveira, D. M., Lima, R. M. F., Clarencio, J., da Silva Velozo, E., de Amorim, I. A., da Mota, T. H. A., Costa, S. L., Silva, F. P., & dos Santos El-Bachá, R. (2016). The classical photoactivated drug 8-methoxypsoralen and related compounds are effective without UV light irradiation against glioma cells. *Neurochemistry International, 99*, 33–41.

de Sá Coutinho, D., Pacheco, M. T., Frozza, R. L., & Bernardi, A. (2018). Anti-inflammatory effects of resveratrol: mechanistic insights. *International Journal of Molecular Sciences, 19*(6), 1812.

Diwu, Z., & Lown, J. W. (1992). Photosensitization with anticancer agents 14. Perylenequinonoid pigments as new potential photodynamic therapeutic agents: formation of tautomeric semiquinone radicals. *Journal of Photochemistry and Photobiology A: Chemistry, 69*(2), 191–199.

dos Santos, A. F., de Almeida, D. Q., Terra, L. F., Baptista, M. S., & Labriola, L. (2019). Photodynamic therapy in cancer treatment-an update review. *Journal of Cancer and Metastasis Treatment*, *5*(25), 10.20517.

Escudero, A., Carrillo-Carrión, C., Castillejos, M. C., Romero-Ben, E., Rosales-Barrios, C., & Khiar, N. (2021). Photodynamic therapy: photosensitizers and nanostructures. *Materials Chemistry Frontiers*, *5*(10), 3788–3812.

Fan, X.-X., Leung, E. L.-H., Xie, Y., Liu, Z. Q., Zheng, Y. F., Yao, X. J., Lu, L. L., Wu, J. L., He, J.-X., & Yuan, Z.-W. (2018). Suppression of lipogenesis via reactive oxygen species–AMPK signaling for treating malignant and proliferative diseases. *Antioxidants & Redox Signaling*, *28*(5), 339–357.

Foti, M. C. (2007). Antioxidant properties of phenols. *Journal of Pharmacy and Pharmacology*, *59*(12), 1673–1685.

Gallardo-Villagrán, M., Leger, D. Y., Liagre, B., & Therrien, B. (2019). Photosensitizers used in the photodynamic therapy of rheumatoid arthritis. *International Journal of Molecular Sciences*, *20*(13), 3339.

Gandhi, Y., Kumar, R., Grewal, J., Rawat, H., Mishra, S. K., Kumar, V., Shakya, S. K., Jain, V., Babu, G., & Sharma, P. (2022). Advances in anti-inflammatory medicinal plants and phytochemicals in the management of arthritis: a comprehensive review. *Food Chemistry Advances*, *1*, 100085.

Gonzalez, M. M., Rasse-Suriani, F. A., Franca, C. A., Diez, R. P., Gholipour, Y., Nonami, H., Erra-Balsells, R., & Cabrerizo, F. M. (2012). Photosensitized electron transfer within a self-assembled nor harmane–2′-deoxyadenosine 5′-monophosphate (dAMP) complex. *Organic & Biomolecular Chemistry*, *10*(47), 9359–9372.

Grzybowski, A., & Pietrzak, K. (2012). From patient to discoverer—Niels Ryberg Finsen (1860–1904)—the founder of phototherapy in dermatology. *Clinics in Dermatology*, *30*(4), 451–455.

Gunaydin, G., Gedik, M. E., & Ayan, S. (2021). Photodynamic therapy—current limitations and novel approaches. *Frontiers in Chemistry*, *9*, 691697.

Gutiérrez, I., Bertolotti, S. G., Biasutti, M., Soltermann, A. T., & García, N. A. (1997). Quinones and hydroxyquinones as generators and quenchers of singlet molecular oxygen. *Canadian journal of chemistry*, *75*(4), 423–428.

Hamblin, M. R. (2008). *Advances in photodynamic therapy: basic, translational, and clinical*. Artech House.

Haukvik, T., Bruzell, E., Kristensen, S., & Tønnesen, H. (2010). Photokilling of bacteria by curcumin in selected polyethylene glycol 400 (PEG 400) preparations. Studies on curcumin and curcuminoids, XLI. *Die Pharmazie-An International Journal of Pharmaceutical Sciences*, *65*(8), 600–606.

Hu, J., Wang, G., Liu, X., Zhou, L., Jiang, M., & Yang, L. (2013). Polo-like kinase 1 (PLK1) is involved in toll-like receptor (TLR)-mediated TNF-α production in monocytic THP-1 cells. *PLoS One*, *8*(10), e78832.

Jendželovská, Z., Jendželovský, R., Kuchárová, B., & Fedoročko, P. (2016). Hypericin in the light and in the dark: two sides of the same coin. *Frontiers in Plant Science*, *7*, 560.

Jiang, Y., Leung, A. W., Wang, X., Zhang, H., & Xu, C. (2014). Effect of photodynamic therapy with hypocrellin B on apoptosis, adhesion, and migration of cancer cells. *International Journal of Radiation Biology*, *90*(7), 575–579.

Karunagaran, D., Rashmi, R., & Kumar, T. (2005). Induction of apoptosis by curcumin and its implications for cancer therapy. *Current Cancer Drug Targets*, *5*(2), 117–129.

Kashef, N., Karami, S., & Djavid, G. E. (2015). Phototoxic effect of hypericin alone and in combination with acetyl cysteine on *Staphylococcus aureus* biofilms. *Photodiagnosis and Photodynamic Therapy*, *12*(2), 186–192.

Kwiatkowski, S., Knap, B., Przystupski, D., Saczko, J., Kędzierska, E., Knap-Czop, K., Kotlińska, J., Michel, O., Kotowski, K., & Kulbacka, J. (2018). Photodynamic therapy–mechanisms, photosensitizers and combinations. *Biomedicine & Pharmacotherapy*, *106*, 1098–1107.

Lançon, A., Frazzi, R., & Latruffe, N. (2016). Anti-oxidant, anti-inflammatory and anti-angiogenic properties of resveratrol in ocular diseases. *Molecules*, *21*(3), 304.

Lee, C.-N., Hsu, R., Chen, H., & Wong, T.-W. (2020). Daylight photodynamic therapy: an update. *Molecules*, *25*(21), 5195.

Li, J.-M., Huang, Y.-C., Kuo, Y.-H., Cheng, C.-C., Kuan, F.-C., Chang, S.-F., Lee, Y.-R., Chin, C.-C., & Shi, C.-S. (2019). Flavopereirine suppresses the growth of colorectal cancer cells through P53 signaling dependence. *Cancers*, *11*(7), 1034.

Li, T., Deng, H., Zhao, J., & Gu, Y. (2015). Elsinochrome A photosensitizers: alternative drugs for photodynamic therapy. *Journal of Innovative Optical Health Sciences*, *8*(01), 1530001.

Liu, Y.-q., Meng, P.-s., Zhang, H.-c., Liu, X., Wang, M.-x., Cao, W.-w., Hu, Z., & Zhang, Z.-g. (2018). Inhibitory effect of aloe emodin mediated photodynamic therapy on human oral mucosa carcinoma in vitro and in vivo. *Biomedicine & Pharmacotherapy*, *97*, 697–707.

Lopes, T. Z., de Moraes, F. R., Tedesco, A. C., Arni, R. K., Rahal, P., & Calmon, M. F. (2020). Berberine associated photodynamic therapy promotes autophagy and apoptosis via ROS generation in renal carcinoma cells. *Biomedicine & Pharmacotherapy*, *123*, 109794.

López-Molina, S., Galiana-Roselló, C., Galiana, C., Gil-Martínez, A., Bandeira, S., & González-García, J. (2021). Alkaloids as photosensitisers for the inactivation of bacteria. *Antibiotics*, *10*(12), 1505.

Luksiene, Z., & de Witte, P. (2003). Hypericin as novel and promising photodynamic therapy tool: studies on intracellular accumulation capacity and growth inhibition efficiency. Medicina, *39*(7), 677–682.

Maheshwari, R. K., Singh, A. K., Gaddipati, J., & Srimal, R. C. (2006). Multiple biological activities of curcumin: a short review. *Life Sciences*, *78*(18), 2081–2087.

Mi, L., Gao, J., Liu, Y., Zhang, N., Zhao, M., Wang, S., & Xu, K. (2023). Photodynamic therapy for arthritis: a promising therapeutic strategy. *Rheumatology & Autoimmunity*, *3*(4), 205–219.

Montaseri, H., Kruger, C. A., & Abrahamse, H. (2021). Inorganic nanoparticles applied for active targeted photodynamic therapy of breast cancer. *Pharmaceutics*, *13*(3), 296.

Montoya, S. C. N., Comini, L. R., Vittar, B. R., Fernández, I. M., Rivarola, V. A., & Cabrera, J. L. (2008). Phototoxic effects of Heterophyllaea pustulata (Rubiaceae). *Toxicon*, *51*(8), 1409–1415.

Morten, A. G., Martinez, L. J., Holt, N., Sik, R. H., Reszka, K., Chignell, C. F., Tonnesen, H. H., & Roberts, J. E. (1999). Photophysical studies on antimalariai drugs. *Photochemistry and Photobiology*, *69*(3), 282–287.

Munier, M., Jubeau, S., Wijaya, A., Morancais, M., Dumay, J., Marchal, L., Jaouen, P., & Fleurence, J. (2014). Physicochemical factors affecting the stability of two pigments: R-phycoerythrin of Grateloupia turuturu and B-phycoerythrin of Porphyridium cruentum. *Food Chemistry*, *150*, 400–407.

Muniyandi, K., George, B., Parimelazhagan, T., & Abrahamse, H. (2020). Role of photoactive phytocompounds in photodynamic therapy of cancer. *Molecules*, *25*(18), 4102.

Nagatani, T., Matsuzaki, T., Kim, S., Baba, N., Ichiyama, S., Miyamoto, H., & Nakajima, H. (1990). Treatment of cutaneous T-cell lymphoma (CTCL) by extracorporeal photochemotherapy. *Journal of Dermatological Science*, *1*(3), 226.

Nakajima, N., & Kawashima, N. (2012). A basic study on hypericin-PDT in vitro. *Photodiagnosis and Photodynamic Therapy*, *9*(3), 196–203.

Nowak-Perlak, M., Ziółkowski, P., & Woźniak, M. (2023). A promising natural anthraquinones mediated by photodynamic therapy for anti-cancer therapy. *Phytomedicine*, *119*, 155035.

Och, A., Podgórski, R., & Nowak, R. (2020). Biological activity of berberine—A summary update. *Toxins*, *12*(11), 713.

Pan, L., Yao, Y., Zheng, H., Yan, S., & Chen, S. (2022). Inhibitory effects and mechanism of action of elsinochrome A on Candida albicans and its biofilm. *Journal of Fungi*, *8*(8), 841.

Panno, M. L., Giordano, F., Palma, M., Bartella, V., Rago, V., Maggiolini, M., Sisci, D., Lanzino, M., De Amicis, F., & Ando, S. (2009). Evidence that bergapten, independently of its photoactivation, enhances p53 gene expression and induces apoptosis in human breast cancer cells. *Current Cancer Drug Targets*, *9*(4), 469–481.

Park, K., & Lee, J.-H. (2007). Photosensitizer effect of curcumin on UVB-irradiated HaCaT cells through activation of caspase pathways. *Oncology Reports*, *17*(3), 537–540.

Park, M.-Y., Kwon, H.-J., & Sung, M.-K. (2009). Evaluation of aloin and aloe-emodin as anti-inflammatory agents in aloe by using murine macrophages. *Bioscience, Biotechnology, and Biochemistry*, *73*(4), 828–832.

Pawłowska, J., Tarasiuk, J., Wolf, C. R., Paine, M. J., & Borowski, E. (2003). Differential ability of cytostatics from anthraquinone group to generate free radicals in three enzymatic systems: NADH dehydrogenase, NADPH cytochrome P450 reductase, and xanthine oxidase. *Oncology Research*, *13*(5), 245–252.

Prasad, P., Verma, S., Surbhi, Ganguly, N. K., Chaturvedi, V., & Mittal, S. A. (2023). Rheumatoid arthritis: advances in treatment strategies. *Molecular and Cellular Biochemistry*, *478*(1), 69–88.

Qi, F., Sun, Y., Lv, M., Qin, F., Cao, W., & Bi, L. (2019). Effects of palmatine hydrochloride mediated photodynamic therapy on oral squamous cell carcinoma. *Photochemical & Photobiological Sciences*, *18*, 1596–1605.

Qi, S., Guo, L., Yan, S., Lee, R. J., Yu, S., & Chen, S. (2019). Hypocrellin A-based photodynamic action induces apoptosis in A549 cells through ROS-mediated mitochondrial signaling pathway. *Acta Pharmaceutica Sinica B*, *9*(2), 279–293.

Quinn, J. C., Kessell, A., & Weston, L. A. (2014). Secondary plant products causing photosensitization in grazing herbivores: their structure, activity and regulation. *International Journal of Molecular Sciences*, *15*(1), 1441–1465.

Quiñonez-Flores, C. M., González-Chávez, S. A., & Pacheco-Tena, C. (2016). Hypoxia and its implications in rheumatoid arthritis. *Journal of Biomedical Science*, *23*, 1–9.

Radu, A.-F., & Bungau, S. G. (2023). Nanomedical approaches in the realm of rheumatoid arthritis. *Ageing Research Reviews*, 101927.

Rai, M., Ingle, A. P., Pandit, R., Paralikar, P., Anasane, N., & Santos, C. A. D. (2020). Curcumin and curcumin-loaded nanoparticles: antipathogenic and antiparasitic activities. *Expert Review of Anti-infective Therapy*, *18*(4), 367–379.

Ratkay, L., Chowdhary, R., Neyndorff, H., Tonzetich, J., Waterfield, J., & Levy, J. (1994). Photodynamic therapy; a comparison with other immunomodulatory treatments of adjuvant-enhanced arthritis in MRL-lpr mice. *Clinical & Experimental Immunology*, *95*(3), 373–377.

Rego-Filho, F. G., de Araujo, M. T., de Oliveira, K. T., & Bagnato, V. S. (2014). Validation of photodynamic action via photobleaching of a new curcumin-based composite with enhanced water solubility. *Journal of fluorescence*, *24*, 1407–1413.

Sarker, M., & Ahn, Y.-H. (2022). Photodynamic inactivation of multidrug-resistant bacteria in wastewater effluent using green phytochemicals as a natural photosensitizer. *Environmental Pollution*, *311*, 120015.

Schett, G. (2007). Cells of the synovium in rheumatoid arthritis. Osteoclasts. *Arthritis Research & Therapy*, *9*, 1–6.

Schweitzer, C., & Schmidt, R. (2003). Physical mechanisms of generation and deactivation of singlet oxygen. *Chemical Reviews*, *103*(5), 1685–1758.

Shah, R., & Verma, P. K. (2018). Therapeutic importance of synthetic thiophene. *Chemistry Central Journal*, *12*, 1–22.

Shi, X., Zhang, C. Y., Gao, J., & Wang, Z. (2019). Recent advances in photodynamic therapy for cancer and infectious diseases. *Wiley Interdisciplinary Reviews: Nanomedicine and Nanobiotechnology*, *11*(5), e1560.

So, K.-K., Chun, J., & Kim, D.-H. (2018). Antimicrobial and antitumor photodynamic effects of phleichrome from the phytopathogenic fungus Cladosporium Phlei. *Mycobiology*, *46*(4), 448–451.

Su, Y., Sun, J., Rao, S., Cai, Y., & Yang, Y. (2011). Photodynamic antimicrobial activity of hypocrellin A. *Journal of Photochemistry and Photobiology B: Biology*, *103*(1), 29–34.

Sulaiman, C., George, B. P., Balachandran, I., & Abrahamse, H. (2022). Photoactive herbal compounds: a green approach to photodynamic therapy. *Molecules*, *27*(16), 5084.

Sun, Y., Guo, X., Gu, X., Qi, F., & Meng, P. (2021). Effect and mechanism of palmatine hydrochloride mediated laser therapy on proliferation and apoptosis in oral squamous cell carcinoma CAL-17 cells. *Academic Journal of Chinese PLA Medical School,* 42(8), 1–7.

Taib, N. M., Hassan, N. M., Kamal, L. Z. M., & Soe, M. K. (2022). Bioautography, combination effects and photo-activated enzymatic restriction inhibitory activity of antimicrobial alkaloids from Glycosmis pentaphylla (Retz.) DC. *Malaysian Journal of Science*, *41*(3), 1–9.

Tan, Y., Sun, H., Lan, Y., Khan, H. M., Zhang, H., Zhang, L., Zhang, F., Cui, Y., Zhang, L., & Huang, D. (2024). Study on 3D printed MXene-berberine-integrated scaffold for photo-activated antibacterial activity and bone regeneration. *Journal of Materials Chemistry B*, *12*(8), 2158–2179.

Tavakkoli Yaraki, M., Liu, B., & Tan, Y. N. (2022). Emerging strategies in enhancing singlet oxygen generation of nano-photosensitizers toward advanced phototherapy. *Nano-Micro Letters*, *14*(1), 123.

Thong, P. S.-P., Watt, F., Ren, M. Q., Tan, P. H., Soo, K. C., & Olivo, M. (2006). Hypericin-photodynamic therapy (PDT) using an alternative treatment regime suitable for multi-fraction PDT. *Journal of Photochemistry and Photobiology B: Biology*, *82*(1), 1–8.

Tønnesen, H. H., De Vries, H., Karlsen, J., & Van Henegouwen, G. B. (1987). Studies on curcumin and curcuminoids IX: investigation of the photobiological activity of curcumin using bacterial indicator systems. *Journal of Pharmaceutical Sciences*, *76*(5), 371–373.

Trauner, K., Gandour-Edwards, R., Bamberg, M., Nishioka, N. S., Flotte, T., Autry, S., & Hasan, T. (1998). Influence of light delivery on photodynamic synovectomy in an antigen-induced arthritis model for rheumatoid arthritis. *Lasers in Surgery and Medicine: The Official Journal of the American Society for Laser Medicine and Surgery*, *22*(3), 147–156.

Ubeda, A., Montesinos, C., Payá, M., Terencio, C., Alcaraz, M. J., & Halliwell, B. (1993). Antioxidant action of benzylisoquinoline alkaloids. *Free Radical Research Communications*, *18*(3), 167–175.

Vandenbogaerde, A., Geboes, K., Cuveele, J., Agostinis, P., Merlevede, W., & De Witte, P. (1996). Antitumour activity of photosensitized hypericin on A431 cell xenografts. *Anticancer Research*, *16*(4A), 1619–1625.

Vaquero, M. J. R., Alberto, M. R., & de Nadra, M. C. M. (2007). Influence of phenolic compounds from wines on the growth of Listeria monocytogenes. *Food Control*, *18*(5), 587–593.

Vittar, N. B. R., Awruch, J., Azizuddin, K., & Rivarola, V. (2010). Caspase-independent apoptosis, in human MCF-7c3 breast cancer cells, following photodynamic therapy, with a novel water-soluble phthalocyanine. *The International Journal of Biochemistry & Cell Biology*, *42*(7), 1123–1131.

Wamer, W. G., Vath, P., & Falvey, D. E. (2003). In vitro studies on the photobiological properties of aloe emodin and aloin A. *Free Radical Biology and Medicine*, *34*(2), 233–242.

Watanabe, Y., Keisuke, E., Kobuchi, H., Kato, Y., Matsuoka, T., Utsumi, T., Yoshioka, T., Horton, A. A., & Utsumi, K. (1995). Photoactivated inhibition of superoxide generation and protein kinase C activity in neutrophils by blepharismin, a protozoan photodynamically active pigment. *Biochemical Pharmacology*, *49*(4), 529–536.

Wimmer, A., Glueck, M., Liu, J., Fefer, M., & Plaetzer, K. (2024). With Blue Light against biofilms: berberine as natural photosensitizer for photodynamic inactivation of human pathogens. *Photonics, 11*(7), 647–657.

Wright, J. S. (2002). Predicting the antioxidant activity of curcumin and curcuminoids. *Journal of Molecular Structure: Theochem*, *591*(1–3), 207–217.

Xu, Y., Wang, D., Zhuang, Z., Jin, K., Zheng, L., Yang, Q., & Guo, K. (2015). Hypericin-mediated photodynamic therapy induces apoptosis in K562 human leukemia cells through JNK pathway modulation. *Molecular Medicine Reports*, *12*(5), 6475–6482.

Yance Jr, D. R., & Sagar, S. M. (2006). Targeting angiogenesis with integrative cancer therapies. *Integrative Cancer Therapies*, *5*(1), 9–29.

Yañuk, J. G., Denofrio, M. P., Rasse-Suriani, F. A., Villarruel, F. D., Fassetta, F., Einschlag, F. S. G., Erra-Balsells, R., Epe, B., & Cabrerizo, F. M. (2018). DNA damage photo-induced by chloroharmine isomers: hydrolysis versus oxidation of nucleobases. *Organic & Biomolecular Chemistry*, *16*(12), 2170–2184.

Ye, Z., Hua, D., Rao, Y., Bai, S., Seeberger, P. H., Yin, J., & Hu, J. (2020). Targeted photodynamic therapy with a novel photosensitizer cercosporin encapsulated multifunctional copolymer. *Colloids and Surfaces A: Physicochemical and Engineering Aspects*, *585*, 124136.

Zhang, K., Gao, S., Guo, J., Ni, G., Chen, Z., Li, F., Zhu, X., Wen, Y., & Guo, Y. (2018). Hypericin-photodynamic therapy inhibits proliferation and induces apoptosis in human rheumatoid arthritis fibroblast-like synoviocytes cell line MH7A. *Iranian Journal of Basic Medical Sciences*, *21*(2), 130.

Zhang, L., Li, D., & Yu, S. (2020). Pharmacological effects of harmine and its derivatives: a review. *Archives of Pharmacal Research*, *43*, 1259–1275.

Zhang, M., Zhang, X., Tian, T., Zhang, Q., Wen, Y., Zhu, J., Xiao, D., Cui, W., & Lin, Y. (2022). Anti-inflammatory activity of curcumin-loaded tetrahedral framework nucleic acids on acute gouty arthritis. *Bioactive Materials*, *8*, 368–380.

Zhao, H., Yin, R., Chen, D., Ren, J., Wang, Y., Zhanga, J., Deng, H., Wang, Y., Qiu, H., & Huang, N. (2014). In vitro and in vivo antitumor activity of a novel hypocrellin B derivative for photodynamic therapy. *Photodiagnosis and Photodynamic Therapy*, *11*(2), 204–212.

Zhou, Y., Wang, J., Yang, W., Qi, X., Lan, L., Luo, L., & Yin, Z. (2017). Bergapten prevents lipopolysaccharide-induced inflammation in RAW264. 7 cells through suppressing JAK/STAT activation and ROS production and increases the survival rate of mice after LPS challenge. *International Immunopharmacology*, *48*, 159–168.

Zhou, Z. X., Mou, S. F., Chen, X. Q., Gong, L. L., & Ge, W. S. (2018). Anti-inflammatory activity of resveratrol prevents inflammation by inhibiting NF-κB in animal models of acute pharyngitis. *Molecular Medicine Reports*, *17*(1), 1269–1274.

Zhu, Q., Sun, Y., Yun, X., Ou, Y., Zhang, W., & Li, J.-X. (2014). Antinociceptive effects of curcumin in a rat model of postoperative pain. *Scientific Reports*, *4*(1), 4932.

Zhu, T., Lin, Z., Tang, X., Liu, J., Zhang, Y., & Zhong, X. (2024). Programmable activation of berbamine and photosensitizers for enhanced photodynamic therapy using emission-switchable upconversion nanoparticles. *International Journal of Pharmaceutics*, *659*, 124202.

Zhu, X., Zeng, K., Qiu, Y., Yan, F., & Lin, C. (2013). Therapeutic effect of emodin on collagen-induced arthritis in mice. *Inflammation*, *36*, 1253–1259.

Zupko, I., Kamuhabwa, A. R., D'Hallewin, M.-A., Baert, L., & De Witte, P. A. (2001). In vivo photodynamic activity of hypericin in transitional cell carcinoma bladder tumors. *International Journal of Oncology*, *18*(5), 1099–1105.

16 Pharmaceutical and Regulatory Considerations and Future Prospects of Phytomedicines and Nutraceuticals

Ejaz Basheer, Shahid Rasool, Muhammad Talha, Muhammad Najeeb Ullah, Ammara Saleem, and Muhammad Furqan Akhtar

WHAT WILL YOU LEARN?

This chapter will provide readers with valuable and comprehensive knowledge about constraints and future prospects of phytochemical products. This chapter will provide information to researchers and pharmaceutical industries about the pharmaceutical, regulatory, and marketing requirements of antiarthritic phytochemical drugs and nutraceuticals and the importance of encouraging mediculture.

16.1 INTRODUCTION

The study of bioactive compounds obtained from natural sources is the basis of the transformation of naturally occurring bioactive compounds into a pharmaceutical product. Today, a wide range of bioactive compounds are available as prescription medicine, ranging from anti-inflammatory and antimicrobials to oncology products, like *morphine*, *silibinin*, *psilocybin*, and *resveratrol*. Moreover, many of the drugs sold today are synthetic modifications of naturally obtained bioactive compounds, for example, *taxol*. After extensive laboratory research and a series of experiments, if a bioactive molecule fails to fulfill subsequent regulatory and marketing requirements, it may lead to just pooling of research data rather than a meaningful scientific contribution toward the well-being of society. Furthermore, if a potential bioactive molecule fails to meet regulatory and marketing requirements, it would be counterproductive in terms of investment and extensive research and could potentially cost billions of dollars.

Currently, in the field of drug discovery, most contemporary scientific endeavors are aimed at the extraction, isolation, and exploration of biological activities of naturally sourced bioactive compounds. The transformation of these bioactive compounds to a patient-useable formulation is an immense challenge, and quality, safety, efficacy, mechanism, drug–drug or drug–food interactions, and contraindications need to be addressed properly. Therefore, compliance with regulatory policies is essential to monitor the safety of these bioactive molecule-based drugs of natural origin. This chapter discusses various considerations, which ensure the transformation of bioactive compounds into patient-useable, safe, efficacious, and quality-compliant medicine.

DOI: 10.1201/9781003535201-20

16.1.1 Regulation of Phytomedicines and Nutraceuticals

Based upon their existence and functionality, plant metabolites can be classified into primary metabolites and secondary metabolites. Primary metabolites serve as a food reservoir and building blocks, essential for plants' daily functioning, growth, and development. Unlike the latter group of metabolites, the former group is not only essential for the daily functioning but also fully capable of eliciting pharmacological and toxicological responses in humans and animals and by virtue of this capability these secondary metabolites are promising candidates for their transformation into therapeutic agents.

In the United States, medicines based on herbal bioactive molecules are termed as botanical drugs and must pass through the regulatory framework for approval by the Federal Drug Administration (FDA). The bioactive molecules of natural origin, for instance, enzymes and vitamins, are attributed as dietary supplements, whereas purified molecules are termed as biodrugs. For either bioactive molecule of natural origin or botanical drugs, the FDA's regulatory policies are not identical to those for nonbotanical drugs like synthetic, semisynthetic, highly purified drugs or drugs modified chemically, and antibiotics obtained from microorganisms. However, like synthetic or semisynthetic drug molecules, botanical drugs/bioactive molecules of natural origin also undergo Abbreviated New Drug Approval or a new drug approval (NDA) (Sunderland, 2020) process to establish quality, safety, and efficacy. FDA's Center for Drug Evaluation and Research (CDER) has already released "Botanical Drug Development Guidance for Industry." According to which, most plant products are considered food supplements. However, in the United Kingdom, herbal medicinal products are regulated under two main categories, i.e., if the applicant claims a medicinal effect, the approval will be granted under medicine legislation, the second type of approval is for the natural compound with strong medicinal or pharmacological activity supported by its in vivo study and considered medicine regardless of the applicant's claim. On the contrary, in Germany almost all herbal products are deemed as medicines. Unfortunately, these different regulatory approaches across jurisdictions negatively impact the authenticity of medicine based on bioactive molecules of natural origin.

FDA (Food and Drug Administration America) and EMA (European Medicines Agency) are among the most stringent regulatory authorities, which regulate and grant approval for herbal bioactive-based medicine and are often compared as equal regulatory counterparts. To understand approval and regulation of herbal medicine based on bioactive compounds of natural origin, it is essential to comprehend approval and regulation of both agencies.

16.2 REGULATORY REQUIREMENTS OF HERBAL MEDICINE

16.2.1 Europe: Regulations on Herbal Medicine Products

To ensure that European Union (EU) residents have access to herbal and traditional medicines, a unified legal framework has been devised. Products with the proper level of quality, safety, and efficacy are able to get market authorization. To grant pharmaceutical businesses access to the market, a specific set of processes is provided. The collection of EU's herbal monographs on herbs and preparations made from them defines a special standard. The development of EU monographs for herbs is an open and transparent procedure, which incorporates suggestions from the scientific strata and concerned stakeholders.

16.2.1.1 European Medicines Agency's Committee on Herbal Medicinal Products

Throughout the EU member states, unified medicinal regulatory law explicitly distinguishing between modern and traditional herbal medicines has been put into effect. The goal is to develop a regulatory framework that considers the specific characteristics associated with herbal medicines. To facilitate market access and harmonize assessment across various member states, standards are being aligned. The standard protocols developed by the Committee on Herbal Medicinal Products (HMPC) and EU herbal monographs, along with quality standards established by the European Pharmacopoeia, serve as an excellent precedent for international harmonization of the regulation of traditional and conventional medicines.

Before going into the details of approval of herbal medicines in the EU, it is necessary to understand the multinational medicine regulatory setup there. The European Commission and the Parliament are responsible for creating European law. Even though "regulations" are established laws that take immediate implementation in the member states, "directives" require national legislation to be put into effect.

The EMA (Amsterdam, the Netherlands), the European Directorate for the Quality of Medicines & Healthcare (EDQM, Strasbourg, France), and other relevant national authorities of the member states (for example, Italian Medicines Agency) are significant key institutions that make up this network. This network is complex and if an applicant wishes to apply for market access, they must be aware of its obligations.

16.2.1.2 The European Directorate for the Quality of Medicines & Healthcare (EQDMA)

The EDQM is an organization that stands for the implementation of quality standards for pharmaceutical products and oversees their use. Following approval by the European Pharmacopoeia Commission, standards are officially permitted to be included in the European Pharmacopoeia. Besides the EU member states, other nations also participate as members of the European Pharmacopoeia Commission, and still more nations participate as observers. The European Pharmacopoeia's quality specifications serve as legally enforceable guidelines for the EU. Reference standards for analyses and tests are also provided by the EDQM. The EDQM also issues certificates of suitability for pharmaceutical substances to certify the compliance with the European Pharmacopoeia's standards and can serve as a reference during an authorization process. To ensure effective quality control of medicine, the directorate also manages a system of laboratories for medicine control.

With regard to herbal medicine, the European Pharmacopoeia stipulates monographs either general or specific to herbal drugs or preparations. In principle, the monographs cover every quality concern associated with plant medicinal products and herbal preparations, including microbial contamination, pesticide residues, assays for standardization, heavy metal content, and aflatoxins. They also cover methods, tests, identification, and assays. The EDQM has constituted expert groups (such as 13 A, 13 B, and Traditional Chinese Medicine (TCM)) tasked with the elaboration of monographs on herbal medications and/or herbal preparations.

16.2.1.3 European Medicines Agency (EMA)

More than 20 years have passed since the EMA was founded. Its primary goal is to organize the network of member states' respective regulatory authorities on medicinal products. There are various procedural options within this framework. While some procedures are managed by the EMA, others are organized by each member state on its own and offer the choice of procedures resulting from a collaboration between the member states. Seven scientific expert committees presently cover various relevant topics for pharmaceuticals.

The scientific expert committees have various sub-apparatuses, such as drafting groups and working parties. Depending upon their specific duties, MA's administrative and scientific employees, national experts, and institutions facilitate advanced levels of evaluation and unified decision-making throughout the EU. Indeed, the idea to share tasks and the duty to protect the public's health is not just applicable to assessments conducted prior to market access. Initial steps like the monitoring and approving clinical trials, risk assessments during post-marketing and execution of actions, and further notable observations and affairs during drug development shared across the EU.

16.2.1.4 Options available for Medicinal Products to Enter Market Within the EU

Prior to final market authorization, herbal medicines or traditional herbal medicines are evaluated for their efficacy, safety, and quality in the EU. The applicant must pick the right procedure to submit a marketing authorization or registration application. The below-mentioned steps are mandatory by law:

16.2.1.4.1 Centralized Process

This application process for marketing approval submitted to EMA is associated with an evaluation supervised by the same. A medicinal product may be allowed to be marketed across the EU after the marketing authorization. This method is used when a specific set of indications (such as neurological/oncological indications) or medicines with particular public health significance are present.

16.2.1.4.2 Decentralized Procedure

In this type of approval, a group of member states led by a reference member state authorizes herbal medicinal products. Member state that takes the lead carries out the evaluation while the rest of member states assess the evaluation done and after a successful process, marketing authorization is awarded for all participating member states.

16.2.1.4.3 Mutual Recognition Procedure

This approval mechanism is suitable for a medicinal product already approved for marketing in a member state. On the basis of an already-established assessment, an evaluation is conducted and on the success of that, an approval for registration is allowed in the concerned member state.

16.2.1.4.4 National Procedure

A proposal may be referred to as an individual regulatory authority of any member state and after fulfilling all the requirements successfully, marketing approval or registration exclusively for the member state involved is awarded.

16.2.1.5 European Union's Unified Standards for Traditional and Herbal Medicines

In the EU, the basics are outlined in Directive 2001/83/EC. The European regulatory apparatus comprehensively defines the traditional herbal medicines, herbal products, herbal ingredients, and herbal remedies as listed in the following sections.

16.2.1.5.1 Herbal Medicine Product

Any medicine that alone has one or more herbal substances, one or more herbal preparations, or one or more of these substances combined with one or more of these preparations as its active ingredients.

16.2.1.5.2 Traditional Herbal Medicinal Products

A natural medicine complying with the requirements of Article 16(a)(1) of Directive 2001/83/EC is reproduced here as under:

> *They have indications exclusively appropriate to traditional herbal medicinal products which, by virtue of their composition and purpose, are intended and designed for use without the supervision of a medical practitioner for diagnostic purposes or for prescription or monitoring of treatment.*

If vitamins and minerals complement the herbal constituent, they may be added.

16.2.1.5.3 Herbal Substances

Herbal substances, sometimes known as "herbal drugs," primarily whole, cut, or fragmented plants, parts of plants, algae, lichen, and fungi in their natural state, often dried but occasionally fresh. Untreated exudates are also regarded as herbal substances. Species, genus, variety, author, plant component used, and the botanical name provide a detailed definition of herbal substances (Table 16.1).

TABLE 16.1
EMA's Scientific Committees

Acronym	Titles
CHMP	Committee for Medicinal Products for Human Use
COMP	Committee for Orphan Medicinal Products
HMPC	Committee on Herbal Medicinal Products
PDCO	Pediatric Committee
CAT	CAT Committee for Advanced Therapies
CVMP	Committee for Medicinal Products for Veterinary Use
PRAC	Pharmacovigilance Risk Assessment Committee

16.2.1.5.4 Committee on Herbal Medicinal Products (HMPC)

The primary responsibility of the committee is to establish EU herbal monographs that discuss the safety and efficacy of herbal preparations or substances made from them. A rapporteur creates a monograph by considering all information available publicly or furnished by the national regulatory authorities. In the course of monograph development, the Working Party on European Union Monographs and European Union List, as well as HMPC, discuss the scientific content of a monograph. To guarantee that the documents are accurate and consistent, a peer review process is also in place. The scientific community and stakeholders can also participate in the development process.

Initially, when a "call for scientific data" is made public, participants can either (i) contribute data or (ii) comment on a draft, which remains public for consultation for three months. After the final approved draft, the EMA publishes a herbal monograph, a reference list, an assessment report, and a summary of feedback collected during the consultation.

16.2.1.6 HPMC Monographs and Guidelines

The HMPC's EU herbal monographs offer a unified stance and make it easier for the EU member states to register traditional herbal medicinal products and provide marketing permissions to herbal medicinal goods. The format of the monographs is similar to the content of a list of product attributes. This contains all necessary details about efficacy and safety. Consequently, the monographs guide the responsible national authorities and lay a foundation for a product-specific evaluation.

16.2.1.7 Concepts

16.2.1.7.1 Well-Established Use

Fundamentally ascertains that a medicinal product has a well-established medical application and efficacy and safety backed by credible bibliographic evidence. This kind of application doesn't essentially require submitting results from own conducted clinical trials, pharmacological and toxicological studies, and other types of testing. Instead, expert reports on nonclinical and clinical data and published scientific literature are considered. The EU requires at a minimum of ten years of authorized medicinal use. The HMPC issued a specific instruction on the clinical evaluation of safety and efficacy. Furthermore, no less than one high-quality controlled study, such as a clinical trial, postmarketing investigation, or epidemiological study, must be carried out.

16.2.1.7.2 Traditional Use

To offer a specific framework for traditional medicines, a specific piece of law called Directive 2004/24/EC was made in addition to the general Directive 2001/83/EC. The European Commission believed that traditional medicines might not be able to satisfy the standards established for all medicinal products due to their unique qualities.

The strategy was to replace specific safety and efficacy standards with the choice termed as simplified registration for traditional medicines with an aged history of medicinal use. A drug must

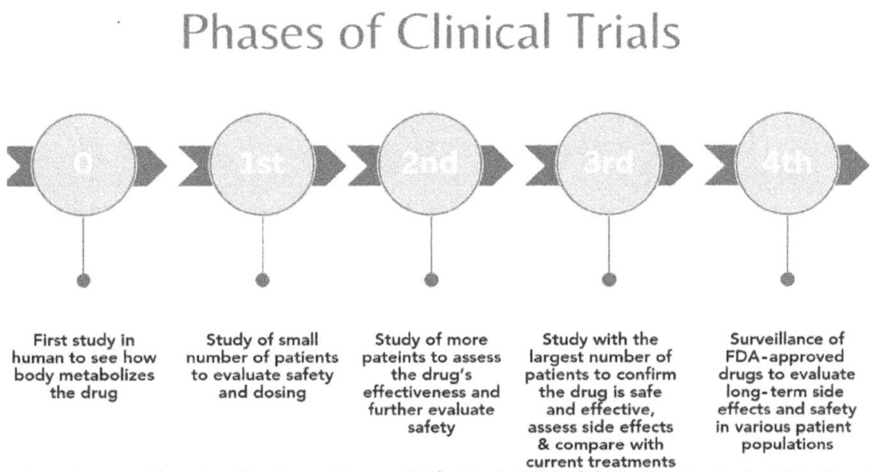

FIGURE 16.1 Phases of clinical trials.

have been used medicinally in the EU for at least 15 of the 30 years required to establish a tradition. The EU or any other region of the world could serve as the second 15 years of therapeutic usage. A set of inclusion criteria was created under the precautionary principle to ensure the security and proper use of conventional herbal medicines.

In the EU, National authorities may ask for more safety information. The standards for quality are the same for all. Traditional herbal medicines must fulfill the same standards as other herbal medicines in terms of quality. The law defines special labeling for traditional herbal therapeutic items. The product must be identified as a traditional herbal drug product used for specific signs and symptoms based on long-term use only. Furthermore, the leaflet in the package must indicate that in case symptoms worsen or persist, the consumer must consult a physician or other qualified healthcare professional. Based on the concept behind a proposal for an herbal medical product, there are varying prerequisites (Figure 16.1).

16.2.2 United States of America: Regulations on Botanical Drugs/Bioactive Molecule of Natural Origin

It is now beyond any shadow of a doubt that the quantum of an international market for herbal products is hovering around one trillion dollars and is increasing by 8%–10% yearly. Currently, herbal medications make up about half of the medicinal supplies available for purchase.

By stipulating guidance for botanical drugs in 2004, the FDA seized the lead for herbal medicine regulation. Consequently, the FDA has authorized 868 semisynthetic compounds derived from natural sources and by 2002, 20 additional anticancer agents also derived from natural sources.

An herbal medicine made from a plant source and used as a drug is what FDA describes as a botanical drug. An FDA-approved conventional medication has a single identifiable active component. However, herbal medicines are found as extracts that contain various chemical components.

According to the Federal Food, Drug, and Cosmetic Act (FD&C Act), botanical goods can be categorized into four segments, i.e., food (such as nutritional supplements), medical devices, pharmaceuticals (including biological drugs), or cosmetics. Before marketing a novel drug product for the intended purpose in the United States, an applicant must request for a Novel Drug Application (NDA). Under a submitted application, the FDA may give an acceptance for a drug product consisting of a drug substance for over-the-counter (OTC) sale. Resultantly, a botanical medication may receive marketing approval for both prescription and OTC use.

The main problem of herbal medicines is their consistency, which can be solved by applying the "totality of the evidence" strategy. This strategy emphasized factors, i.e., biological assay, production control, quality control, and raw material control. To ensure the quality of the raw materials involved in drug production, agriculture and collection practices are used as part of raw material control. Spectroscopic and chromatographic techniques are used in quality control to analyze the medicine and find any impurities or contaminants. Process validation is a component of manufacturing control, which is used to guarantee that the medicine is produced consistently. Finally, biological experiments are employed to reveal the drug's mode of action.

To accomplish approval for a botanical drug from the FDA, the regulations that must be satisfied are often referred to as the protocols mandatory for the herbal drug development program. These standards and procedures are created to guarantee that the medication is effective and safe for its claimed purpose. Data such as the chemical composition of the medicine, its pharmacological activity, clinical safety, and efficacy are mandatory for qualification. Formal examination of the data requires CDER to review the data to make sure the drug abides by the standards and guidelines necessary for the botanical drug development program.

Furthermore, although botanicals have been used for centuries, there is no strong evidence to support either their clinical efficacy or safety. This indicates that there is insufficient safety and efficacy data for the intended use of botanicals. To ensure safety and efficacious use of the drug, it becomes crucial that the criteria and procedures mentioned above are followed.

16.2.2.1 Definitions

As per the Food and Drug Authority (FDA):

Botanicals mean products that include plant materials, algae, macroscopic fungi, and combinations thereof. Not including products consisting of

- Animals or animal parts (insects and annelids) and/or minerals.
  Exception: Present as a minor component in a traditional botanical preparation (traditional Chinese and Ayurvedic medicines).
- Materials obtained from botanical species genetically modified (by recombinant DNA technology or cloning) for the purpose of producing a single molecular entity.
- Fermentation of yeast, bacteria, and plant cells including plants used as substrates or other microscopic organisms with fermentation intended to produce a single molecular entity (antibiotics, amino acids, and vitamins).
- Highly purified substances that can be either derived from a naturally occurring source (paclitaxel) or chemically modified (estrogens synthesized from yam extracts).

16.2.2.2 Types

Botanical products, as per different conditions, may be classified and regulated as drugs, dietary supplements, foods, or cosmetics.

- A botanical product would generally be regulated as a drug, if it's aimed to be used for treating a disease.
- A botanical product would generally be regulated as a dietary supplement, if it's ingested, labeled as a dietary supplement or for use that affects the structure or functions of the body.
- A botanical product would generally be regulated as a food or a food additive if it's used raw or dried to flavor food.
- A botanical product would generally be regulated as a cosmetic if it contains botanical ingredients intended to be used for moisturizing the skin.

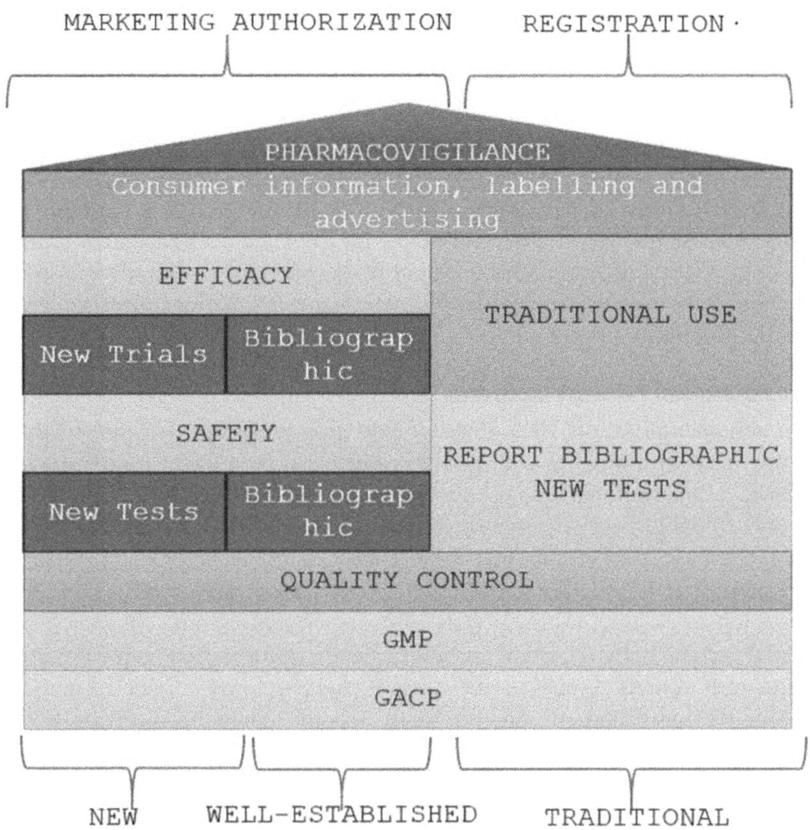

FIGURE 16.2 Pharmacovigilance.

16.2.2.3 Botanical Drugs

A botanical product used in diagnosing, mitigating, curing, or treating disease will be defined as a drug under Section 201(g)(1)(B) of the FD&C. A botanical product intended specifically for the prevention of a disease will be defined as a drug under Section 201(g)(1)(B) and be regulated as a drug.

The Center for Drug Evaluation and Research (CDER) has published the *Botanical Drug Development Guidance for Industry*, which accentuates the development plans for botanical drugs in particular:

- Botanical drugs to be submitted in new drug applications (NDAs)
- Submitting investigational new drug applications (INDs)
- Botanicals for the future NDA submissions
- OTC drug monograph for botanical drugs
- Note: Doesn't provide guidelines for botanical drugs marketed as per biologics license applications (Figure 16.2).

16.2.3 Regulatory Process

Botanical Drug Development Guidance for Industry covers everything from preparing your IND documents to resolving botanical-specific issues such as seasonal variation in chemical composition and the presence of multiple active ingredients. In case of botanicals, the FDA may require some fundamental details about the medicine and its products during phases I and II investigations. Most of the chemical tests and manufacturing control for the product, including the toxicological information at this point, is optional for botanical products. However, if the drugs have an excellent set of data on safe human use, the above-mentioned data and animal toxicity data may be of minimal use or not even necessary for these phases. However, data should still be gathered during this time even if they are not necessarily needed in these phases because they must be submitted before phase III studies can commence. Phase III is a nearly sudden transition to rigorous regulations, where even previous human use is not enough evidence of safety for FDA. One advantage of botanical pharmaceuticals is that combination drugs receive a waiver from individual component assessments, which is required for drugs with a known fixed combination or single component drugs.

It consists of the following stages:

16.2.3.1 Investigational New Drug Status

It is essential to demonstrate to the FDA that the new drug proposed does not likely pose a threat to the patients involved in a clinical trial and there exists a belief that it will work for its stated purpose. It includes a description of the product, chemistry, manufacturing, & control information, pharmacological & toxicological information, an investigator's brochure, a clinical protocol for the proposed clinical trial, and previous human experience.

It is mainly submitted for FDA authorization to conduct human trials. The IND application obtains authorization to conduct clinical trials using an investigational substance from the FDA. This application includes a compilation of all of the existing information on the botanical substance. The FDA has this requirement to ensure that clinical trials are *only* conducted using products that are reasonably safe for human patients and are likely to achieve beneficial results. For pharmaceutical companies, reaching this step requires conducting extensive laboratory research both in petri dishes and in animal models. However, most botanicals already have a lengthy history of human use and a substantial amount of information in the literature. For botanically based drugs, sufficient information for a satisfactory IND application can usually be sourced from previously conducted studies and the botanical drug company does not have to conduct new preclinical research before human trials. The FDA has 30 days to review and respond to IND applications.

There are two possible responses: either authorization to conduct the trial or Clinical Hold, i.e., substantial, such as finding that human subjects would be subjected to unnecessary and serious risk of injury, or minor, such as requiring adjustments to the clinical protocol.

16.2.3.2 Phase II Clinical Trials

Phase II clinical trials evaluate whether or not the product actually works when compared to a suitable placebo. For laboratory-developed drugs, this stage does not occur until after extensive laboratory tests and phase I clinical trials. Most botanical products, however, are able to bypass preclinical testing and phase I testing, entering the process at phase II clinical trials. These typically require 50–250 human patients. The exact number of patients required is determined by a statistical analysis of the existing evidence on the ingredient's actions. Products with larger effects require fewer patients while products with milder effects require more patients. There is no fixed number of studies required for FDA authorization or to advance to phase III studies. Rather the FDA considers the overall body of evidence about the product. In rare cases, a single trial may be sufficient, while other cases may require multiple studies. The total number required to advance to phase III clinical trials depends on the product. While studies are billed per patient, it is important to remember that studies with too few patients are more likely to fail to capture the beneficial effects of a product.

16.2.3.3 Phase III Clinical Trials

These trials study large numbers of human subjects to ensure the product works in diverse groups of people, i.e., different ages, sexes, races, and regions. Identify all of the rare side effects or adverse effects that take place. Once the FDA determines that the quality and quantity of evidence obtained in phase II studies are sufficient, a drug candidate can advance to phase III studies. The total amount of evidence required depends on the effect size of the botanical product and how much information was available in the literature at the start of the process. These studies typically require 500–5,000 participants, depending on the evidence collected prior to this stage. The end goal of this step is to have a completed New Drug Application (NDA) for the FDA to review. Until preliminary findings from phase II studies are available for review, it is difficult to predict how many studies and patients will be required to fulfill the requirements in this final step. Because these studies are larger and require more diversity in human patients, they are the most expensive studies to conduct. For a botanical product, these may range from $300k to $1 million. Cost is determined by the total number of human participants required, study duration, the severity of the condition being treated, and similar factors.

16.2.3.4 New Drug Application and Approval

New Drug applications mainly consist of the following components:

- Product description
- Documentation of previous human experience
- Tests for quality
- Raw material
- Botanical drug substance and drug product
- Nonclinical safety assessment
- Clinical pharmacology
- Clinical evidence for safety
- Clinical evidence for efficacy
- Therapeutic consistency
- Postmarketing considerations

The New Drug Application (NDA) is the formal request for FDA approval to market a product as a drug in the United States. The NDA must include all existing data (and analyses of the data) compiled in the previous steps. When the NDA is complete, it is submitted to the FDA to enter the review process. FDA focuses on evaluating the benefits and risks of the product. A drug is approved when the benefits outweigh the risks, based on an assessment of high-quality evidence.

During this process, a multidisciplinary team of experts from the FDA reviews:

- Clinical evidence compiled in the studies conducted under the IND
- Statistical analysis of these studies
- Toxicology
- Pharmacology
- Chemistry
- Proposed drug label

Within the first 60 days, you will know if the FDA has accepted the application for filing.

Assuming it is accepted, you will receive a "74-day letter" confirming the type of review for your product, any action items requiring an immediate update in the application, and the action date. Over the next several months, the FDA will send information requests. These vary from questions on where information can be found in the NDA to requesting additional analyses of data. During the final weeks, the wording on the label is negotiated. This is typically the time of greatest activity and rapid turnarounds are expected. The entire process from application to label negotiation takes approximately 8–12 months.

16.3 REGULATORY STATUS OF ANTIARTHRITIC NUTRACEUTICALS

Before the advancement of a conventional medicinal system, herbs or herbal-based medicines were the only means of treatment. Medicinal plants and herbs were the only allies of ailing humanity. Before and during the 19th century, Folk remedies were exclusively derived from natural product extracts, particularly those with botanical origins. But as chemistry advanced in the latter part of the 19th century, the isolation of physiologically active compounds began to take center stage in the realm of medicine. Furthermore, comparatively pure forms of organic molecules with biological activity have emerged as the preferred medication for various ailments. For example, the aspirin precursor salicylic acid was extracted from willow bark in 1874.

There is encouraging research supporting the use of a few natural remedies to treat arthritis. Additionally, data indicates that these herbal remedies may also aid in lowering NSAID usage. It is clear that certain herbal remedies when used alone do not perform inferior to methotrexate, and when used in conjunction with it, the results were better than when taken alone (Sundy et al., 2011). In addition to this data, it is noteworthy here that, contrary to popular assumption, herbal medicine can have negative effects, albeit modest ones (Soeken, Miller, & Ernst, 2003). Many synthetic medications that are thought to be the standard for care for RA have so many adverse effects that they are not as widely accepted as they could be. These details might significantly raise the likelihood that herbal remedies will be used to treat arthritis. Many herbal medicines have been used for a long time to treat rheumatoid arthritis, according to the traditional medical literature. Many therapeutic substances that have been identified from plant sources and can regulate inflammatory mediators and show action against these substances include quinones, alkaloids, flavonoids, anthocyanins, catechins, terpenes, and anthoxanthins; they are all well known for having anti-inflammatory properties (Khanna et al., 2007).

Currently, several plant extracts, most often in the form of decoctions, are employed in traditional medicine to treat arthritis. Moreover, OTC medications including phytochemicals and nutraceuticals are accessible for the treatment of arthritis; for example, liniments, balms, oils, creams, patches, and ointments. The toxicity, quality, and efficacy of these preparations have yet to be determined. These formulations may generally be divided into groups based on the dose type and intended use. These assertions have the potential to expand into conventional claims and established applications. In any scenario, marketing permission can be obtained by contacting the appropriate regulatory body whose jurisdiction encompasses the marketing region.

16.4 CHALLENGES WITH PHYTOCHEMICAL STANDARDIZATION AND BIOAVAILABILITY

Complex characteristics of herbal drugs are attributed to diverse combinations of phytocomponents in herbal products. As a result, they raise significant obstacles for themselves to satisfy quality and safety standards. To gain the trust of communities and integrate plant-based products into the mainstream healthcare system, manufacturers, researchers, and regulators must employ rigorous scientific methods to ensure the standardization and quality of traditional herbal products. The excellence and consistency of the products typically depend on the quality control of the source materials because the identity of the finished products is not clearly defined and because the production of plant-based medications essentially does not involve any purification steps (Awasthi et al., 2014).

In clinical practice, determining the efficacy of phytochemical and nutraceutical products is difficult because of inadequate quality control and accessibility. Additionally, difficulties with preparation, taste, odor, and storage are shown to be drawbacks and have an impact on the pharmacological effectiveness of conventional herbal dosage forms. Patient compliance has decreased as a result of the more emphasis on herbal medication's convenience of use rather than its dosage form (Diamond, 2000).

Therapeutic doses can be administered in various ways, but even for conventional herbal medications, the solid dosage form is the most preferred. Oral dosage forms are more convenient to use and provide a precise dose, enabling patients to self-administer and improving compliance. Consequently, it is thought to be the primary means of administering phytochemicals to achieve systemic effects (Orsat, Yang, Changrue, & Raghavan, 2007).

The primary goal of phototherapeutic development for phytochemicals and nutraceuticals is to provide a product that is readily available, technologically practicable, and both chemically and physically stable. The majority of the time, a plant extract is processed to create a solid dosage form (Bonati, 1991). In addition, there are technical problems with the pharmaceutical dosage form formulation of herbal extract since the pharmaceutical characteristics are not appropriate for the type of extract that was obtained. Various technological hurdles may be seen in herbal extracts, such as their high hygroscopic nature that impacts flow characteristics, uniformity and stability, high cohesiveness, poor compressibility, and so on. Changes to the physical condition of the extract may be beneficial in trying to address these problems (Orsat et al., 2007). Further contributing factors to the complexity of polyherbal dosage forms are the large number of solvents used in the preparation of these concentrated formulations without adequate information, as well as the least known physico-technical details of the crude drugs and their interaction and relationship with the excipients being used. Additionally, the plant extract granules will have different dry extract concentrations based on the inherent variations in the dry mass of the decoction; as a result, they occasionally need to standardize up to a specific concentration of decoction per granule. To reduce the variation in dry mass during decoction, concentrate standardization is essential. This reduces variability in its qualities and establishes a relationship between volume and components (Zhang et al., 2010). Nutraceutical and phytochemical products are often heterogeneous complex mixes that either antagonistically or synergistically demonstrate their medicinal properties. It mostly depends on the kind of dose form or herbal medicine formulation. If the phytochemicals responsible for the medicinal effect are widely recognized and when compared to a chemically defined drug, the notion of primary similarity is highly relevant. Nevertheless, for extracts lacking known active biomarkers, the information on known chemical ingredients is useful but insufficient to achieve bioequivalence. In this case, pharmacological or bioassay investigations that assess the applicability of therapeutic efficacy are helpful. For the sake of bioequivalence, phytochemical and nutraceutical products may be compared head-to-head with innovator preparations under the following conditions:

i. Pharmaceutical standardization or equivalence
ii. In vitro or biopharmaceutical equivalence
iii. Bioequivalence with different endpoints (in vitro model, animal model)
iv. Or clinical studies

As is evident from above list one cannot overlook the importance of clinical studies in phytochemical and nutraceutical goods. The only way to demonstrate the efficacy and safety of any herbal medicine product or allopathic therapy is through a randomized clinical study. Clinical trials for phytochemical and nutraceutical products follow the same procedures and objectives as those for allopathic drugs (single component). Herbal formulations have already been used in several randomized, double-blind, placebo-controlled trials. They have previously completed clinical trials to determine the efficacy and safety of phytochemical and nutraceutical goods. Although the use of polyherbal medications has a long history, it is crucial to validate and rationalize their effectiveness and safety through randomized clinical trials. This is because they have known negative effects and should be used with caution.

In light of the foregoing, it is possible to harness the potential of herbal medicinal plants to generate pharmaceutical products that meet quality, safety, and efficacy requirements by transforming evidence-based ethnopharmacological knowledge.

16.5 GLOBAL AND LOCAL MARKETING OF PHYTOCHEMICAL DRUGS

Even though the market for herbal medicines is growing steadily worldwide, there are still many hurdles to overcome to market these medicines. For example, in the United States alone, on the FDA's behalf, there are 500 pre-IND meetings and IND applications, but none have succeeded in moving forward to meet the finalization of the NDA stage.

This could be a result of the following reasons:

16.5.1 Inadequate Proof of Efficacy

As per regulatory guidelines to demonstrate and document a new drug candidate's safety and efficacy, "adequate and well-controlled" multicenter clinical studies are mandatory. The highest degree of inspection is also ensured before approving. Nevertheless, the studies need to have a predetermined target group, stringent experimental controls, adequate outcome measures, independent and fair monitoring, and precise analysis. Therefore, the most frequent cause of failing to advance to the NDA stage is the absence of statistically significant proof of a drug's efficacy and clinical relevance.

16.5.2 Unrealistic Expectations and Goals

Another major reason for the slow approval process is that the drug sponsors frequently set expectations that are too high, particularly when they lack experience.

- The regulatory authority evaluates botanical drugs less rigorously than conventional synthetic drugs because the initial phases of IND development are comparatively less stern, which gives sponsors false hope.
- Lack of botanical regulatory standards that are standardized globally.
- A large number of non-US botanical drug sponsors or non-EU applicants who run drug research operations in the United States or EU are unaware of the actual distinctions between United States, EU, and foreign market regulations (Table 16.2).

TABLE 16.2
Difference in Herbal Drugs in Europe and United States

Europe	US
Definitions	
Herbal Drugs Both the terms: herbal drug as in European Pharmacopeia and herbal substance as in European Union's legislation are synonymous to each other. **Herbal Medicinal Products (HMPs)** A medicinal product constituted with one or more herbal substances or preparations or herbal substance mixed with herbal preparations, as active constituent. **Herbal Preparations** Herbal substances when subjected to processes such as extraction and expression, purification distillation, concentration, fractionation or fermentation, herbal preparations are obtained. Examples: Herbal substances as powders, herbal extracts, essential oils, tinctures, juices obtained via expression and processed exudates. **Herbal Substances** Same as Herbal Drugs as in European Pharmacopeia. Example: Namely whole or cut plants, parts of plants and algae.	As per **Food and Drug Authority (FDA);** *Botanicals mean products that include, plant materials, algae, macroscopic fungi, and combinations thereof.* • A botanical product would generally be regulated as drug, if it's aimed to be used for treating a disease. • A botanical product would generally be regulated as a dietary supplement, if it's taken by mouth, labelled as a dietary supplement or for use which affects the structure or functions of the body. • A botanical product would generally be regulated as a food or a food additive, if it's used raw or dried as an ingredient to flavor food. • A botanical product would generally be regulated as a cosmetic, if it contains botanical ingredients intended to be used for moisturizing the skin.

(Continued)

TABLE 16.2 (*Continued*)
Difference in Herbal Drugs in Europe and United States

Europe	US
Requirements	
Herbal Substance • Botanical Characteristics • Macroscopical & microscopical characterization • Biological variation • Geographical differences • Conditions for cultivation/harvesting/drying • Pre- and post-harvest chemical procedures • Stability profile of the constituents **Herbal Preparations** • Description of herbal preparation • Quality of herbal substance • Preparation method • Constituents • Other constituents • Drying conditions • Stability profile of the constituents • Microbial purity during storage • Batches tested in pre-clinical and clinical testing **Herbal Medicinal Products** • Quality • Manufacturing • Stability profile of the active substance or formulation in the packaging • Batches tested in pre-clinical and clinical testing	**Investigational New Drug (IND) Application** • Description of Product • Chemistry, Manufacturing, & Control Information • Pharmacological & Toxicological Information • Investigator's Brochure • Clinical Protocol for Proposed Clinical Trial • Previous Human Experience **Clinical Trials** • Phase-1 • Phase-2 **New Drug Application and Approval** • Product description • Documentation of previous human experience • Tests for quality • Raw material • Botanical drug substance & drug product • Non-clinical safety assessment • Clinical pharmacology • Clinical evidence for safety • Clinical evidence for efficacy • Therapeutic consistency • Post marketing considerations

16.6 GOOD MANUFACTURING PRACTICES (GMP) FOR HERBAL MEDICINE

GMP encompasses quality, safety, and efficacy. To ensure GMP for herbal medicine is a holistic approach, which starts right from identification of appropriate herb, its collection if obtained from wild source, cultivation, harvesting, if it grows on a commercial level, then its storage and transportation till final formulation of the herbal medicine.

Major issues in the implantation of GMP standards for herbal medicine manufacturing is variation in their source, active constituent variation of single herb from multiple sources, different cultivation, or collection practice for the same herb from different geological sources and the sorted nature of formulation and its manufacturing method. In addition, the registration process is also varied with different regulatory authorities. For instance, EU deals with herbal medicine as an herbal medicinal product; however, Food and Drug Administration USA (FDA) regulates the same as dietary supplements. Moreover, other major stakeholders of herbal medicine like China practice herbal medicine as TCM and regulate it accordingly.

The role of the World Health Organization (WHO) is pivotal in harmonizing GMP guidelines for herbal medicine across different countries. The "WHO guidelines on Good Manufacturing Practices (GMP)" (World Health Organization, 2007) and "WHO updated GMP: supplementary guidelines for the manufacture of Herbal Medicines" outlined a robust framework for the implementation of the quality, efficacy, and safety of phytochemical-based products.

16.6.1 WHO Guidelines on Good Manufacturing Practices (GMPs)

The guidelines provide a set of principles and practices that should be followed in the manufacture, processing, packaging, labeling, and storage of herbal medicines to meet appropriate quality standards included the following.

1. Quality management and documentation
2. Premises and equipment requirements
3. Personnel training and hygiene
4. Sanitation and pest control
5. Procurement and quality control of raw materials
6. Manufacturing processes and controls
7. Packaging, labeling, and storage
8. Quality control testing and stability monitoring
9. Complaints and product recalls

These guidelines are focused on developing quality systems and helping manufacturers ensure good practices to manufacture herbal medicines without variation of quality, effective, and safe for use.

16.6.2 WHO Supplementary Guidelines for the Manufacture of Herbal Medicines

These amendments address natural healthcare items, traditional herbal medicines, and herbal medicines. The WHO's revised GMP supplemental recommendations for the production of herbal medicines are summarized here.

16.6.2.1 Quality Management

Emphasizes the importance of a comprehensive quality management system.

Ensures consistent quality, safety, and efficacy of herbal medicines.

16.6.2.2 Herbal Starting Materials

Outlines specific guidelines for choosing, gathering, identifying, and storing herbal beginning materials.

16.6.2.3 Packaging and Labeling

Contains certain guidelines for labeling and packaging that guarantee the traceability and integrity of the product.

16.6.2.4 Quality Control

Outlines testing and quality control measures, including macroscopic, microscopic, and phytochemical identification methods.

16.6.2.5 Validation

Demands that crucial production procedures, such as drying and extraction, be validated.

16.6.2.6 Personal Training

Emphasizes the importance of adequate training and competence of personnel involved in herbal medicine manufacture.

16.6.2.7 Premises and Equipment

Establishes specifications for the construction and upkeep of manufacturing buildings and machinery to guarantee quality and avoid contamination. The revised recommendations seek to improve the legal environment and guarantee effectiveness and safety.

16.7 FUTURE PROSPECTS OF PHYTOTHERAPY AND PHYTOMEDICINES

The use of phytomedicine for medicinal purposes has been witnessing a significant rise, and this trend is projected to continue and even expand in the future. Some promising future prospects are as follows:

16.7.1 Market Growth

It is anticipated that the worldwide market for phytomedicines will expand at a rapid rate. The global market for herbal medicines was estimated to be worth USD 83.0 billion in 2019 and is projected to increase at a compound annual growth rate of 18.9% between 2020 and 2027, according to a research by Grand View Research (Size, 2020).

16.7.2 Growing Demand

Due to customer preferences for natural therapies and the perception that they have fewer adverse effects than synthetic pharmaceuticals, there is a growing market for natural and organic health goods. This pattern is driving the market for phytomedicines to expand.

16.7.3 Research and Development

Discovery of new phytomedicinal substances is the result of advances in biotechnological developments and phytochemical research. This is playing an important role in improving phytomedicines' safety and efficacy and leading to their increased acceptance in traditional medicine.

16.7.4 Regulatory Support

As mentioned above, advanced countries have already set up an extensive framework of guidelines for phytomedicine regulation, which is motivating other nations to put guidelines for safety and effectiveness in place. This is anticipated to increase market expansion and customer confidence.

16.7.5 Combined Therapy

Using phytomedicine in conjunction with traditional medical procedures is becoming increasingly common. Potential benefits of this integrated approach include more individualized and comprehensive healthcare treatments.

16.7.6 Economic Outlook

The impact of phytomedicine is multifaceted, and it will affect various sectors in multiple ways.

It will not only create new opportunities in the agriculture sector but also lead to increased income for farmers through the cultivation of high-value medicinal plants and sustainable agriculture will promote conservation and biodiversity research. In the pharmaceutical sector, search for new phytomedicinal compounds will drive investment in R&D and new products will generate revenue for pharmaceutical companies and contribute to economic development. Healthcare sector will benefit from cost-effective treatments, and complementary therapies will become an integral part of it. The increasing consumer preference for natural and organic products will boost the retail sales and online sales of phytomedicines. Countries rich in medicinal plant biodiversity, such as India and China, can generate substantial export revenue from phytomedicine by exporting the phytochemicals and phytomedicinal plants as a raw material to other countries.

REFERENCES

Awasthi, M. K., Pandey, A. K., Khan, J., Bundela, P. S., Wong, J. W., & Selvam, A. (2014). Evaluation of thermophilic fungal consortium for organic municipal solid waste composting. *Bioresource Technology*, *168*, 214–221.

Bonati, A. (1991). How and why should we standardize phytopharmaceutical drugs for clinical validation? *Journal of Ethnopharmacology*, *32*(1–3), 195–197.

Diamond, A. (2000). Close interrelation of motor development and cognitive development and of the cerebellum and prefrontal cortex. *Child Development*, *71*(1), 44–56.

Food, U., & Administration, D. Botanical Drug Development Guidance for Industry. 2016.

Khanna, D., Sethi, G., Ahn, K. S., Pandey, M. K., Kunnumakkara, A. B., Sung, B., Aggarwal, A., & Aggarwal, B. B. (2007). Natural products as a gold mine for arthritis treatment. *Current Opinion in Pharmacology*, *7*(3), 344–351.

Orsat, V., Yang, W., Changrue, V., & Raghavan, G. (2007). Microwave-assisted drying of biomaterials. *Food and Bioproducts Processing*, *85*(3), 255–263.

Size, M. V. M. (2020). Share & trends analysis report. https://www.grandviewresearch.com/industry-analysis/automated-guided-vehicle-agv-market

Soeken, K., Miller, S., & Ernst, E. (2003). Herbal medicines for the treatment of rheumatoid arthritis: a systematic review. *Rheumatology*, *42*(5), 652–659.

Sunderland, J. J. (2020). The academic NDA: justification, process, and lessons learned. *Journal of Nuclear Medicine*, *61*(4), 480–487.

Sundy, J. S., Baraf, H. S., Yood, R. A., Edwards, N. L., Gutierrez-Urena, S. R., Treadwell, E. L., Vßzquez-Mellado, J., White, W. B., Lipsky, P. E., & Horowitz, Z. (2011). Efficacy and tolerability of pegloticase for the treatment of chronic gout in patients refractory to conventional treatment: two randomized controlled trials. *JAMA*, *306*(7), 711–720.

World Health Organization. (2007). *WHO Guidelines on Good Manufacturing Practices (GMP) for Herbal Medicines*. World Health Organization.

Zhang, M., Brewer, A. C., Schröder, K., Santos, C. X., Grieve, D. J., Wang, M., Anilkumar, N., Yu, B., Dong, X., & Walker, S. J. (2010). NADPH oxidase-4 mediates protection against chronic load-induced stress in mouse hearts by enhancing angiogenesis. *Proceedings of the National Academy of Sciences*, *107*(42), 18121–18126.

Index

Note: **Bold** page numbers refer to tables and *italic* page numbers refer to figures.

Q

R

S

T

U

V

W

Z

For Product Safety Concerns and Information please contact our EU representative GPSR@taylorandfrancis.com
Taylor & Francis Verlag GmbH, Kaufingerstraße 24, 80331 München, Germany

www.ingramcontent.com/pod-product-compliance
Ingram Content Group UK Ltd.
Pitfield, Milton Keynes, MK11 3LW, UK
UKHW052219190926
475702UK00006B/1111
9781032860862